Jan Harmsen, André B. de Haan, Pieter L. J. Swinkels

Product and Process Design

Also of Interest

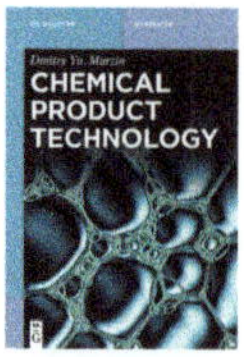

Chemical Product Technology.
Murzin, 2018
ISBN 978-3-11-047531-9, e-ISBN 978-3-11-047552-4

Chemical Process Synthesis.
Connecting Chemical with Systems Engineering Procedures
Bezerra, 2018
ISBN 978-3-11-046825-0, e-ISBN 978-3-11-046826-7

Process Synthesis and Process Intensification.
Methodological Approaches
Rong (Ed.), 2017
ISBN 978-3-11-046505-1, e-ISBN 978-3-11-046506-8

Process Engineering.
Addressing the Gap between Study and Chemical Industry
Kleiber, 2016
ISBN 978-3-11-031209-6, e-ISBN 978-3-11-031211-9

Jan Harmsen, André B. de Haan,
Pieter L. J. Swinkels

Product and Process Design

Driving Innovation

DE GRUYTER

Authors
ir. Jan Harmsen
Harmsen Consultancy B.V.
Hoofdweg Zuid 18
2912 ED Nieuwerkerk aan den IJssel
The Netherlands

prof. dr. ir. André B. de Haan
Delft University of Technology
Department of Chemical Engineering
Section Transport Phenomena
Van der Maasweg 9
2629 HZ Delft
The Netherlands

ir. Pieter L. J. Swinkels
Managing Director PDEng programmes
Faculty of Applied Sciences
Delft University of Technology
Departments of Chemical Engineering and Biotechnology
Van der Maasweg 9
2629 HZ Delft
The Netherlands

ISBN 978-3-11-046772-7
e-ISBN (PDF) 978-3-11-046774-1
e-ISBN (EPUB) 978-3-11-046775-8

Library of Congress Control Number: 2018934236

Bibliographic information published by the Deutsche Nationalbibliothek
The Deutsche Nationalbibliothek lists this publication in the Deutsche Nationalbibliografie; detailed bibliographic data are available on the Internet at http://dnb.dnb.de.

Cover image: Bart van Overbeeke Fotografie
Typesetting: le-tex publishing services GmbH, Leipzig
Printing and binding: CPI books GmbH, Leck

www.degruyter.com

Preface

We, the authors, have extensive experience in designing and developing innovative products and processes, for a broad variety of companies. Some of these companies are strongly product driven, while others are strongly process driven, and some by both products and processes. The companies belong to various industries, such as oil and gas, consumer products, chemicals, life sciences, food, and engineering, procurement, and construction.

We also have extensive experience in teaching design methodology for products and processes both in universities and at companies. Coming from these combined industrial and academic experiences we felt a strong need to write a book covering all relevant design methods and heuristics focused on innovative solutions to establish breakthrough as well as adapted innovations. This book can be used as a textbook for teaching in academia and industries but also for inspiring industrial product and process engineers to create novel designs.

From our industrial experiences we have gleaned that industrial innovation can be tremendously accelerated by incorporating design activities throughout the innovation process to drive projects from early-stage idea generation to commercial implementation. Our academic experience has taught us how to unlock and develop the design skills of students and enable them to generate novel concepts for products and processes.

These experiences, complemented by recent insights into design and innovation in the literature, have been used to create this book. During this process, a lot of discussion took place regarding the content, the connections between the chapters, and the format and structure of the book. We felt privileged that such a large group, with different knowledge and experiences but with the same conviction and motivation, were in a position to cooperate with us to complete this book.

We would like to acknowledge the people who contributed to this book: Johan Grievink for writing Chapter 9 and contributing to Chapters 4 and 8; Ido Pat-El, for writing Chapter 11 on health and safety, and Section 5.7 on designing for human factors; Jos van Reisen, for writing Section 5.8, on designing for process energy efficiency; Christhian Almeida Rivera, for contributing to Chapters 4 and 8 and proofreading Chapter 9; Rob van der Lans for contributing to the characteristic times of Chapter 8; Gabrie Meesters for contributing to Chapters 3 and 8; Mark Toonen, Innovation Manager at Cosun, for providing information on innovation management in practice; Tom Brooijmans of Cosun Biobased Products for reviewing Chapter 10; Gijsbert Korevaar, Asst Prof. of energy, services and systems, TU Delft, for reviewing Chapter 12; Maarten Verkerk, Professor at TU Eindhoven, for reviewing the use of modal aspects in Chapter 3; Yvette van Herwijnen, of AMI consultancy, for providing input on Chapter 3 on project management.

https://doi.org/10.1515/9783110467741-201

We thank Giljam Bierman for many years of contribution in design projects and design methods at the (post-MSc) PDEng level at TU Delft, and for contributing this expertise in Chapter 4. We thank the following people for their inspiration and cooperation in product and process design at TU Delft over the last 25 years:

Hans Wesselingh, Peter Appel, Gert Frens, Johan Grievink, Cees Luteijn, Giljam Bierman, Peter Jansens, Jacob Moulijn, Freek Kapteijn, Ernst Sudholter, Michiel Kreutzer, Ruud van Ommen, Ger Koper, Margot Weijnen, Paulien Herder, Peter Bongers, Cristhian Almeida Rivera, Gabrie Meesters, Rafiqul Gani, Kees van Overveld, Adrie Huesman, Henk van den Berg, Jan Dijk, Jeu Lambrichts, Peter Daudey, Urjan Jacobs, Henk Nugteren, Peter Hamersma, Jos van Reisen, Sorin Bildea, Tony Kiss, Karel Asselbergs, Luuk van der Wielen, Adrie Straathof, Mark van Loosdrecht, Sef Heijnen, Marcel Ottens, Maria Cuellar Soares, Geert Jan Witkamp, Henk Noorman, Peter Verheijen, as well as the many, many MSc students in chemical engineering, PDEng trainees in process & equipment design, bioprocess engineering, chemical product design, and PhD students in process systems engineering over the last 25 years.

We thank the photographer Bart van Overbeeke, DuPho, Dutch Professional Photographers, for his photograph on the book cover and David Vervloet (PhD student), and his supervisors Ruud van Ommen and Freek Kapteijn, for allowing their experimental reactor setup to be photographed. We thank Marjo Dekker, Moments of Memory fotografie, for her photograph of Jan Harmsen and we thank Harm P.W.L. Swinkels for his photograph of Pieter Swinkels.

Contents

Part C: **Design optimization**

Authors' biographies

ir. Jan Harmsen

Harmsen Consultancy BV
Hoofdweg Zuid 18, 2912 ED,
Nieuwerkerk aan de IJssel, The Netherlands
Email: jan@harmsenconsultancy.nl

Jan Harmsen is currently a consultant for sustainable process innovation. He provides advice to and teaches courses for industry and academia. After graduating in chemical technology from Twente University in 1977, he joined Shell. There he held professional positions in process research, notably on three-phase reactors and biotechnology, in process development, in reaction engineering, process concept design, process implementation, and finally in process intensification until 2010. He became, part-time, Hoogewerff Professor of Sustainable Chemical Technology in 1997, first at Delft University of Technology and later at Groningen University until 2013. He authored the following books: Jan Harmsen, *Industrial Process Scale-up: A practical innovation guide from idea to commercial implementation* (Elsevier, 2013); Gerald Jonker and Jan Harmsen, *Engineering for sustainability: A practical guide for sustainable design* (Elsevier, 2012); Jan Harmsen, Joseph Powell (eds.), *Sustainable Development in the Process Industries: Cases and Impact* (John Wiley & Sons, 2010).

prof. dr. ir. André B. de Haan

Delft University of Technology (TU Delft)
Faculty of Applied Sciences,
Department of Chemical Engineering
Van der Maasweg 9, 2629 HZ Delft,
The Netherlands
Email: a.b.dehaan@tudelft.nl

Prof. André de Haan currently combines a position as principal technologist at the Cosun Innovation Centre with a position as part-time professor at Delft University of Technology. After finishing his PhD, he held various positions at DSM (1991–1999), held the chair of separation technology at the University of Twente (1999–2006) and the chair in process systems engineering at the Technical University of Eindhoven (2006–2010). He worked for Corbion–Purac as a corporate scientist in process technology (2010–2016), initially in combination with a part-time position at the Technical University of Eindhoven (2010–2013), which he exchanged for a post at Delft University of Technology in 2014. Since 2002 he has been editor for the journal, *Separation and Purification Technology*. He authored the books **Industrial Separation Processes** and **Process Technology**, both published by De Gruyter.

ir. Pieter L. J. Swinkels

Delft University of Technology (TU Delft)
Faculty of Applied Sciences,
Department of Chemical Engineering
Van der Maasweg 9, 2629 HZ Delft,
The Netherlands
Email: p.l.j.swinkels@tudelft.nl
Photo by: H. (Harm) P. W. L. Swinkels

ir. Pieter Swinkels is director of the post-MSc PDEng designer programs in chemical and biochemical product and process design at Delft University of Technology (TU Delft). After obtaining his ir. (MSc) degree at Technical University Eindhoven, with specialization in bioprocess engineering at TU Delft, he was employed by Unilever. He was a member of the R&D team that invented novel detergent powder formulations and manufacturing processes (1986–1991). Within Unilever's specialty chemicals businesses (later part of National Starch & Chemical Company and ICI) he worked in various positions in manufacturing, product/process design & development, and as divisional process development manager (1992–2001). In 2001, he moved to TU Delft as Asst. Professor of product and process design and engineering, and became PDEng program director. He develops chemical and biochemical product/process design methodologies, which he teaches and applies in MSc and PDEng design projects in cooperation with companies from his extensive industrial network.

prof. em. ir. Johan Grievink

Delft University of Technology (TU Delft)
Faculty Applied Sciences,
Department of Chemical Engineering
Van der Maasweg 9, 2629 HZ Delft,
The Netherlands
Email: j.grievink@tudelft.nl

Johan Grievink is a chemical engineering Professor Emeritus at Delft University of Technology. He was educated as a chemical engineer and applied mathematician at Twente University, after which he joined Shell Research, Amsterdam in 1971. There, he contributed in various positions to the rapidly evolving field of mathematical modeling and computing for development, design, and operations of chemical processes. In 1992 he transferred to TU Delft, to teach process modeling, design, and optimization of operations as well as to develop educational policies for MSc and PDEng programs. The focus of his research was on processes for conversion of energy carriers (e.g., natural gas-to-liquid fuels) and for industrial crystallization. His research gradually broadened from process engineering to the interactions between product and process engineering. Having retired in 2008, he continues to perform some academic duties and to offer industrial consultancy in process modeling and optimization.

ir. Ido E. Pat-El, MSc.

Royal HaskoningDHV
Industry & Buildings
Contactweg 47, 1014 AN Amsterdam,
The Netherlands
Email: ido.pat-el@rhdhv.com

ir. Ido Pat-El graduated from the Delft University of Technology in 2007 with a Masters in chemical engineering. During his college years he performed safety-related projects for organizations such as TNO Defence Security and Safety, the Dutch State Supervision of Mines, and the Center of Industrial Fire Safety of the Rotterdam–Rijnmond Port District Fire Services. He is currently employed as a process safety specialist at the engineering and consulting company, Royal HaskoningDHV The Netherlands B.V. In that role he performs health, safety and environmental assessments for the oil, gas and chemical industry. He has been involved as Lead Technical Safety Engineer in the design of various offshore oil and gas facilities. Ido has written papers about his field of work and has given presentations on process safety related subjects in national and international forums. He provides technical safety guest lectures at the Delft University of Technology and the Eindhoven University of Technology.

dr. ir. Jos L. B. van Reisen, PDEng

CB&I
Engineering & Construction
Prinses Beatrixlaan 35,
2595 AK The Hague, The Netherlands
Email: jvanreisen@cbi.com

Jos van Reisen holds the position of senior process engineer at CB&I, engineering & construction group, located in The Hague. As guest lecturer, he teaches energy efficient process design at Delft University of Technology since 1993. He has a MSc (ir) degree and a professional doctorate in engineering (PDEng) in chemical engineering from Delft University of Technology. He has also received a PhD in chemical engineering from Delft University of Technology for his research on a design method for the application of multistream heat exchangers in the energy saving retrofit of chemical processes. He worked for ABB Lummus Global from 1998 until its merger with CB&I and worked on a wide range of projects from basic to detailed engineering and construction supervision for clients throughout the energy chain both up- and downstream.

Part A: **Innovation and industry**

1 Goal, scope, and structure

1.1 Goal

The world requires novel products and their processes to increase human health, to reduce the burden on the environment, and to increase prosperity, particularly in deprived areas, as defined in the United Nations Global Grand Challenges and subscribed to in a declaration by the presidents of the World Congress Chemical Engineering (WCCE), the European Federation of Chemical Engineers, the European Society of Bioengineering, the IACChE, and the APCChe, at the WCCE10 in Barcelona, during October 1–5, 2017 [1]. These combined requirements call for the radical innovation of products and their processes.

These needs are necessary for innovation but are not sufficient. The need for innovation has to also be identified by industrial companies, and then projects must be started to generate the novel products and technologies and to successfully implement them in the market.

Moreover, the innovation projects must be effective and efficient. Effective means that a need is fulfilled. Efficiency means that the novel product and/or technology require minimal resources and that the innovation project is carried out with minimal effort, time, and cost. Even more importantly, the novel technology developed in the company must be able to be competitive with existing technologies in the markets.

These innovation projects can only achieve all these goals by having trained and experienced people skilled to execute innovation projects. Trained people are recruited from universities and have completed the appropriate programs and courses.

The focus of this book is, therefore, two-fold. It is written as a textbook for design courses and as a guideline for industrial innovators to generate better products and processes.

To be precise, the authors had the following goals and readers in mind:
- To educate student engineers for their future careers in designing new products, processes, and systems that address the needs of people, save the environment, and create prosperity locally and globally
- To inspire educators with novel methods in educating students in the art of design
- To inspire industrial researchers to invent, design, and develop novel product and processes
- To support start-up entrepreneurs in organizing their endeavors with the right mix of people in terms of education, experience, and behavior, through cooperation with companies and institutes in open innovation projects
- To assist managers of research and development departments in getting a good mix of people with specific skills and providing them with innovation methods that have been proven to work

https://doi.org/10.1515/9783110467741-001

- To help companies organize and communicate an integrated approach to R&D, design, manufacturing, and marketing of novel products and processes for successful launches
- To inform organizations who want to enter the market of process-product industries how these industries are structured and how their innovation is organized

This last point may need some clarification. The authors have noticed that what is going on in this set of industries is not known nor understood by people outside this set of industries. Even for people working in these industries but not directly involved in the actual manufacturing process, it is still hard to understand how they function.

There are several reasons for this lack of understanding:

(1) The first reason is that most people have never been in contact with these industries, because the manufacturing sites are at locations remote from cities and office buildings.
(2) A second reason is that these industries relate primary resources via long supply chains of intermediate products to the final consumer. People only see that final product, but not the intermediate products and their processes.
(3) A third reason is that these industries are very complex. They have many system levels.
(4) A fourth reason is that there are many industry branches related to processes and products. Often people within a certain branch are not aware of the existence of most other branches.

For companies wanting to enter this set of industries, for companies wanting to change their supply chain, and for people wanting to work in these industries, this book should also be useful. The focus of the book is, therefore, on methods, guidelines, and heuristics for concurrent novel products, novel process design, and the development process that takes an idea to the market in which business case development and experimental data generation are systematically integrated.

Because major and decisive effects on final products and processes occur in the concept stages, the most emphasis is on those stages. However, the other stages to commercial implementation are also treated.

The content of this book is partially derived from the tacit knowledge of the authors, shaped by experiences in design and innovation. This tacit knowledge has to the be transformed into coded knowledge using published structures on design steps and innovation stages and based on findings and categories in design knowledge as explained in Chapter 4.

The content of this book is validated as much as possible, where available scientific literature is used to back up the knowledge of the authors. In many parts, the information is made plausible by engineering reasoning. In several parts, the content is based on one or a few cases. For the education parts, the validation is that students using the theory were able to generate novel and practical designs.

1.2 Scope

The scope of this book provides general knowledge on generating novel products and their processes through a combination of designing and experimentation. The content of the book is limited to generic knowledge and methods of design and innovation.

The book does not provide detailed descriptions of specific process technologies or specific process steps such as unit operations. For that the reader is referred to other books, notably to *Process Technology* by de Haan [2], and for descriptions of specific processes, to the book by Moulijn [3].

1.2.1 Products classification and description

The class of products treated in this book is homogeneous down to the micron scale and to molecular scale. Typical products belonging to this class are: transport fuels (petrol, diesel, kerosene) chemicals, plastics, resins, medicines, healthcare products, detergents, food products, feed products, and ceramics.

A product is called a 'chemical product' when it is formed by molecular scale transformations by means of reactions and, possibly, by creating multiscale physical structures, such as dispersed or stacked thermodynamic phases. All such products are manufactured by the chemical process industry and related manufacturing sectors, such as food, energy, and electronic materials.

A special subclass is a set of products that are homogeneous down to the micron scale but become heterogeneous below this scale and have two or more phases. These are called 'structured products'. An example is mayonnaise, which is a stable dispersion of water in an oily phase. Another example is a heterogeneous catalyst. Designing these types of products requires special knowledge as described in the section on the design of structured products.

A special class of products are in fact microprocesses that operate on chemical engineering principles (reactions, separations, fluid flow) and have information as the primary output and not a chemical product or energy as such. The information concerns features of the feed material(s) being processed. Examples of such microprocess products are integrated labs-on-a-chip where the purpose is rapid screening of many samples and finding the ones with best performance or generating medical diagnostics from samples of body fluids.

The design of discrete products and devices such as cars do not belong to the core of this book, but many design and innovation methods provided in this book are also useful for designing such products and devices.

1.2.2 Description of process industries

Industries that produce homogeneous products are often called process industries. There are many process industry branches. Here is a list to get an impression: oil and gas (petroleum), chemicals, plastics, rubber, pharmaceuticals, food, beverages, ceramics, base metals, coal, textiles, coating and paint, paper and pulp, electricity (power), and (drinking) water.

Most process companies are large and have their own research and development, design department, manufacturing processes, marketing and sales and a head office with supporting departments such as legal and public affairs. Nowadays the research and development is combined with the design department into a technology center so that design works more closely with R&D and can be better integrated into the whole innovation trajectory. Apart from these manufacturing companies there are many other companies and institutes relevant to innovation in the field of process industries. Those are described in detail in Chapter 3.

1.3 Book structure

The structure of the book follows a systematic top down systems approach.

- Part A is about the what, why, and how of industrial innovation. It provides context for any innovation and design by a description of all system levels in nature, down to the smallest molecular scale. It also describes innovation structures, methods, and collaboration options with other organizations and elaborates on the innovation stage-gate approach with all major steps included. These stages are referred to throughout other parts of the book. This section also contains management guidelines for governing innovation portfolios and innovation projects.
- Part B is about the why and how of concept design for innovation. This section focuses on generating novel ideas and concepts. It provides, in a central chapter, a comprehensive method for concurrent product and process design for each innovation stage. This chapter will be particularly useful for experienced designers and innovators to generate, plan, and manage complex radical innovations. This section also contains other chapters that systematically guide students and less experienced designers from design scoping through to design execution. These chapters focus on discovery and concept stages.
- Part C is about how to optimize designs through modeling, evaluations, and reporting.
- Part D is about how to teach design for innovation. It describes established education engineering programs and courses at Delft University of Technology.

References and further reading

[1] Negro C, Garcia-Ochoa F, Tanguy P, Ferreira G, Thibault J, Yamamoto S, Gani R. Barcelona Declaration – 10th World Congress of Chemical Engineering 1–5 October 2017, Chemical Engineering Research and Design, 129, 2018, A1–A2.

[2] De Haan AB. Process Technology An Introduction, Berlin: De Gruyter, 2015.

[3] Moulijn JA, Makkee M, van Diepen AE. Chemical Process Technology, 2nd edn, Hoboken: J. Wiley, 2013.

2 Systems relevant to design for innovation

2.1 Introduction to all relevant system levels

Table 2.1 provides an overview of all system levels relevant to product-process innovations, as derived from Allenby [1] and Hartmann [2]. The wording of each level has been chosen such that it should be easy to understand for all involved in innovation.

Each system level is connected to the neighboring levels via mass and energy streams. Changes inside a system level thereby affect all other system levels. Products and their manufacturing processes are, therefore, not islands, but are connected to larger and smaller systems.

The purpose of this systems approach is to:

- Help innovators to get the big picture of how the object of innovation is connected to other systems
- Assist designers to include higher and lower system levels in the design scope
- Provide managers a framework to communicate with other parties and stakeholders involved or affected by the innovation. It shows for instance the connections between industry, society, and nature. This should help to generate innovative solutions that are attractive to society and have no negative impact on the environment.
- Facilitate effective and efficient communications, in particular by using block flow diagrams to represent the systems

Tab. 2.1: All system levels. Derived from Allenby [1] and Hartmann [2].

System Level	Size (m)	Description; example
Earth	40×10^6	World system of nature and society
Society	10×10^6	All human activities
Value chain	10×10^6	Cradle to grave chains
Industrial symbiosis network	40×10^3	Industry + societal
Industrial complex	10×10^3	Factories connected at industrial site
Factory	1×10^3	Processes at same location and owner
Process	400	Manufacturing set transforming feed to product
Process step	100	Subprocess to intermediate product
Unit operation	10	Generic rules defined process step; distillation
Main equipment	1	Mechanically defined artefact; pump
Characteristic subprocess	0.1	Defined exact geometric boundaries; tray
Microelement	10^{-3}	Elementary element; catalyst particle
Elementary	10^{-9}	Characteristic principle; chemical reaction

https://doi.org/10.1515/9783110467741-002

2.2 Earth system level

The earth system level and its main system elements are schematically represented in Figure 2.1. Mass and energy flows connect all these elements. The whole earth system is approximately closed in terms of mass.

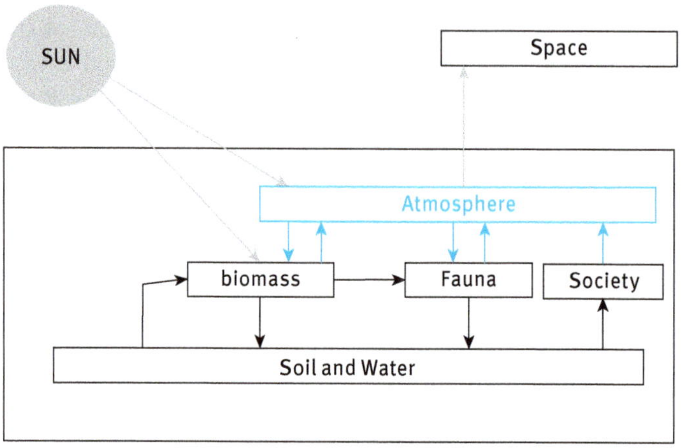

Fig. 2.1: Earth system: elements, energy and mass flows.

In terms of energy, it is in a dynamic near equilibrium. Sunlight energy, thermodynamically a high-quality energy input equivalent to a temperature of 6000 K, enters the system and is radiated back from the earth's surface (black mass radiation temperature) at 256 K. That radiation is absorbed by gases in the atmosphere, notably water vapor, carbon dioxide, methane, and other components. These in turn radiate heat back into space at a temperature of 288 K [3]. The difference between black body radiation from the earth surface and radiation from the earth's atmosphere of 32 K, is called global warming.

The system element of biomass stores a tiny fraction of the solar radiation by the photosynthesis of water and carbon dioxide into carbohydrates (and some other components). The system element of fauna converts carbohydrates back to carbon dioxide and water, while using the released high quality energy for muscle activities.

For the last ten thousand years, up to the industrial revolution in the nineteenth century, this system level has been dynamically stable with no significant accumulation of materials in any of the system elements [3].

This earth system level is an inspiration for engineers, as it shows that a nearly steady state on a global scale is obtainable for mass transformations driven by solar energy. This inspiration is an important part of design for industrial ecology and is described in Chapter 5.

Presently, due to human activities, represented in Figure 2.2 by the system "society," some system elements are no longer in a dynamic steady state. A major disturbance is the burning of fossil carbon fuels, by which carbon dioxide is accumulated in the atmosphere.

This accumulation increases the absorption of black body radiation, by which global warming increases. This in turn means that more water evaporates from the ocean, which condenses elsewhere, causing more wind and more rain and thereby changing climate behavior in various parts of the earth. These changes are summarized in the term "climate change" [3].

2.3 Society system level

Figure 2.2 shows the society system level and its system elements of culture, politics, and the technosphere.

The technosphere contains all human activities of creating and using technologies. This creation and use of technologies all cause mass and energy flows to and from the earth system.

Cultural and political system elements influence the technosphere via information flows and legislation. Major global actors include the United Nations (UN), climate change conferences, nongovernmental organizations (NGOs), the World Business Counsel for Sustainable Development (WBCSD), and universities world-wide.

Examples of these influences are the Paris Agreement on climate change, set up by the United Nations, through which 195 country governments agreed to take measures to reach zero net global warming gas emissions by 2050 [4] and the formulation of Sustainable Development Goals (SDG) by the UN and the WBCSD [5].

These global agreements are turned into policies by governments and by industries that stimulate innovation to create technologies that are embedded in the earth

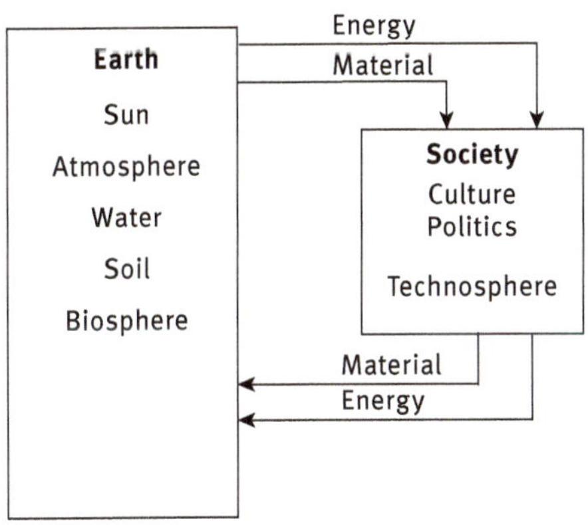

Fig. 2.2: Earth and society systems: their elements and connections.

system, and structure society in such a way that they keep these surrounding systems intact, while creating prosperity.

On lower society levels, stakeholders also interact with and influence the technosphere. Governments provide laws and budgets for research institutes and universities to generate knowledge and novel technologies. Governments in general also provide budgets for education, which in turn generate the next generation of skillful people to develop society.

Chapter 3 provides more information on important parties in society and their relations to innovation in the technosphere.

2.4 Technosphere system level

The technosphere level is a subsystem of society and is the collective activities of mankind creating and using technology. The technosphere is connected by material and energy flow to the earth system.

As an example, take the carbon cycle between the technosphere and nature. Large amounts of carbon are being extracted from the soil in the form of fossil resources, coal, crude oil, and natural gas, which are burned and put into the atmosphere in the form of carbon dioxide. This is a major cause of carbon dioxide concentration increase in the atmosphere, causing additional global warming.

Through innovation, increasingly renewable sources of energy are being developed, notably water, wind, solar, and biomass, in such a way that in the end the technosphere will be connected to nature via mass and energy in a steady state flow, hence,

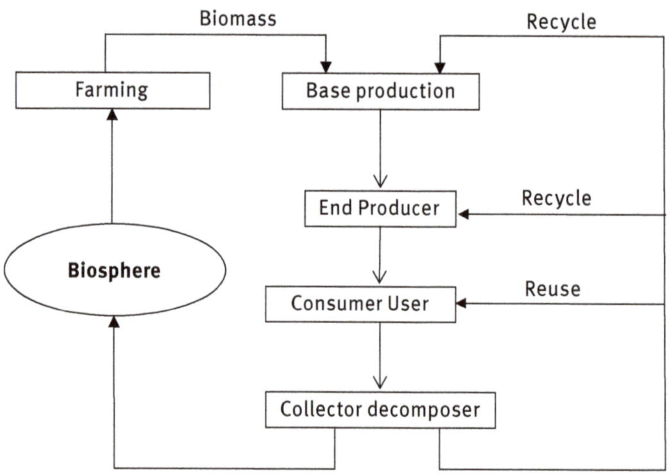

Fig. 2.3: Circular Economy mass flows of technosphere and biosphere. Derived from Ellen MacArthur Foundation.

with no accumulation of carbon in one subsection and depletion in a different subsection of nature. This is shown in Figure 2.3.

All major stakeholders involved in innovation in the technosphere are described in Chapter 3.

2.5 Value chain system level

The value chain system level is about the economic value increase of material flows from primary natural resources transformed into consumer products via industrial upgrading steps. Value chains are a subset of society and are connected to the technosphere via mass and energy flows. Value chains are also called supply chains, by companies further down the value chain.

A major current value chain for the chemical industry is shown in Figure 2.4. It starts from fossil sources of crude oil or natural gas and ends in consumer products.

A growing supply chain starts from biomass such as sugar beet, corn, or sugar cane, to glucose and then to food, plastics (polylactic acid), and also to pharmaceutical products, as indicated in Figure 2.4.

This growing supply chain from biomass shows that in innovation not only fossil resource based value chains have to be considered but also these renewable value chains. Through breakthrough innovation, more products are likely to be produced from renewable resources at prices competing with products from fossil sources.

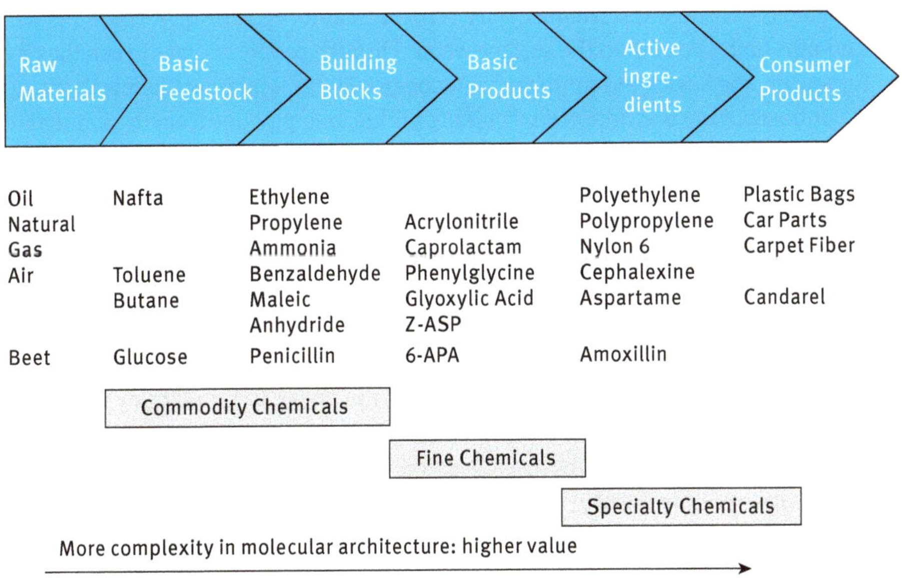

Fig. 2.4: Chemical industry value chain.

Chapter 4 shows how in product and process design driving innovation, novel value chains are taken into account.

2.6 Industrial symbiosis system level

The industrial symbiosis system level is about connections between process factories and other society elements such as cities. District heating by industrial waste heat is an example of industrial symbiosis.

It is also the system level where different engineering disciplines meet. Sustainable housing design for example, means that architects and process engineers must interact to get an integrally optimized housing system where washing, waste reuse, heating and lighting and the use of sunlight as energy source are to be combined. Industrial symbiosis is, therefore, a level in which innovation can flourish. A good example showing the power of industrial symbiosis thinking in saving cost and reducing the environmental impact is the municipal waste water conversion to boiler steam water for Dow Chemical in Terneuzen, the Netherlands.

For decades, Dow Chemical took salty surface water from the Westerschelde and converted that into boiler feed water. The most energy intensive and costly process step was the desalting using ion-exchange resin beds. The central Dow organization embraced industrial symbiosis as a useful method. A methodology was developed and local factories were invited to come up with industrial symbiosis solutions.

Dow Chemical Terneuzen picked up the idea by looking around for options and identifying the waste water stream of city Terneuzen as an opportunity, since this water contains far less salt than the surface water. Through cooperation between the city of Terneuzen, the Zeeuws-Vlaanderen water board, Elides, and Dow Chemical, the regional infrastructure was adapted by making a 6-km long pipeline from the municipal waste water treatment plant to the Evides pumping station. The Reverse Osmosis (RO) plant for treating brackish estuary water was partly changed to biologically-treated waste water. The implemented solution meant the reuse of 2.5 million km^3 of municipal waste water; rather than treating this as an effluent, the waste water replaced the surface water, causing 65% in energy savings, and a reduction of chemical intake for cleaning the RO membranes. In 2007 the novel solution obtained the EU Responsible Care Award [6]. This new industrial symbiosis system is illustrated in Figure 2.5.

This industrial symbiosis case demonstrates the power of thinking in more than one system level. Only by going to a higher system level than the industrial complex could the solution be found. It is also an example of social innovation. Chapter 3 also describes various other types of innovation. Chapter 5 shows a design approach based on industrial symbiosis.

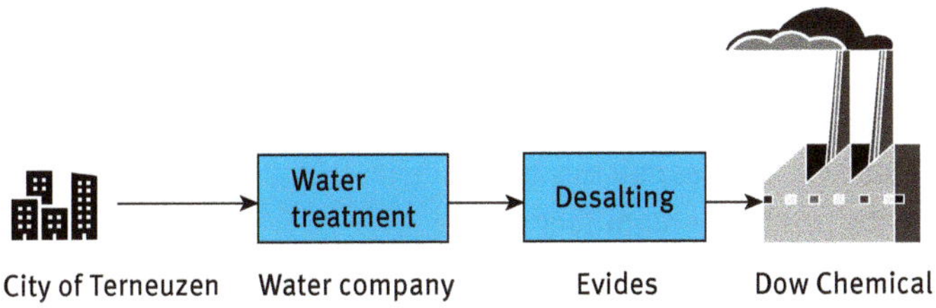

Fig. 2.5: Turning municipal waste water into industrial boiler feed water: industrial symbiosis by the city of Terneuzen and Dow Chemical – winner of the European Responsible Care Award [6].

2.7 Industrial complex system level

An industrial complex contains many factories connected by networks of many different mass and energy flows. Often certain shared utilities such as steam generation and waste water treatment are common. In Germany, these complexes are often called Verbund.

Designing a process to be incorporated in an existing industrial complex, is called designing for a brown field, while designing a process as a standalone with no common utilities is called designing for a green field. Designing for a brown field situation has many advantages as the major facilities such as utilities and infrastructures for all kinds of energy such as electricity, steam, base chemicals, and waste water treatment are available.

Designing a process for a so-called green field situation, so for a location with nothing there, means that the infrastructure must also be designed. This means that a concept for the whole complex has to be generated. This on the other hand also provides many opportunities for designing a very efficient and environmentally friendly complex, where industrial symbiosis level design thinking is applied.

2.8 Factory system level

A factory consists of one or more manufacturing processes of the same owner at the same location. It also includes storage facilities for feeds and products. In most cases it is a subsystem of an industrial complex. It converts input materials into higher value output products. Other outputs such as byproducts and energy may be sent to other parts of the industrial complex.

In designing a factory, it is always beneficial to take the higher system levels into account. Often facilities can be shared and byproducts used in other factories of the industrial complex.

2.9 Process system level

A process is part of a factory. It is a set of process steps to transform input materials into a market product. Some companies call this a (chemical) plant. Often the process (or plant) is named after the product it manufactures.

Process design is one of the core topics dealt with in this book; notably in Chapter 4–13.

2.10 Product system level

The term 'product' means in general something that is manufactured and sold to customers. If it is sold to industrial customers, it is often called an intermediate product. If it is sold to consumers, it is called a consumer product. This book focuses on homogeneous products (see Chapter 1). These are for instance intermediate chemicals, such as solvents and olefins, and consumer products such as food products, detergents, and plastics. Chapter 1 provides examples of product types from various industry branches.

The majority of these products have a special function in consumer use. Most of these products react chemically or physically when used. Examples are: food products that react inside humans and animals to other components that are used to grow and maintain a healthy body, as well as detergents that interact with dirt and clothing during washing, such that the dirt is separated from the clothing.

Because of these special functional purposes novel product design is a complex act, as the behavior of the client with the novel product and the product behavior have to be considered. The product design is even more complex, as design of the manufacturing process belonging to the product also has to be taken into account, and the higher system levels, such as the supply chain, have to be considered. Chapter 4 provides an elaborate method for doing so.

2.11 Process step system level

The process step system level needs some explanation. It is a subsystem of a process. It is a cluster of process unit operations that perform a desired action. Examples are feedstock pretreatment, chemical conversion, separation, product purification, and product packaging.

This system level is very useful when designing a large novel process or an industrial complex with several large processes. It keeps the representation of the whole design simple with a limited number of block flow diagrams, in which each block is a process step. Connection changes between process steps, for instance, are easily made. Communicating the whole design is then also easy.

2.12 Unit operation system level

A unit operation is a subprocess step characterized by the same physical and chemical rules as for any application. Thus, the same unit operation can be applied to many different applications using the same rules. This system level was defined by Arthur D. Little in 1916 [7]. Its definition resulted in enormously more efficient process designs because it allowed for a) educating students about unit operations applicable to any process design, b) improving the generic rules through research, and c) improving actual process designs by applying rules (prior to this unit operation invention, only field-specific technologies existed such as sugar technology, coal technology, crude oil technology, etc. Knowledge generated was only applicable for a particular field).

Process design based on unit operations still dominates the design method of chemical processes. New unit operations such as membrane separation appear and find their way into many applications. Design based on function identification and function integration is a novel process design method that facilitates breakthrough process designs. Both ways of process design are treated in this book, in particular in Chapter 4–10.

2.13 Main equipment system level

Main equipment denotes all individual major apparatuses required for the processes. A unit operation distillation for instance will consist of the following main pieces of equipment: a column with internals, a top condenser with a heat exchanger, a top vessel to separate vapor from liquid and to act as a holding tank, a bottom boiler heat exchanger, and a pump to feed liquid reflux back to the column. Disciplines involved are mechanical engineering, electrical engineering, and materials engineering.

2.14 Characteristic process subsystem level

A characteristic process subsystem is a subsection of a unit operation or main equipment. It is characterized by a few well described physical and/or chemical phenomena and clear boundary conditions. A characteristic process subsystem, with well-defined geometric boundaries, is a tray in a distillation column, or a static mixer element in a static mixer. In general it is a very useful entity in design, because it allows for the dedicated modeling of the subunit and connects it to all subunits. This type of modeling is then often standardized as modules. The modules of the process subsystems are then connected by simple energy and mass flows. Computer modeling packages for these process subsystems and their connections that form integral models are for sale.

2.15 Microelement system level

A microelement, such as a catalyst particle or a droplet in a liquid extraction unit, is governed by boundary conditions (such as interphase equilibrium), basic heat transport, and mass transport rules. The innovation on this level can involve catalysis, microbiology, interface chemistry, particle technology, and chemical engineering reaction engineering.

The conditions at the interface are part of microelement modeling. In this way the microelement can be studied separately from the hydrodynamics of the fluids surrounding the microelement. It can, therefore, be studied prior to the selection of the process equipment using dedicated laboratory equipment with controlled uniform conditions outside the microelement.

2.16 Elementary system level

The elementary system level is uniform over its scale and governed by basic physical and chemical properties and chemical reactions. All these properties and reactions are independent of the equipment type and scale, and hence can be studied separately from scale dependent phenomena, by experts in chemistry, kinetics, and physical properties.

References and further reading

[1] Allenby B. Industrial Ecology: Policy Framework and Implementation, Saddle River New Jersey: Prentice Hall Inc., 1999.
[2] Hartmann K. Analysis and Synthesis of Chemical Process Systems, Hoboken: Elsevier, 2013.
[3] Boeker E, an Grondelle R. Environmental physics, 2nd edn, Chicester: J. Wiley, 1995.
[4] UN. Paris Agreement, United Nations Treaty Collection, 8 July 2016, Sourced from: https://treaties.un.org/pages/ViewDetails.aspx?src=TREATY&mtdsg_no=XXVII-7-d&chapter=27&clang=_en. Last access date: 10 January 2018.
[5] UN. Transforming our world: the 2030 Agenda for Sustainable Development, Resolution adopted by the General Assembly on 25 September 2015, accessed on 5 May 2017. Sourced from: http://www.un.org/sustainabledevelopment/sustainable-development-goals/. Last access date: 10 January, 2018.
[6] Wu Q. Industrial Ecosystem principles in industrial symbiosis: by-product synergy, in Harmsen J, Powel J (eds), Sustainable Development in the process Industries, Cases and Impact, Hoboken: J. Wiley, 2010.
[7] Little AD, et al. Chemistry in history, Chemical Heritage Foundation, Retrieved 13 November 2013. Sourced from https://en.wikipedia.org/w/index.php?title=Unit_operation&oldid=709344969. Last access date: 15 January 2018.
[8] De Haan AB. Process Technology: An Introduction, Berlin: Walter de Gruyter GmbH, 2015.
[9] Moulijn J, et al. Chemical Process Technology, 2nd edn, Hoboken: J. Wiley, 2013.

3 Managing innovation

3.1 Overview

3.1.1 Innovation terms

The word innovation in this book is used in three ways. First, it means the successful introduction of a new product or process into the market. Second, it means having management methods and systems in place to always have a consistent portfolio of innovation projects in progress. This is called innovation portfolio management. Third, it means the management of individual innovation projects into successful market introductions. This is called innovation project management. The overall meaning of this chapter is to show that innovation pays off and is even more beneficial to companies when managed by proven portfolio and project methods.

The historic business trends in product and process innovations leading to concurrent product-process innovation, including the supply chain and the product market, are shown in Section 3.2. This section also provides business motives for innovation.

Various types of innovation are described in Section 3.3. Innovation partners of industrial innovation are described in Section 3.4. Portfolio management of innovation inside enterprises is described in Section 3.5. The remaining sections all deal with project management in the innovation stages. An overview of stages and key activities is shown in the next section.

This chapter treats the management of innovation in a limited way. Far more information on innovation management is found in dedicated textbooks such as those written by Goffin [1] and Cooper [2]

3.1.2 Stage-gate approach

Innovation project management, as described in this book, is based on a stage-gate approach. Table 3.1 shows the names of the stages and some stage characteristics. The same stages may have different names in companies, but the activities during each stage will be nearly the same and the sequence of stages will be similar. Gates are placed in between the stages. At these gates, the decision is taken to stop or to continue the project.

This stage-gate approach ensures that with minimum effort in the early stages, only the most critical aspects are investigated so that a project can be stopped timely. A full rationale of the stage-gate method is provided in Section 3.6. Sections 3.7–3.12 provide descriptions of the stages.

https://doi.org/10.1515/9783110467741-003

Tab. 3.1: Overview of Innovation Stages.

	Innovation Stages					
	Discovery	Concept	Feasibility	Develop-ment	Detailed engineering	Implementation
Purpose	Idea generation	Screening	Concept feasibility	Validation	Engineering, procurement, construction	Product launch, process commissioning, and start up
Invested %	0.01	0.1	2	10	94	100
Stage-gate decision	Strategy fit	Business case	Business case	Business case	Top mgmt.	
Duration (years)	0.3–1	1–2	1	1–5	1–2	0.1–3
Design scope %	1	10	50	90	100	
Experimental	Proof of principle	Proof of concept	Product Prototype	Pilot plant	–	Start up

3.1.2.1 Short description of stage characteristics

The discovery stage comprises the generation of ideas. In some companies this is also called the ideation stage. Very little design is carried out. The novel product and the process will be sketched. The investment in a project will be a very small percentage of the overall investment. The main critical question to pass the gate is: Does the idea fit the business strategy? Successful ideas that pass the gate to the concept stage will be followed up by design and experimentation to such a level that a crude estimate of the commercial scale investment cost can be made. In the feasibility stage a commercial scale design will be made by process engineers, in which the scope of the design will be 50% of the total scope. This type of design is called front-end loading 1 (FEL-1) in America and Asia and may be called "50% concept design" or "50% concept engineering" or, in Europe, "50% base engineering."

The investment estimate will be based on the design inside the scope and on an estimate of the elements outside the design scope using correlations. The design and cost of a downscaled pilot plant will also be part of the feasibility stage. If the business case is attractive enough, a decision to go into the development stage will be taken.

In the development stage new product testing and pilot plant testing will be carried out, followed by a commercial scale design based on the product testing and the pilot plant testing. The design scope will now cover 90% of all items of the commercial scale plant. The investment cost estimate has thereby an uncertainty of maximally 10 percent, but is likely to be only a few percent. This is called front-end loading 3 (FEL-3). For very large investments a design with a scope of 80% definition will be made, as this requires less design cost to be made and is still accurate enough to take a decision to continue or not). This is called front-end loading 2 (FEL-2).

If the decision is made in favor of the commercial scale, an Engineering Procurement Construction (EPC) contractor will be asked to make a detailed engineering design, and execute the procurement of equipment and construction of the process. When the construction is finished, the new plant will be commissioned and will be launched in start up. The new product will then be launched into the market.

The duration of each stage strongly depends on the project innovation class (see Section 3.3). Breakthrough innovations of Class 6 can take up to 19 years. Class 0 may take 1 year. The figures provided are indications only.

3.2 Business focus and motives for innovation

3.2.1 Business focus trends in product and process innovation

The trends in business focus on chemical product and process innovation of the last 100 years are summarized in Table 3.2. The information is from Arora [3].

In the period between 1920–1950, the innovation focus was on novel products such as plastics. Process design became more systematic by the introduction of the concept of unit operations. Between 1950–1980, the focus shifted to far larger processes to improve efficiency and reduce investment cost per ton of product.

In the next two periods of 1950–1980 and 1980–2015, global competition and concerns about the environment (acid rain, ozone layer depletion, and global climate change) induced process innovation with process intensification as the major leading principle. Computer modeling greatly facilitated this innovation [4]. This is also the reason that Chapter 8 and 9 on product and process modeling are included in this book.

Tab. 3.2: History of product and process innovations.

Period:	1920–1950	1950–1980	1980–2015
Business Driver	Novel products	Global scale-up	Global competition Global environment
Design	Unit operations	Unit operations	Function integration
Methods	Monographs	Phenomena models	Integrated models Simulations
Scale-up	Empirical	Pilot plants	Pilot plant validation

Source Business Driver: Ashish Arora, Implications for energy innovation from the chemical industry, NBER, 2010.

In all past design methods, product design and process design were carried out separately. In some companies, their research and development were also carried out separately. The product was developed under the assumption that the process would consist of available standard unit operations and that no interaction between product development and process development was needed. The textbook *Product and Process Design Principles* by Seider, shows how these designs can be made [5].

Recently a simultaneous product and process design method based on simulation and optimization has been proposed. However, the method requires a complete mathematical description of all physical and chemical phenomena of the product (and of potential processes), so can only be used in limited cases [6].

Almeida and others developed a so-called anticipating reciprocal approach to product and process design, which facilitates optimum combined design of product and process. His method is the base for our book as found in Chapter 4, 6, 7, 8, and 9. The advantages are briefly [6]:

(1) It incorporates a larger solution space than sequential approaches, by which more optimum combined solutions of product and process design can be found.
(2) It allows for qualitative and quantitative design approaches.
(3) It has been successfully applied in industry.

3.2.2 Business motives for innovation

Companies in general differ in their motivation to innovate. The sections hereafter describe most encountered innovation drivers.

3.2.2.1 Competition as innovation driver

Arthur D. Little carried out a global innovation benchmark study of many companies, covering both processes and consumer products. The bench study revealed that innovation is regarded by all companies as the major pillar for staying in business. For most companies, innovation is regarded as the major pillar for sustaining competitive advantage and the long term increase of company value [7].

Top innovators actively try to open up new business areas and product developments are significantly important to them. The following statistics of these top innovators indicate their importance [7].

Top innovators:

- Realize up to 2× higher sales from new products/services
- Have up to 2× higher earnings from new products/services
- Have half the time for the breaking even of ratio of earning to innovation cost

What happens if companies lag behind in innovation? That is indicated by the statistical results of a large worldwide benchmark study on energy efficiency of chemical

Tab. 3.3: Energy efficiency ratio of processes in various world sectors [9].

World sector	Average Energy Efficiency Ratio %
Western Europe	176
North America	207
Rest of the world	224
Asia	230

processes by PDC. They gathered data of different processes in commercial operation for the same product, and did so for 100 different chemical products. The results are shown in Table 3.3 [8]. For each product, the most efficient process is set at 100% energy efficiency. Each energy consumption per ton of product (for the same product) is calculated and divided by the figure for the most efficient process and expressed as a percentage called the Energy Efficiency Ratio (EER). The average EER per world sector has also been calculated and is shown in Table 3.3. It shows that the average EER is a factor 2 times higher than the best processes. It also means that a lot of processes are even less energy efficient. These far less efficient processes thus lag in innovation and will struggle to survive in a competitive world.

3.2.2.2 Learning curve as innovation driver

Learning curves (also called experience curves, or Boston learning curves) are plots of the market price of a product (vertical axis) plotted against all cumulative sales of that product (horizontal axis) and taking for both axes the logarithmic value, so taking a log-log plot. In most cases a down sloping straight line is obtained and a slope is determined. These learning curves thus quantify the rate at which the market price decreases with increasing cumulative sales [9].

These experience curves often hold for many decades and become a self-fulfilling prophecy as companies keep innovating to reduce the cost, knowing that the competition also does so. A good example of such a learning curve is the case of the photovoltaic solar cell, as provided by Swanson. It has been determined from 1979 until 2005 and shows a cost reduction of 20% per doubling of installed capacity over that period [10].

The beginning of a learning curve is when commercial scale production starts and the product enters the market. The time prior to that beginning can be very long. The basic principle of a solar cell has been known since the 18th century. The breakthrough technology for commercial production arrived in the seventies with the silicon photovoltaic technology of both the product and its required manufacturing process. That technology breakthrough resulted in a breakthrough innovation. Commercial scale implementation was then enormously successful with a rapid increase of installed solar cells. This success in turn caused a strong drive to reduce the manufacturing cost. This cost reduction follows a learning experience curve of 20% cost reduction for every doubling of installed capacity, price expressed in ($/w) from 1979 until the present, called Swanson's law [10].

The main implication of innovation for products for which learning curves are available is that the innovation must aim at the moving future cost target. A book about learning curves has been written by Jaber [9].

3.2.2.3 Circular economy as innovation driver

Contributing towards a circular economy is for many companies an innovation driver. It means that products and processes are invented that fit the circular economy concept. Chapter 5 contains a section on the definition of a circular economy.

The World Business Counsel for Sustainable Development (WBCSD) provides the following five innovative business models for this purpose [11]:

- Circular supplies: meaning that all inputs are biobased or fully recyclable and that only renewable energy is used for driving the processes.
- Resource recovery: meaning that useful resources are recovered from used materials, byproducts, and waste.
- Product life extension: meaning that product lifecycles are extended by repairing, upgrading, and reselling, as well as through innovative product redesign.
- Sharing platform: meaning that product users are connected to one another and are encouraged to share its use and access or have ownership to increase product use.
- Product as a service: Meaning moving away from product ownership and offering customers paid access to products, allowing companies to retain the benefits of circular resource productivity or ownership to increase product use

The WBCSD also provides three disruptive technologies facilitating circular economy [11]:

- DIGITAL TECHNOLOGIES such as Internet of Things (IoT), big data, blockchain, and RFID help companies track resources and monitor utilization and waste capacity.
- PHYSICAL TECHNOLOGIES such as 3D printing, robotics, energy storage and harvesting, modular design technology, and nanotechnology help companies reduce production and material costs and reduce environmental impact.
- BIOLOGICAL TECHNOLOGIES such as bio-energy, biobased materials, biocatalysis, hydroponics and aeroponics help companies move away from fossil based energy sources.

3.2.2.4 Sustainable development as innovation driver

The UN general assembly of September 25th, 2015, defined 17 specific measurable targets, named the Sustainable Development Goals (SDGs), to be achieved by 2030. These are listed in Chapter 13.

Chakravorti mentions three reasons for adopting the sustainable development goals of the UN for business [12]. "First, the global goals campaign represents a significant new opportunity for companies that view emerging and frontier markets as their source of long-term growth. According to estimates from McKinsey, consumers in these markets could be worth $30 trillion by 2025, a significant step up from the 2010 value of $12 trillion. Since 2011, as emerging markets have suffered from slower growth and fresh social unrest, that $30 trillion prize seems more distant. Acting on the global goals could help address several of these obstacles that are giving rise to "trapped value" in the emerging markets.

Second, with the public declarations by many companies to help with the goals, there is likely to be competitive pressure. Some companies could get a jumpstart in their industry in organizing partnerships and even positioning themselves as leaders in sustainable development using the goals as a branding anchor. Being slow to act could lead to the risk of being left out of these relationships. It then could form a competitive disadvantage from a brand equity perspective.

Third, the goals cannot be realized without business participation. The price tag for accomplishing these global goals is estimated to be up to $3 trillion a year, for 15 years. For most governments, financing the global goals campaigns will be a stretch; governments have already committed in the past for similar targets."

There are special links between the business driver for sustainable development and value chains. For the environmental impact of products, life cycle assessment is now common practice. The value chain is a subpart of the life cycle chain, so inevitably other companies of the value chain must be involved to reduce the environmental impact. For renewables such as feedstock even completely new value chains must be established.

For the social dimension of sustainable development, partners along the value chain have to also be involved. This is the case when the early part of the value chain is in developing countries. A major part of the value should be created in those countries and negative impacts, such as child labor, must be avoided.

3.2.2.5 Corporate social responsibility as innovation driver

In the business realm, sustainable development is often called Corporate Social Responsibility (CSR). It is founded on the United Nations description of sustainable development with environmental, social, and governance (ESG) dimensions. One motive for firms to endorse a CSR driven strategy is to obtain capital at a lower cost.

In a large and recent empirical research on firms' abilities to access finance in financial markets, Cheng proves that companies with some better CSR performance face indeed lower capital constraints.

Cheng investigated data from 2,439 public listed firms during the period 2002 to 2008. As a measure for CSR performance he used a panel dataset with environmental, social, and governance (ESG) performance scores obtained from Thomson Reuters

Tab. 3.4: Corporate social responsibility ASSET4 performance information structure [14].

Economic	Environmental	Social	Corporate governance
Client loyalty	Resource reduction	Employment quality	Board structure
Performance	Emission reduction	Health & safety	Compensation policy
Shareholder loyalty	Product innovation	Training & development	Board functions

ASSET4; a Swiss-based company that specializes in providing objective, relevant, auditable, and systematic ESG information and investment analysis tools to professional investors, who built their portfolios by integrating ESG (nonfinancial) data into their traditional investment analysis. It is estimated that investors representing more than €2.5 trillion assets under management use the ASSET4 data, including prominent investment houses such as BlackRock [13].

The CSR measurements of Thomson Reuters ASSET4 involve 220 criteria grouped under four pillars each with main categories as shown in Table 3.4 [14].

Corporate Social Responsibility (CSR) is a process for companies to integrate social, governance, environmental, and supply chain sustainability into operations and corporate strategy. This subject is beyond the scope of the book. However, assessing a novel technology on contributions to sustainable development is a necessary step in incorporating the technology in the CSR strategy of the company.

3.2.2.6 World problems as driver for the product-process industries

Specifically, for the product-process industries the results of the Delft skyline debates on world problems and innovation technology area are of interest. In the Delft Skyline Debates, 75 scientists and engineers from several disciplines, but the majority

Tab. 3.5: World scale problems and innovation areas [15].

Subject	Aspects
World scale problems:	
Shortage	Water, food, energy materials
Pollution	Air, water, soil
Human Health	Diseases
Innovation areas	Food, health, living, transport
Constraints	No waste, renewables, solar based
Technology Areas:	
Solar cells and electricity storage	Small scale, local
Fuel cells + manufacturing	Small scale, local
Molecular controlled reactions	Personalized medicines, tissues, organs
Membranes	Safe affordable water
Process Intensified biomass conversions	Small scale, local

from chemistry and chemical engineering, from all over Europe, discussed world scale problems and innovation areas for solutions [15]. The results are summarized in Table 3.5.

The purpose of this table is to stimulate thinking on solving these world scale problems. Focusing on innovation in these areas for both products and process innovation is likely to provide extra business in the future and will also contribute to the Sustainable Development Goals.

3.3 Innovation classes and types

3.3.1 Innovation classes

There are many classifications for innovations along similar definitions. Verloop [16] provides a classification originally made by Gaynor with the terms, breakthrough, new, and incremental. McKinsey includes (lack of) knowledge of the market as an additional classification element [17]. We add as an additional parameter, a new value chain up front for the product (and process), as a new feed and a new supplier have additional risks. Combining all these elements led us to the innovation class definitions of Table 3.6.

The data of the last three columns of Table 3.6 are obtained from McKinsey [13]. They analyzed 118 chemical company innovation results. Their definition of time to market is from the formal project initiation, which is likely to be the same as the start of the development stage, to the time where the cumulative product sales equal the R&D cost. Their success rate is defined as the portion of projects in an innovation class

Tab. 3.6: Innovation classification for combinations of markets, supply chains, products, and processes.

Innovation class	Market	Product	Supply chain	Process	Time to market years	Success %	Extra Margin %
6 Breakthrough radically new	New	New	New	New	8–19	15–20	0–60
5 Radically new	Existing	New	New	New			
4 New	Existing	Modified	Modified	New	6–15	30–40	0–10
4 New	Existing	Existing	New	New			
3 New	Existing	Existing	Existing	New			
3 New	Existing	Modified	Existing	Modified			
2 Novel	Existing	Existing	Existing	Modified	2–5	40–50	0–5
1 Conventional	Existing	Existing	Existing	Increased capacity			
0 Conventional	Existing	Existing	Existing	Existing			

that created positive returns on investment, expressed as net present value, using the cost of capital with no risk adjustment.

Here are short descriptions of each innovation class.

Innovation class 6

'Breakthrough radically new' involves a new market, a new product, a new supply chain, and a new process. The new market may be nonexistent or new to the company. For homogeneous new products, a market nearly always exists, so an even higher innovation class was not defined. With the new product, a radically new product is meant, which also involves a new supply chain, as new ingredients are needed, and of course the process is new.

Examples are: Lab-on-a-chip products for medical personal diagnostics and large household sized batteries for balancing local solar energy production (i.e., Tesla 'Powerwall').

Innovation class 5

'Radically new' involves an existing market known to the company, but the product, supply chain, and process is new. So apart from existing market knowledge about the product, it has the same degree of novelty because of the combination of the new product, new supply chain, and new process. The time to market, success rate, and extra margin will be in between Classes 6 and 4.

Examples are:

- Hydrogen as transportation fuel (next to diesel and gasoline)
- Fischer–Tropsch produced gasoline and diesel, with superior fuel efficiency, and no sulfur components emissions
- Biobased soft drink bottles with improved barrier properties (PEF)
- Water based paints (solvent-free)

Innovation class 4

'New' involves a modified product, a modified supply chain, and a new process. A modified product means a product that has nearly the same application with a change in relative amounts of ingredients or one or two new ingredients. In product innovation, this sometimes is called a platform derivative. This class also includes an existing product for a new supply chain and a new process.

Examples are:

- Bio-ethanol containing gasoline (bio-ethanol process and supply chain is new)
- Biobased polyethylene (monomer sourcing part modified)
- Low-VOC water based paints (improved postreaction or VOC removal processing)

- Ice cream with improved properties (caloric value and sensory properties; flavor delivery and mouth feel; improved product formulation and structure through new processing)

Innovation class 3

involves a radically new process for an existing product.

A radically novel process means that novel principles for mass movement, reaction, separation, and heat exchange are applied. These principles are also combined in novel ways in novel equipment.

A past example is the Fluid Catalytic Cracking Process that converted refinery fractions into gasoline. The process, developed during World War II, is entirely based on gas-solid fluidization in catalytic cracking and in coke combustion, while the heat exchange is carried out by the solids flow. It allowed very large production capacities at very low investment and processing cost.

At that time fluidization was not practiced in the process industries and for important elements of the process such as the standpipe between the combustor and the catalytic cracker riser, precise knowledge of the flow phenomena was not known. This standpipe has two functions: a) fast, reliable catalyst transport from the combustor to the cracker by gravity and b) a seal for gasoline vapor to prevent flow to the combustor. Due to the enormous need for gasoline, the oil company (Exxon) took the risk and constructed the commercial process with enormous commercial scale success. The process is still in operation in refineries [18].

A more recent example is reactive distillation. It combines reaction and distillation in a single column. Eastman Chemical wanted to replace a conventional design with 11 unit operations with a single column combining extractive stripping and absorption with reaction. Design knowledge for these combined functions was minimal and it required several pilot plants before the basic principles were fully understood. The commercial scale process was an enormous success, with reduced investment and operation cost by a factor of 5, while increasing the production safety, health, environment, and reliability [19].

This type of innovation requires an enormous design effort, not only in creativity, but also in developing new design methods, modeling, and simulation tools.

Innovation class 2

Involves a modified process as only novel aspect. It contains no new processing functions in relation to the product manufacturing, but only fringe novelties, such as a heat integration with additional heat exchangers; improved process control; optimized operating conditions; alternative utility sources; and retrofitting a heat exchanger network. It requires on average 2–5 years for commercialization [17]. The average success rate of only 40–50% may come as surprise. Section 3.5.5 provides the reasons for this low success rate.

Innovation class 1

Involves a process with a higher capacity than previous commercial scale processes already in operation. Examples are capacity increase by expansion or incremental debottlenecking, or a new plant with the same technology with existing processes for the same product.

A word of warning is needed here. Often a process is considered conventional when the applied unit operations are conventional, and the project is placed in this category. However, if those unit operations have not been applied for the particular product or streams, that process is new and should be treated as new. It still may need a pilot plant to remove major uncertainties and reduce the risks to an acceptable level. Section 3.10 describes this risk of not considering a process as new.

Innovation class 0

Involves a carbon copy of an existing process in commercial operation. It may come as a surprise that projects in this class still sometimes fail. Section 3.5.5 of this book reports the reasons for the large percentage of failure even if the project involves no apparent novelty.

Overall remarks

Not all possible combinations are covered in these classifications. The reader, however, can easily place his project in an innovation class, by interpolation using the degree of novelty for the following parameters: market, product, supply chain, and process.

3.3.2 Innovation by serendipity

The New Oxford Dictionary of English defines serendipity as the occurrence and development of events by chance in a satisfactory or beneficial way, understanding chance as any event that takes place in the absence of any obvious project (randomly or accidentally) that is not relevant to any present need, or in which the cause is unknown.

Innovations presented as examples of serendipity have an important characteristic: they were made by individuals able to "see bridges where others saw holes" and connect events creatively, based on the perception of a significant link. Walpore [20] defined the term as: "Serendipity is finding by accident and by sharpness of mind something not looked for." So, it combines an accident and cleverness.

A recent example of serendipity is the discovery by my former Shell colleague Einte Drent, of a novel polyketone polymer. He was trying to synthesize solvents from synthesis gas. He saw in one of his experiments that white flocks had been formed. Rather than discarding the result as a failure, he investigated the chemical composition of the flock and found that it was a polyketone. This novel product was further developed and appeared to be an engineering polymer with several properties superior to nylon [21].

Companies who want to be innovative foster a climate and have employees by which serendipity is stimulated. The results of serendipity innovations will, in the end, fall in the categories of the previous section.

3.3.3 Social innovations

In some innovations, no new technology is involved and yet they can be called innovations. Some can be called social innovations. These are novel combinations of stakeholders, such as described in Chapter 2 for system level industrial symbiosis. The case is cooperation between the city of Vermeulen, the waterboard, and Dow Chemical to take municipal waste water and use that as feedstock for boiler feed water production.

These social innovations require a lot of communication and discussions, but involve low innovation project costs and low investment costs. The benefits are often large economically, ecologically, and socially.

3.4 Innovation partners

3.4.1 In-house versus open innovation

Some companies perform all innovation inside their organization. This is called in-house innovation. Others perform innovation with partners. This is called open innovation. The advantages and disadvantages are summarized in Table 3.7

Tab. 3.7: Advantages and disadvantages of in-house and open innovation.

Parameter	In-house innovation	Open innovation
Speed	Rapid	Slow; takes extra contracts and organization
IP	Intellectual property easily protected	IP protection needs negotiation and attention
Quality	Can be limited by narrow view	Can be high due to outside expert knowledge
Cost	High	Low by sharing cost between partners
When	Strong in-house knowledge base	Breakthrough concept knowledge is outside

When more than one company is involved in an innovation project then that innovation project is called an open innovation project. If several projects with a common theme are clustered and developed with the same partners then this may be called an innovation platform. The advantages of these latter types of collaborations are that the best partners for each knowledge element can be selected and the costs are shared. The disadvantages are that:

(a) Intellectual property contracts should be made and signed. Often these take a considerable time, before the actual project can start and
(b) Organizing and leading the project is more complex. Each partner has his or her own agenda and interest. To define a common goal and stay focused on this goal takes effort and requires a project leader experienced in open innovation.

3.4.2 Innovation partners

Potential innovation partners are depicted in Figure 3.1 together with main information flows.

Each potential partner type of an open innovation project is briefly described in the subsequent sections.

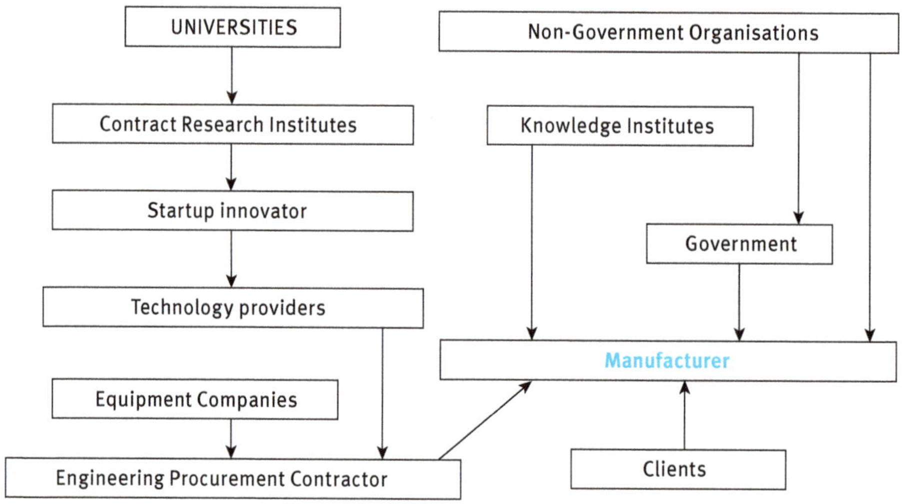

Fig. 3.1: Innovation stakeholders and connections.

3.4.3 Universities

Universities produce educated students and research results. University education programs focused on homogeneous product and process education and research are first of all technical universities and polytechnics. They have in most cases a chemical engineering department and/or a biotechnology department and/or a food technology department. General universities also have in some cases chemical engineering, biotechnology, and food processing departments. Mining departments also may have a section on ore processing.

University education programs are in general for BSc and MSc students. Some universities also have programs for PDEng students. For PhD research students, university departments often also offer education programs. Chapter 14 illustrates the content options for designing each of these education programs.

University research characteristics include a focus on the new elements of a technology and not on a complete integral solution, and on describing and publishing experimental results. The research is in general carried out by PhD students with no technical experience. The research papers and the PhD thesis in most cases have a Technology Readiness Level (TRL) of 0 (TRL is defined in Section 3.6). Most academic research stays at this bottom level.

Because of the enormous number of university research groups worldwide and the publications in scientific journals, readable by all, a large number of inventions from universities nevertheless move up in technology readiness levels and become successful innovations.

Academic research also produces design methods and modeling methods. These methods facilitate the innovation velocity and facilitate optimized solutions. Modeling methods for reactive distillation and extractive distillation, for instance, have shortened the development time and have resulted in more optimal designs.

3.4.4 Contract research organizations

Contract Research Organizations (CROs), sometimes called knowledge institutes, carry out research for a customer. Often these customers are small and medium enterprises with a small research and development department, or no such department.

CROs also carry out research and development on their own budget, and then later when the results are positive and protected by patents and other means, they sell the results to customers, often via a license agreement.

CROs in general pick up ideas and research results from universities, and help Small Medium Enterprises (SMEs) to innovate. In this way, they play an important role in the general innovation level of a country. Because of this role often governments subsidize these organizations and often a government customer, such as a military, logistics, and safety department, is a customer of a CRO.

3.4.5 Startup innovators

Startup innovators are companies that start from nothing with an invention. The leader is a person who strongly believes that their product or technology will change the world and goes for it. Often the product is ill-defined and the innovator lacks marketing skills. By trying to sell it she learns the hard way that she must modify her ways to make a product successful and how to position it in a market.

A spectacular example of a chemical product-process startup is Synthon. It was started in 1991 by a Dutch chemist Jacques Lemans. In 1993, the first generic product, dobutamine, a sympathomimetic drug used in the treatment of heart failure and cardiogenic shock, was launched, and quickly became a commercial success. Synthon is now a global company with over 1600 employees, producing generic drugs at affordable prices [22].

3.4.6 Technology providers

Technology providers provide unique knowledge about a specific technology to their customers, so that the customer, an end user, or an Engineering-Procurement-Contractor, can complete the design of a product or process with that knowledge. Often technology providers also have an engineering department so that they can deliver the complete engineering package to purchase and construct the technology. Their unique knowledge is often applied to a specific design problem and then tested in their pilot plant facilities.

3.4.7 Equipment providers

Equipment companies manufacture and sell equipment. The design knowledge about the equipment is retained for the larger part inside the company. They also have specialized manufacturing facilities for their equipment. Some equipment companies are very innovative with new equipment for existing and new problems.

3.4.8 Engineering procurement construction contractors

Engineering procurement construction (EPC) contractors perform in general nearly all activities for staged detailed engineering. They take the front-end loading design of the product manufacturer as input. With that and the manufacturer requirements, they make a detailed design, purchase the equipment and materials, and construct the process.

Their work force consists of engineers of all disciplines and of a variety of experience levels. They are often specialized in process design for one industry branch, such as petrochemicals, food, pharmaceuticals, or electricity production.

For a product manufacturer, it is of paramount importance to select an EPC firm that has experience in the industry branch. For EPC activities, the motto is: the devil is in the details. If for instance an EPC is chosen for a food production plant, but has never had such experience, then it is likely that the process will not clean properly for decontamination, the most important criterion for a food production facility.

3.4.9 Tolling manufacturers

Tolling manufacturers have processes that can be hired for producing an amount of product. Tolling manufacturers for instance have several small-scale processes suitable for reaction, distillation, and crystallization. Often the process operation range is wide and the construction material is such that many different media can be applied without causing strong corrosion. The toller's knowledge about what can be processed varies enormously. Some tollers leave the process operation to some extent to the operators of the hirer. Some tollers have enormous knowledge and experience and provide not only the process equipment but also their knowledge on how to best operate the process for the new product.

The advantages of involving a toll manufacturer are that: the product is quickly produced, no pilot plant has to be designed and constructed, and knowledge of the toller is obtained for the operation mode.

3.5 Portfolio innovation management

3.5.1 Objectives of portfolio management

Portfolio management is about managing all innovation projects to meet strategic business objectives. This well-stated definition is obtained from the Project Management Institute (PMI) as shown by Bakker [23]. The objective of portfolio management is to determine the optimal mix and sequence of projects in their innovation stages. It is, therefore, a dynamic decision-making activity, as some projects will stop at a stage gate, while others proceed, needing more resources.

There are four major rules for portfolio management [23]:

Rule 1: All projects of a portfolio represent planned investments by the company.

Rule 2: These projects are all aligned with the company's strategic objectives.

Rule 3: A large portfolio will be divided in clusters of projects of similar characteristics.

Rule 4: All projects are quantifiable so that they can be measured, ranked, and prioritized.

Ad Rule 1: Often companies also have small projects that are not part of the portfolio. These loose projects can be damaging to portfolio management, as these are not discussed in the portfolio management meetings, so are not properly managed. An investigation by Blichfeldt and Eskerod in 2008 of 128 companies showed that these small projects together often tie up considerable resources originally allocated to the portfolio projects [23]. In this way, they damage projects inside the managed portfolio.

This problem can be avoided by having a cluster, called discovery or significant change, where all small embryonic projects are placed and an annual budget for that cluster is allocated. In this way, these projects no longer consume budget in an uncontrolled way. Management of such an ideation cluster can be further improved by having experienced researchers part-time allocated for that management, facilitating ideas popping up during the year by advice, and providing a budget (within the overall budget for that cluster).

Ad Rule 2: This rule can only be applied if the company has a clear innovation strategy.

Ad Rule 3: This clustering on similar characteristics is not easy. Several options are available, along business division lines, or along technology type lines, but these all have the danger of being locked in thinking, meaning that novel ideas cannot be considered. It is better to first have a high-level innovation matrix clustering as described in the next section. Further clusters underneath each main heading of the innovation ambition matrix can then be formed. Section 3.5.2 proposes a very practical main clustering, from which further subclustering can be done.

Ad Rule 4: Metrics applicable to the whole portfolio and to individual innovation projects for ranking and prioritizing are very important. Wrong metrics will cause wrong outcomes. Metrics for companies can be developed or enriched using the generic criteria list of Chapter 4.

3.5.2 Ambition matrix for innovation portfolio management

For large established companies, the innovation ambition matrix described by Nagji can be very useful. It groups all innovation projects of a company into three clusters: core, adjacent, and transformational [24]. There are similar clusters described in literature, based on the terms in the innovation classes of Section 3.3, but the advantage of Nagji's clustering over others is that it has generic names and clear definitions applicable to all product and process innovation oriented companies.

3.5.2.1 Core cluster

In this cluster, existing products for existing customers and existing processes are further developed. The development of these minor changes of products will often be carried out in departments supporting existing products and processes, often close to the manufacturing facilities.

The projects of this cluster fall in innovation Class 3; see Section 3.3. Their time to market is short. The success chances of introducing these modified products are high, and the expected internal rate of return sufficient. Management is simple, as it uses in-house innovation, by experienced employees with known methods for product development.

A ranking of potential projects and the best selected with a total budget that meets the set budget can be done according to the following criteria. First, for Safety, Health, and Environment, clear questions regarding acceptability are to be answered by yes or no. A no means that project will not be part of the ranking procedure. Ranking of the remaining potential projects are mainly through Economics (forecasted additional sales, profit, and return on investment), Technical feasibility and Social (market) acceptance. The capitalized criteria words in this paragraph form the acronym SHEETS, described in Chapter 4 in detail.

3.5.2.2 Adjacent cluster

In this cluster, new businesses for adjacent customers and adjacent markets are developed. Here it is important to recognize that new customers and new markets create additional risks. These risks can be mitigated by generating new knowledge about those customers and markets. Often adjustments in products must be made to meet the requirements of those new customers. This in turn means adjustments to the manufacturing processes.

Projects of this cluster fall into innovation Class 4; see Section 3.3. The time to market is much longer than for the core cluster. For chemical products, the expected internal rate of return is 13–18%.

Management of this cluster is critical to success. It is very important for these projects to go through all the innovation stages starting at the ideation stage, so that alternative options are generated and evaluated. If the ideation stage is neglected and a project is jumpstarted in a stage further up in the innovation funnel, then the best idea may not be pursued. When the new product enters the market, competitors may already have found a better way of designing and producing it. The ideation stage can, however, be quickly executed and most of the emphasis and budget will be allotted to the follow-up stages.

Ranking of projects and selecting the best (in such a way that the total budget stays within the total set budget for this cluster) is more difficult. Ranking criteria may be fixed by the company and experienced managers can use their judgment to perform this ranking. Table 3.8 is used inside a product-process innovation company for this purpose.

Tab. 3.8: Scoring table for economic attractiveness of novel products.

Criterion	Score 4	Score 3	Score 2	Score 1
Market size	Very large	Large	Moderate	Limited
Market growth	Very strong > 10%	Strong > 5%	Light 2%	None
Sales potential after 10 years	Very large	Large	Moderate	Small
Profitability after 10 years	> 40% gross margin	> 25% gross margin	> 10% gross margin	<10% gr. margin

Tab. 3.9: Scoring table for potential innovation projects using risks criteria.

Criterion	Score 4	Score 3	Score 2	Score 1
Entry barrier for us	None	Overcome < 1 y	Overcome < 2 y	Overcome > 2 y
Entry barrier for competitors	Overcome > 2 y	Overcome < 2 y	Overcome < 1 y	None
Scale-up financial risks	Easy, can be outsourced	Moderate; cannot be outsourced	Hard; cannot be outsourced	Unclear, risks unknown
Our competitiveness	Very strong	Strong	Light advantage	No advantage seen
Own IP position	Strong for long time	Limited for short time	Unlikely	Unclear

For each score, the company has to first quantify what they mean by the terms. This will strongly depend on the company and their branch. Some companies consider 500 million euros very large, others consider 500 billion euros very large.

The scoring is carried out by experienced marketing employees. The total for all criteria can then be added up, so that each potential project gets a simple economic dimensionless value.

Attractiveness of opportunity including risks with the following criteria and their scoring is suggested with Table 3.9. It is used inside a company with innovation in product-process combinations.

The entry barrier is an important criterion about being legally allowed to produce and sell the product. For instance, for a novel food product the entrance barrier is food approval by a government body.

'Scale-up financial risk' means the risk involved in investing in production processes (for the new product). It has to do with the novelty of the process. If the process consists of classic unit operations, can be completely defined such that an EPC contractor can design it, and a pilot plant validation is not needed (see Section 3.11), then a score of 4 is obtained.

The scoring is carried out by experienced managers and the individual score results as well as the total result are used for evaluating projects for the concept stage gate to feasibility and for the next stage gate.

3.5.2.3 Transformational cluster

In this cluster, breakthrough products are generated for new (or even not yet existing) markets targeting new customer needs. New manufacturing processes for these novel products are also generated.

Projects of this cluster fall into innovation Classes 5 and 6. They have high risks and high rewards. The risks are reduced by following the stage-gate process.

For this cluster, it is important that sufficient numbers of new ideas are generated and are nurtured to reach sufficient definition so that the best options can be selected for follow-up. Because new competences need to be acquired for these novel products for novel markets, it is also important to form new basic research groups. These can generate basic design data and knowledge so that the novel products can be embedded in the company. Chapter 6 describes in detail for each innovation stage how basic design data are generated.

Roadmapping to identify and link basic design data generation, skill development, supporting technologies, manufacturing technologies, and the products is very useful for this cluster. Goffin provides a generic picture of such roadmapping [1]. Chapter 4 on concurrent product-process design will help to plan the design and research effort, with the Delft Design Map (DDM) and the fingerprint tables for innovation Classes 5 and 6.

This transformational cluster may also be subdivided into technology platforms to serve specific (new) business divisions. DSM for instance has three technology platforms: [25]:

- **DSM Biomedical:** Innovative materials that deliver more advanced clinical procedures and improved patient outcomes
- **DSM Biobased Products & Services:** Advanced enzymes and yeast platforms: enabling advanced bio-energy and biobased chemicals
- **DSM Advanced Surfaces:** Smart coatings and surface technologies to boost performance in the solar industry

For potential projects in the transformational cluster the strategic fit is the key criterion. Table 3.10 is used in an innovative product-process company and can be of help in evaluating these ideas and potential projects, notably in the discovery and concept stages.

Tab. 3.10: Strategic fit score for breakthrough ideas and projects.

Criterion	Score 4	Score 3	Score 2	Score 1
Input	Existing to us	Existing, new to us	New, proven far away	Emerging
Process Technology	Existing to us	Existing, new to us	Proven elsewhere	Emerging
Product	Variants of existing to us	Existing, new to us	New, variant of existing	Radically new
Market	Existing to us	Existing, new to us	Immature, high entrance barrier, niche product	None, unknown

Clarification of some terms: ‘to us’ means to the company involved. ‘Proven far away’ means proven in a world continent, where the company has no manufacturing presence.

3.5.2.4 Budget distribution over clusters

The best distribution of innovation budget over the three clusters will depend on the business strategy of the company. Ngai provides a rule of thumb for the distribution of budget over the three clusters [24].

Core:	10%,
Adjacent:	20%
Transformational:	70%

The total corporate budget for innovation comprises for most companies the stages discovery, concept, and feasibility.

The budget for individual projects in the stage development, detailed design, and startup is decided by the relevant business section or by a corporate decision.

3.5.3 Risk adjusted value for innovation management

Risk adjusted value for innovation management is still not a well-developed field. Academic research based on industrial practice is sparse. There is a variety of methods used by companies, but the most used method is judgment by management [23]. Here are some plausible guidelines for the innovation stages.

For the early stages discovery, concept and feasibility, the total spending is limited and the only risk is that all money spent is lost, because the projects do not all pass the stage gate to development. If this total spending can be carried by the company, the risk is very low. The total budget spent on these early innovation stages can be related to the typical spending of the industry in the particular branch. Most companies chose to go for the average value of a branch. If they should catch up with their competitors they may spend more. If they are ahead of the competition then they may keep the spending the same as in previous years, in particular when the overall return of investment on innovation is high.

The temptation to cut innovation portfolio budgets for the early stages is always there when the going is tough. The immediate effect on business performance is positive, as cost is reduced. In the longer run, however, market share is lost to competitors, which then appears to be nearly impossible to get back as innovations take time. The author experienced this when in the eighties, research for a particular polymer product was stopped. Within 10 years market share was lost and could never be recovered.

3.5.3.1 Risk adjusted value in development stage and beyond: crossing the valley of death

For the development stages, often major investments on prototyping tests and pilot plant tests are needed. In general, the maximum budget for the whole development stage of the portfolio is set by top management. The individual projects are first evaluated using risks and expected added value. Table 3.10 will be useful in this respect. Then all projects for the development stage can be presented in the same way and ranked.

Special attention is paid to the so-called 'valley of death'. When the projected cumulative profit versus time of an innovation project is made, it will first show an ever deeper negative value until the lower minimum is reached and then losses will decrease as sales start to generate money and finally a net profit will be made. The negative part of this projected curve is called the valley of death.

The projected profit curve can only be made with some meaning for the first time at the end of the feasibility stage, when the cost of the commercial scale process, the prototype, and the pilot plant are known. The deepest point of the profit curve, so the maximum projected loss, is the most critical part. If that projected loss is larger than what the company can bear, or is willing to bear, then the project will stop, or be modified such that the valley of death is less deep.

The deepest point of the profit curve should also be compared to the long run profit when more than one manufacturing plant will be built and the total sales become large. Chapter 10 and Figure 10.1 provide a detailed description of the projected profit curve and the valley of death. Further reading on this topic is provided by Markham [26]. Cooper provides further reading material into innovation portfolio management [27–29].

3.5.4 Innovation management guidelines for small enterprises

3.5.4.1 Innovation guidelines for small enterprises

The main success factors for innovative small companies and institutes gathered from numerous literature sources have been summarized by Conquest [30] and are found in Table 3.11.

3.5.4.2 Guidelines for breakthrough innovation companies

If the innovation concerns breakthrough technology that can alter the company completely then the intuition of top management plays a very important role. Verganti, who worked inside very innovative successful companies applying design driven innovation found no formal ranking method, but instead found that the only criterion was: do the CEOs find the idea sufficiently exciting and sustainable. If the CEO asked for a financial analysis of the value of the design, then that meant that the project would be stopped [31].

It is also worth considering whether the CEO's intuition has been built by years of experience in successful breakthrough innovations. If that is the case then follow-up is warranted. If, however, the CEO has no experience with successful breakthrough innovations then failure is likely to happen. Kahneman, the Nobel prize winner de-

Tab. 3.11: Management success factors for innovation in small enterprises [30].

Aspect	Description
Leadership	
Visionary	Top management has innovation vision and shares it
Long term commitment	Improvements are judged by contribution to long term
Risk acceptance	Top accepts that not all innovations are successful
Strategy	
Positioning innovation	Improvements and renewal are an important pillar of the strategy.
Signals from environment	Our company can pick up signals, transform the strategy accordingly, and make clear what the impact needs to be in the present business process structure.
Cooperation in chain	Our company shares actively strategic information and business drivers with chain partners to innovate services, products, and processes.
Steering	
Tasks and responsibilities	Innovation tasks and responsibilities are explicitly delegated.
Innovation protection	Innovation initiatives are given room to grow.
Rewards	Steering and rewards for innovation are explicitly embedded in steering and reward structure.
Culture	
Felt need to innovate	Employees' hearts and minds know the purpose of needing to innovate.
Learning climate	There is an open communication learning climate in which mistakes and problems are shared.
Tolerance	Novel ideas are always appreciated, even when they are contrary to common wisdom.
Knowledge and competence	
Personnel diversity	Our company has a large variety of people regarding knowledge and competencies.
Knowledge sharing	It is common to share knowledge of procedures and processes.
Innovation ownership	People responsible for innovation have the knowledge and competence to create a collective atmosphere to sell innovation.

scribes in his book that intuition that has been formed by experience in a regular environment can on average be trusted [32]. A breakthrough innovative company may form such a regular environment.

3.5.5 Project failures and their causes

This section deals with project failure rates and their causes. As the causes are also related to portfolio management the subject is treated here.

3.5.5.1 Innovation failure statistics

We treat here specifically process innovation failures, so failures where only the process is new (innovation Class 3) and the product is conventional. The practice of the last 30 years of introducing novel processes is that more than 50% of those projects are disasters. Disaster is defined as having more than 30% cost growth beyond the budget and more than 38% schedule slippage. These statistics have been obtained by Independent Project Analysis from over 12,000 projects in oil and gas, petrochemicals, fine chemicals, and pharmaceutical sectors [23]. This means that the risk of failure is large.

3.5.5.2 Causes of project failures

Here is a list of causes of project failures in the stages up front of detailed engineering gathered by IPA and published in Table 1.2 of [23].

- Lack of adequate analysis of potential solutions
- Schedule is agenda driven rather than data driven
- Optimistic forecasting
- Overly aggressive appraisal strategy
- Lack of front-end loading (feasibility and development stage deliverables)
- High degree of complexity/innovation
- Unrealistic plan
- Unrealistic budget
- Lack of understanding of stakeholders' aspirations

In addition to this list an experienced innovation manager of a product-process innovation company observed in his career following additions causes:

- Lack of strategic fit
- Lack of freedom to operate
- Lack of understanding of customer needs and profit

All causes can be avoided by sound project management.

3.5.5.3 Megaproject failures

In addition to causes reported by Bakker for all types of projects, Merrow reports specifically failures of mega-industrial process projects [33]. He found seven major causes of failures derived from the data of mega-industrial process projects in the IPA database. These megaprojects contain also more novel technologies than the other projects in the IPA database, so the conclusions are very interesting regarding innovation. Here are the major causes of megaprocess failures.

(1) Too little money spent in innovation stages, often due to greedy lead sponsor of budget
(2) Too ambitious schedule, not taking into account good industrial practices and time required

(3) Not a good deal is arranged between the budget provider and the project executor early in the project
(4) Too little effort spent on the early innovation stages up front of detailed engineering
(5) Reduce the established budget for the innovation project by 20% by the CEO (lead sponsor)
(6) Transfer the risk to the Engineering Procurement Construction Contractor by lump-sum contract
(7) Beating up (threatening to fire) project managers for an overrun-on cost

The first four causes are mainly related to the early innovation stages up front of detailed engineering, where novel technologies are developed.

The last three causes are mainly related to the detailed engineering and implementation stages.

This again leads to the conclusion that sound project management in each innovation stage is needed to increase the success rate of innovation projects.

3.5.5.4 Project failures in detailed engineering stage

Here is the list of reasons for project failure from IPA for this stage as reported by [23]:
- Lack of clearly defined scope
- Incomplete design and errors
- Wrong contracting strategy
- Wrong contractors
- Lack of adequate resources (skills and numbers)
- Personnel changes
- Personality clashes
- Inadequate risk management
- Lack of adequate change control

3.6 Project management of innovation

3.6.1 Objectives of innovation project management

Project management is about individual projects. It is needed to ensure that ideas result in a successful novel product and process market implementation. A project in general is a temporary organization with a defined beginning and end brought to life to deliver a unique scope within given boundary conditions (schedule, cost, and quality). Management of these projects is about controlling the execution of the project so that it reaches its goal and stays within its constraints.

An innovation project is an endeavor in which human, financial, and material resources are organized in a novel way to undertake a unique scope of work of given specification, within constraints of cost and time, to achieve beneficial change defined by quantitative and qualitative objectives [23].

In the early innovation stages of ideation and concept when the number of people and disciplines involved is limited to a few, this management is limited to ensuring alignment to business objectives and to ensure exploring and analyzing many potential solutions. In these early stages of discovery and concepts, the project manager will only spend a fraction of his time on this management. Arranging meetings between project team members to monitor progress and to ensure that the project is focused on the goal and stays within its constraints is the main task.

In the next stages feasibility and development when more people, more disciplines, and more stakeholders are involved, this management becomes a full-time job. The factors critical to success for this project management are:

- Making sure what the deliverables for each stage gate need to be
- Making sure that sufficient time and effort is spent to produce these deliverables and that no shortcuts are taken to jump over the stage gates, without the deliverables

3.6.2 Project management by stage-gate system

3.6.2.1 Product and process innovation stages: general aspects

A major method to innovate at the lowest cost and time has been the introduction of the stages and gates approach. The basic reasoning is that in the early part of innovation, major risks of failure and major risk on nonoptimum concepts, are reduced quickly and at low cost, by exploring the whole field of options, selecting the best, and then performing proof of principle experiments for the best option.

At each gate between two stages a decision is taken to continue, or abandon the project. Some companies also include other options, such as putting the project on hold, or recycling the project, i.e., sending the project team back to provide more or better information, or give a conditional go to continue, depending on specific future events happening [29].

The purpose of the stages and gates method is to avoid spending large sums of money on wrong ideas for a long time and spending those large sums of money on good ideas and only in the last stages when the risks have already become smaller, because more reliable information is available.

Cooper initiated a stage-gate method for product innovation in 1990. Many product innovation companies quickly adopted his method. His method has also been validated in the end by investigating many companies, including health care, chemicals, and polymer companies. Companies that had a formal stages and gates method in place and those that also had committed gatekeepers did far better than companies with no formal stages and gates method [28]. A reference book on the stages and gates method is by Cooper [27].

This stages and gates approach to innovation was applied by many process focused companies, in particular oil and gas companies even before the product innovation oriented stages of Cooper. They used, however, different names for the stages. Bakker provides an overview of different names for the same or similar stages [23].

In this book, we provide a uniform nomenclature for all product and process focused industry branches for each stage's major activities, as shown in Table 3.12. It is partly based on the product innovation stage definitions of Cooper and partly based on process focused companies, and Engineering Procurement Construction (EPC) companies [23]. We condensed these descriptions into a single stage-gate description to fit the concurrent innovation of products and processes.

Table 3.12 is structured from top to bottom on general personnel management, business, design and experimentation. This table will be helpful in the overview of steps from idea to implementation, and the mutual understanding of employees in the R&D, engineering, and business departments in their use of terms, and in understanding their differences in attitude.

Table 3.13 indicates for each stage which disciplines are involved to what extent. In combination with Table 3.12 and Belbin team roles as described in Appendix A3, teams can be composed for each stage.

This stage-gate approach to industrial innovation ensures that unfeasible ideas are quickly discarded with minimal effort and that good ideas are pursued. When combining the stage-gate approach with the systematic design approach of this book then good ideas are further enriched so that both the company and society greatly benefit from the novel products and processes.

3.6.2.2 End-of-life stage

We do not show in the overview the end-of-life stage for product and process, as for that stage no generic design methods are available. In the design of products and processes in the earlier stages considerations for end-of-life will have to be considered, so that product and process recycling are made easy. This foresight aspect of design for end-of-life is treated in the design chapters of this book and in Life Cycle Assessment part of Chapter 12.

3.6.3 Project entries to stages by technology readiness level method

Technology transfer from a university, institute, or technology provider to a company, such as an end-user company is not easy. The readiness level for commercial scale implementation should be established, so that is clear what still needs to be done.

Tab. 3.12: Key items stages and gates for product-process design driven innovation.

Items	Innovation Stages Discovery	Concept	Feasibility	Development	Detailed engineering	Implementation
General management						
Purpose	Idea generation	Screening	Feasibility concept	Validation	Engineering Procurement Construction	Launch Startup
Required attitude	Optimism	Optimism	Realism	Realism	Pessimism	Pessimism
Project management	Nurture part-time	Research part-time	Project part-time	Project part-time	Project full-time	Project + Operation
Open innovation	Low	Academic	Knowledge Institute	T. provider toll manuf.	EPC	–
Funding	Game changer	Annual R&D budget	Annual R&D budget	Specific Project	Specific Project	Business
Business						
Risk of failure %	100	100	50	30	10	10
Invested %	0.01	0.1	2	10	94	100
Capex estimate uncertainty %		200	50	30, 10	0	
Market items		Demand opportunity	Size estimate	Size determination		Launch
Business question	What starting?	Full range explored?	Best Selected?		Everything in place?	What did we learn?
Business case			Initial	Confirmed Value, Risks		
Stage-gate decision	Strategy fit	Business case	Business case	Business case	Top mgmt.	
Duration (years)	1–2	1–5	1–2	0.5–1	1–2	0.2–0.5
Design						
Product design	Sketch	Concept	Prototype	Complete	Detailed	Description
Process design	Sketch	Concept	Unit Operations	Main Equipment	Detailed	Startup plan
Main deliverable	Sketch concept	Concept Economic potential		Basis data Design, Test results	Design books	Plan, Post startup review
Project engineering			FEL-1	FEL-2, FEL-3	Procurement package	
Experimental	Proof of principle	Proof of concept	Product Prototype	Pilot plant	–	Startup

Tab. 3.13: Typical involvement company parts in product and process innovation. As percentage per stage.

	Discovery and Concept	Feasibility and Development	Detailed engineering	Product Launch and Process startup
Science	90	10	5	5
Engineering	5	75	75	5
Operation		5	5	75
Marketing		5	5	5
Business	5	5	10	10

Technology Readiness Level (TRL) assessment is very useful in this respect. It is about quickly assessing of a novel technology where it is in the innovation stage. It is particularly used for novel technologies resulting from academic research or technologies provided by startup companies. End-users then use the result of that assessment to decide to pursue the technology (partly) in house or to leave it for further development outside the company.

There are several technology readiness level (TRL) tables available from NASA [34], ESA [35], DOE [36], and EC [37]. However, these are all defined by government organizations and do not everywhere use terms familiar to the industry. They also do not consider the (mass) manufacturing of the technology, but only focus on the product technology itself. To fill this gap in addition a Manufacturing Readiness Level (MRL) table for novel manufacturing technologies (for existing products) has been defined [38]. Recently the European Association of Research Technology Organizations (EARTO) developed a TRL level description for both novel products and processes [39]. From these tables, we derived a TRL table for novel homogeneous products and for their novel manufacturing processes using terms familiar to these industries, shown in Table 3.14. We added the level "0" so that also ideas, which have not reached TRL 1 can be categorized.

The table is particularly useful to assess novel technologies from universities and from technology providers. It is also being useful in open innovation projects with various partners to help to establish consensus on the technology status and what needs to be done for commercialization of various process options [40].

A warning remark must be made here. The TRL table suggests that novel technologies at TRL 1–4 are outside the industrial realm. Also in industry, ideas are researched from TRL 1 to 4.

Tab. 3.14: Technology readiness level definitions for novel products and processes.

TRL Level	Description	Innovation Stage
0	Idea of novel product or process stated	Discovery
1	Experimental proof of key novel element of product or process; called proof of principle	Discovery
2	Concept design of product and process	Concept
3	Experimental proof of product concept elements and process elements	Concept
4	Experimental validation of integrated product (prototype) and integrated process (lab scale)	Feasibility
5	Industrial professional's techno-economics assessment product and process	Feasibility
6	Novel product prototype tested by industry Novel process technology demonstrated by industrial pilot plant	Development
7	Novel product launched in market First commercial scale demonstration of process plant in operation	Implement
8	Full launch product to market Learning points from demo plant incorporated in commercial process design	Implement
9	Commercial process in operation and evaluation report available	Implement

3.6.4 Team formation

3.6.4.1 Team member behavior characteristics required over the stages and teams

Team members should be chosen for their skills but also whether they could work together. The Belbin group formation method is very successful in forming groups that cooperate well.

3.6.4.2 Group design with Belbin team roles

Project groups are mostly designed based on disciplines required only. However, if groups must cooperate closely for the desired results then they should have complementary personal skills as well. In addition to discipline design the Belbin team roles can be used to design the group or team. Appendix 1 shows how to apply this method.

Desired behavior in different stage

The desired behavior of people in the various innovation stages is also important. Optimism should prevail in the early stages of discovery and concept, as most ideas will not make it to the commercial implementation and the lack of knowledge should not hamper people to believe in the success of the project. In the feasibility and development stages realism should prevail, as now it is important to show and validate that the new product and process indeed work. In the detailed engineering and startup stage pessimism should prevail, as now all details of the design and the startup also have been right. Pessimistic people are very good in finding detailed concerns why the novel product and process will not work. All these concerns should be turned into specific modifications and actions to still get the product and process working.

3.6.5 Intellectual property creation and protection

Intellectual property is something that is new, useful in commercial practice, is documented and protected from use by others. Intellectual property can be created by a design, as a design always has new elements, and is nearly always useful when exploited in practice. It is always documented. Protection from commercial use by others is then the last step. There are many ways to protect intellectual property. One way is to keep it secret. For small companies this may be doable. For larger companies this will be hard. A second way is to file patents. Now the information shown in the patent is made public but cannot be used by the competitors for business purposes. A third way is to file copy rights for essential documents containing the innovation. A fourth way is to file a design right. A fifth way is to file a trade mark.

The main error made on this topic that the novelty is presented orally or in writing in the open. Open means any public meeting place of any publication in any form. Even bilateral discussions with no formal nondisclosure agreement are public. Once the novelty is public no patent protection is possible anymore.

Literature on intellectual property generation and protection is vast. Verloop argues that intellectual property is a form of intellectual capital, and should be used defensively, i.e., preventing other companies from commercially exploiting the innovation and offensively by exploiting it in various ways [16].

3.7 Discovery stage

3.7.1 Purpose discovery stage

The purpose of the discovery stage is to obtain ideas and to define the ideas to such a level that it becomes clear what they are about so that at the end of the discovery stage it is clear whether they fit to the company strategy can pass on to the concept stage, or not.

3.7.2 Managing projects in discovery stage

The discovery stage can be compared to hatching of an egg. The only essential conditions for success, provided by the hen, are protection and warmth. The actual development inside the egg is invisible and autonomous.

The same holds for ideas generation and development in the discovery stage. Companies highly dependent on a continuous stream of new ideas to feed their innovation funnel, protect researchers in their early idea development. Some, like IBM and 3M, do it by letting researchers spending 10–15% of their time on pursuing their idea, without requiring a formal budget and reporting.

Other companies have a formal organization such as a Game Changer organization where researchers can ask for a budget to pursue their idea, any time of the year, with a minimal description of the idea. A member of the game changer team will then help to describe the idea such that it fits in the company strategy and will provide a budget to perform the most critical test to demonstrate the feasibility of the idea. Often this is a proof of principle experiment for a novel product or process step.

If the business case looks sufficiently attractive but the idea involves many innovation aspects then the idea may also be analyzed for what specific additional innovations will be required. Once the idea has been proven in some way and the business case has been sufficiently developed, the researcher will propose the idea for taking up in the annual research budget round, see also the table generic criteria for stage-gate decisions.

3.7.3 Creativity methods

In the discovery stage creativity is needed to generate many and good ideas. The literature on creativity methods for innovation is vast. A good introduction to further information is by Nijstad [41] and Hermann, [42].

Brainstorming is a good method to generate unusual associations and new perspectives to the problem at hand. The most important rules are:
(a) No critical remarks on ideas
(b) Associate and enlarge ideas

Brainwriting is another useful method. It works well for introverted people, as they can write ideas down on their own. Process engineers in general are introverted; hence this method is mentioned. The complete method can be found in the Appendix A3.

3.7.4 Discovery stage-gate decision evaluation criteria

What in general needs to be done in the discovery stage is expressed here under elements of the reporting for the stage-gate decision.

3.7.5 Idea description

If the idea is a product then a functional description of the purpose of the product is needed. Options for product composition to fulfill the product functionality may be given. If the idea concerns a process then a block flow diagram with feedstock and product streams is required.

3.7.6 Proof of principle

The result of a proof of principle experiment for the main critical novelty of product or process is needed. The challenge for the researcher is to define an experimental setup that is simple, quick to construct and such that the test result can convince others that the novel principle works. In general it requires several attempts before the proof of principle is successful.

3.7.7 Business case

The business case is the business justification of a project [43]. It is very clear from this definition that it specifically belongs to a specific company. The business case cannot be established by others outside a company. Or in other words the company owns the business case. Even better a specified person inside the company owns the business case. This is not the project leader.

The business case development of an idea in the discovery stage may be done by answering the following questions [43]:
(1) Does it create value for the company?
(2) Does it fit to the strategy of the company?
(3) Does the company need the idea?
(4) Can the company (together with others) develop the idea?

For a project to pass the discovery stage gate these four questions of the business case may be used. If positively answered then the preliminary business case for the idea is established.

3.7.8 Creating value

The question of creating value for the company through radically novel products may be hard to answer. It may be answered in stating, yes, we believe in its capacity to create value for the company and society.

A simple economic potential estimate may be provided on the potential economic contribution per year of the novel product or process, when commercialized. It may be done by just estimating a sales price per ton and subtracting from that the feedstock cost per ton of product and estimating the total sales in ton/year.

3.7.9 Strategic fit

The fit of the novel product or process with the company strategy will be indicated. If it cannot be made fit to the strategy and a strategy change of the company is not foreseeable then the idea should not be pursued. However, some champions of an idea are not easily stopped and because of their stubbornness and creativity, may find ways to pursue the idea.

3.7.10 Necessity of the idea

Even if the novel product and/or process fits the company strategy, still the question remains of whether the idea is needed. If so, then it will be pursued. If not, it can be stored and may be retrieved when circumstances change and the idea is needed. This, however, requires an idea storage and retrieval system that most companies will not have.

3.7.11 Idea development doable (with others)

The question of whether the idea can be developed by the company with help of others is also often hard to answer. It requires knowledge of what technology providers and academic research groups are available to solve some of the foreseen knowledge gaps.

3.8 Concept stage

3.8.1 Purpose of concept stage

The purpose of the concept stage is to obtain a full picture of what the goal of the novel concept is, what knowledge is needed, what knowledge is available, and which most critical success factors are to be investigated in the concept stage so that at the end of this stage, a decision can be taken, based on all criteria, to pursue the concept or not.

For a novel product, this involves developing a customer value proposition, technical requirements for product and process definition, identification of critical to quality variables and experiments to proof the concept.

3.8.2 Customer value proposition in concept stage

For a novel product concept, it means, first, that a customer value proposition is made. This is a marketing statement that describes why a customer should buy a product. For a business to business product it consists of the sum of total benefits that a vendor promises a customer will receive in return for the customer's associated payment [44]. Table 3.15 contains its main elements.

Tab. 3.15: Customer Value Proposition elements.

Customer value proposition elements	Description
Product type	Single product, or product family, product platform
Technical value proposition	Technical advantages over existing product
Product technology	Product components
Manufacturing process technology	Process options for manufacturing of new product
Materials technology	Input materials for the new product

This customer value proposition requires a strong interaction between the product developers and the business and marketing staff of the company. Design thinking and design models help these conversations enormously by having drawings and pictures of the product design, by having both a functional product description (what it does) and by having several physiochemical alternatives fulfilling those functions. By also making basic process concept designs with material balances, first indications of feedstock cost per alternative are available. Following a checklist derived from human senses may be useful to make sure that all relevant aspects of the product requirements are covered:
- Feel to relevant body parts
- Visual appearance
- Smell
- Taste
- Sound

3.8.3 Technical product requirements in concept stage

The primary consumer perception requirements should be translated into technical requirements. Other functional goals and requirements from the customer value proposition also should be translated into technical requirements. This translation may involve several steps, before a list of quantified technical specifications of the product is obtained. Quality function deployment and house of quality methods may be used for this translation of customer needs into technical requirements. The summary of the technical product requirements is called critical-to-quality requirements.

3.8.4 Process concept generation

The new method of process concept generation comprises designing a functional block-flow diagram with its material inputs and outputs for the chosen product concept. If several product concepts have been generated then at least one functional block-flow diagram for each product alternative will be made. Major energy inputs and outputs will also be determined. For each functional (task) block, feasible unit operations will be selected, or even better, functions will be combined and feasible functional integrated concept solutions will be defined.

The dominant method of the past, designing with unit operations, does not deliver breakthrough innovation anymore. Process design by task integration is emerging now as the dominant design method, as the advantages in investment and operating cost, lower environmental impact and inherent safe design have been proven for hundreds of commercial scale implementations [8]. The design method is described in detail in Chapter 6 and 7.

3.8.5 Proof of process concept

All key issues are derived from the process concept design and a lab-bench scale research program is defined and executed to remove these key issues and establish information for a better process concept design.

3.8.6 Selection of best process alternative

The best process alternative will be selected based on SHEETS criteria (see Chapter 4), with often a high weight factor for lowest feedstock cost and lowest investment cost. Selection methods are provided in Chapter 7.

3.8.7 Process concept design

For the best alternative, a process concept design will be made containing:
- Feedstock specifications
- Product specifications and quality performance requirements
- Catalyst type and amount
- Byproduct stream specifications
- Stream compositions for each stream
- Heat balances for each process unit
- Selected unit operations with pressures, temperature, and residence time
- Selected equipment types and main sizing aspects
- Choices of construction materials
- Main control strategy is indicated for sensitive process aspects
- Auxiliaries
- Utilities needed

3.8.8 Basic design data generation

Basic design data will be generated for the concept design. Chapter 6 provides details on what data are to be generated and how this can be done efficiently.

3.8.9 Concept stage gate evaluation

Business and research management will be informed about the results of the concept stage and about the planning and cost of the feasibility stage. For all criteria information will be provided such that a decision to pass the stage gate or stop the project can be made.

3.9 Feasibility stage

The purpose of the feasibility stage is to verify the feasibility of the product concept and the process concept for all critical aspects for success.

3.9.1 Concurrent product-process design and testing

The technical product feasibility is proven by experimentally testing the integrated product. Often this integrated product is called the prototype. Here unexpected effects of integration may cause the product to fail. Often the failure can be repaired by analysis of the cause of the failure. Because the tests are carried out at laboratory scale, modifications are fast and at low cost.

Through the direct customer and end-consumer feedback, both the product design and the optimal process design can be fixed. The product design includes not only the optimal composition and structure, but also will contain tolerable composition levels (including impurities). Also, the option to have multiple material suppliers validated, will help later in the negotiation of raw materials purchasing and supply conditions. Finally, the tolerable key product attributes and their relationship with composition and processing should be very clearly reported for testing in the next stage: the manufacturing stage.

3.9.2 Microplant in feasibility stage

The best process design concept is integrally tested experimentally at laboratory scale. This test unit is best called a microplant.

A properly designed microplant contains all major process units in the same line up, recycle structure, and commercial scale design. It contains all reactions and separations and it can operate in the commercial design window of process conditions.

Its purpose is to study the reactions and separations under the design conditions, with recycling. Reaction conversion rates and byproduct formation under recycle conditions are determined to improve reaction kinetics. Separation performance results are used to validate and improve physical equilibrium models for the separation steps.

Including the recycle flows in the microplant is also very useful. Heavy boiling species for instance can build up in it, affecting the reactions and the separations. Knowing this is of enormous value for improving the commercial scale design and reducing its risk of failure.

Because of its small scale, the construction and operation cost are low. Often is it made from glass, by which problems such as foaming and fouling are also quickly localized and the process is quickly modified to solve them. It can be used to test various process design options [45].

A downscaled microplant is a very cost-effective element to reduce risks and provide data for optimization of the commercial scale design. It is the smallest scale of an integrated version of the commercial scale design concept including recycle flows. The typical production capacity range is 1–100 gram/h and the equipment diameter is typically 1–5 mm [45].

The emergence of microscale equipment with well-defined residence time distribution and uniform temperatures has helped enormously to boost the use of microscale units.

For new pharmaceutical products, these microplants increase the speed of process development and thereby reduce the time to market. In some cases, the laboratory scale process is run for several weeks to quickly provide product for clinical testing.

In fact, a whole new industry branch of very small-scale devices has emerged [46] with over 274 companies providing these devices for the pharmaceutical industry [47]. A new term, microfluidics, is often used for these devices. The term refers to devices, systems, and methods for the manipulation of very small fluid flows, as small as micro, nano, pico and even femtoliter/hr flows [48]. Other terms used are: lab-on-a-chip and flow chemistry.

The application of microflow systems is increasing also for fine chemicals. In this industry branch, most of the production is carried out in a batch operated mechanically stirred reactor. Cooling is provided by having a solvent that evaporates from the reaction liquid, which in turn is cooled by an overhead condenser. The cooled liquid solvent is returned to the reaction liquid.

This reactor and its operation have large scale-up uncertainties, which are solved by an experimental program on the batch reactor. Byproduct formation is also often far more than the amount under ideal conditions, due to the far longer residence time than required for the reaction itself. Also, the solvent after its use is becoming a waste byproduct, similarly with the cleaning liquid in between batches.

Boodhoo describes the following microprocess technologies in some detail [50]: micromixers, microheat exchangers, and microreactors. Of the latter class, multichannel reactors and spinning disk reactors are described in detail. Günther provides an overview of multiphase microfluidics for chemicals production at laboratory scale [51] and Kenning covers microfluidic unit operations: extraction, absorption/desorption, distillation, as well as particle or droplets separation [52].

If successful, the process design is completed with process instrumentation and control and the product packaging method is designed. The degree of design definition for an economic assessment is typically 50%. This means that 50% of the investment is based on the process design and 50% is based on correlations for cost of plot preparation, concrete structure, and outside battery limits.

Remaining uncertainties regarding process integration and scale-up will be addressed in a pilot plant in the development stage. The design and cost of the pilot plant and cold-flow model, however, belong to the feasibility stage. It may well be that the cost of the pilot plant is so high that the project stops at the feasibility stage gate.

The feasibility of the product and the process design are assessed at the stage gate using all criteria of Chapters 10–13 and focused on validation of its feasibility for all those criteria. Details on the feasibility stage gate evaluation will be found in these evaluation chapters.

3.9.3 Scale-up strategy and information

The first question to answer is what scale-up strategy should be followed. For novel products with limited patent protection and high added value, such as for pharmaceuticals, a short R&D time is very important. Equipment cost is in general a minor

Tab. 3.16: Scale-up methods available for various process equipment. Scale-up methods: Brute Force (B.F.), Correlation based (Cor.), Modeling based (Mod.).

Equipment	Residence time distr.	Micromixing	Mass transfer
Trickle bed	B.F., Cor. + dilution	B. F	B.F., Cor. + wetting
L, G Packed bed	B.F., Cor.	B.F.	B.F., Cor.
Static mixer L, G.	B.F., Cor., Mod.	B.F., Mod., Cor.	B.F., Cor., Mod
Static mixer G/L	B.F., Cor., Mod.	B.F.	B.F., Cor., Mod.
Tray column	Cor.		
Plate column	Cor.		
G/L Bubble column	Cor., Modeling	Unknown	Cor.
Cross flow bub. col.	Cor., Modeling	Unknown	Cor.
Mech st. tank L, G	Cor., Mod.	Mod.	–
Mech st. tank G/L	L Cor.; G unknown	Unknown	Unknown

item in the overall cost. So, easily scalable equipment requiring only one validation step are preferred. Millimeter flow reactors for which the brute force method can be applied have here a large advantage.

For bulk products for which reliable production to satisfy the customers and to minimize the startup cost of the commercial scale process is very important, the scale-up risk needs to be low. The innovation class table with its risk data of Section 3.3 can be used to assess the risk involved.

Table 3.16 provides for some, often applied, process equipment available scale-up methods.

For reaction, the residence time distribution (r.t.d.) is always relevant. For two-phase reactors both the r.t.d. and the mass transfer is always relevant. If the reactions are fast and multiple reactions are involved often micromixing is also relevant.

For crystallization, the micromixing and the residence time distribution and the mass transfer is relevant.

For gas/liquid phase separations such as in distillation and absorption, both the mass transfer and r.t.d. are important.

A special warning must be given for the scale-up of two-phase mechanically stirred reactors. These reactors have a highly nonuniform turbulent intensity profile. Near the impeller high shear and energy dissipation occurs. There the dispersion is made very fine. Further away the turbulent field is less intense and there coalescence of the dispersed bubbles or droplets can occur. At scale-up the areas of high and low turbulent intensity change and often the ratio of high and low intensity changes. This makes scale-up notoriously uncertain. In emulsion polymerizations carried out in mechanically stirred tank reactors where the dispersion quality also determines the final product quality, scale-up is still a high-risk affair. Choosing other equipment such as static mixers would avoid this scale-up problem.

3.9.4 Scale-up information equipment

For all process equipment scale-up information will have to be provided, so that the commercial scale design base and its associated risk, is known for each process equipment. In some cases, the risk may be reduced by mitigation measures.

3.9.5 Business case feasibility stage

In the feasibility stage, the business case is much further developed. Also, here the business case values of all projects are established and so the value of the whole portfolio of innovation projects [23].

3.10 Development stage

3.10.1 Purpose development stage

The purpose of the development stage is to reduce the risk of introducing the novel product and process to acceptable levels. This means that at the end of the development stage the chance of a successful product launch is high and that the commercial scale process will meet the product requirements and process capacity and that it can be started up within the average industry startup time.

3.10.2 Product testing in development stage

For product performance at the client level, different product versions are produced at a processing scale that is large enough to perform an accurate scale-up of the process and/or is large enough to produce sufficient amounts for testing the product in real conditions at the direct customer and/or end user of the final product with the new ingredient/component. Customer evaluations of the product should reveal the appreciation for, and the value of the product for them. Modifications of the product are still possible and needed, not only to verify the product functions, but also to maximize the business opportunity by tuning the added value parts of the product and reducing the costs (raw materials, processing) of producing the product.

Often a miniplant is run to produce different product versions based on raw materials (different suppliers, quality levels, process settings, etc.) and to validate the product in the industrial customers processes and in the final consumers' application. For this purpose, sufficient material for testing should be produced, and this often also determines the size of the mini- or pilot plant. If this scale is not large enough and a pilot plant will be necessary, that plant will be constructed and operated in the development stage.

Also, critical logistics processes, storage, filling, packaging, transportation, unloading, etc. at direct customers may be identified and verified with the product batches. Again, if major investment cost is involved then that investment will be part of the development stage.

3.10.3 Process testing in development stage

3.10.3.1 Miniplant purpose and design

The miniplant ideally is also a downscaled version of the commercial scale design including all recycling, but also now containing the same type of reactors and separators with the same number of mass transfer units, residence time distribution, and the same construction materials as the commercial scale. It typically has a production capacity of 0.1 kg/h [45].

With this miniplant, further validation of the design concept and of the process models are established and problems are quickly identified and solved. Potential interaction between the construction materials and the components such as catalysis by construction material and corrosion of construction material can now also be observed as these plants operate continuously and for a long period (several months). Additionally, corrosion test material samples are placed in all streams to optimize for the commercial scale the construction material choices. Automatic process control and data storage is purchased from a reliable supplier of the distributed control system.

In some cases, a miniplant avoids the need of a large pilot plant and then saves a large investment. The pilot plant section demonstrates when a larger pilot plant is still needed.

The following design documents for a miniplant are required [45]:

- A process flow sheet
- An engineering flow diagram
- All equipment types and sizes
- Objectives of instrumentation
- Specifications for insulation of equipment and piping
- List of measuring points
- Safety concept and logic: how sensors are linked to actuators and safety procedure

3.10.3.2 Pilot plant purposes and design

The first question to answer is: is a pilot plant needed? The pilot plant can be needed for:

(A) Producing a large amount of product for quality and market testing
(B) Identifying scale-up problems

(C) Validating process models and providing input for mass and energy balance updates
(D) Training operating personnel of the future commercial plant
(E) Improving the estimates of catalyst life times
(F) Testing construction materials for corrosion
(G) Validating process control and operating procedures

A pilot plant is also very useful in getting reliable stream compositions. This increases the stream knowledge factor *F* of the startup time correlation from nearly 0 to 1 (see the section on startup). The difference in startup time of those two *F* values is on average 3.2 months. The amount of money lost by not having a specification product for that period in general easily outweighs the cost of a pilot plant.

3.10.3.3 Pilot plant as downscaled version commercial scale design

If a pilot plant is needed then it should always be a downscaled version of the commercial scale plant with all unit operations and all recycle flows. Also, the commercial scale process conditions such as temperatures, pressures, and residence times should be inside the pilot plant operating window. Only in this way the tests will be representative of the previously optimized commercial scale plant.

If the pilot plan is not designed as a downscaled version of the commercial scale then this means that the commercial scale plant must contain the same process type units and recycling as the pilot plant. In some cases, this means that multiple units must be selected in parallel for the commercial scale, because the selected equipment for the pilot plant appeared not to be scalable. It will be clear that this way of scale-up means a higher investment cost for the commercial scale process. So it is highly undesirable, although it is still a practice in some companies. If the commercial scale design also differs from the pilot plant in process type and recycle structure, then the risk of failure increases enormously. According to Merrow this is one of the major causes of commercial scale failure, meaning that the startup time is 40% longer than budgeted for (see startup time prediction) with a 40% additional cost to remedy the failure [33].

A: Pilot plant for producing large amount of product for testing

Often a large amount of product is needed for performance tests for one or more clients. This testing often is desired by the client even if the product is promised to be the same and only the process has been modified. This even holds for intermediate bulk products. Trace amounts not detected by analysis can still affect the product performance at the level of the client; for instance, when the new trace components degrade the catalyst in the client process.

If the product is novel, large amounts are often needed for all kinds of market launch tests [45].

B: Pilot plant to identify scale-up process problems

A pilot plant is needed to identify scale-up process challenges if either [53]:
- More than one new process step is involved
- The process step contains a new step and has a recycle stream over more than one step
- The process contains a novel solids processing step
- The feedstock is a crude source, whose composition knowledge is incomplete

If the pilot plant is needed but is still omitted, startups in most cases lead to failure, which means that the process as such cannot be operated and/or meet the design specifications on product quality and capacity [54]. An additional budget of over 30% of the investment is often needed to solve the problems [23]. These costs far exceed the pilot plant cost, which are in most cases less than 10% of the commercial scale investment cost.

C: Pilot plant to validate process simulation models

Process simulation models are now the main source of detailed information of process streams composition knowledge everywhere in the process including the final product output stream. This information is used to select the construction materials, to select and size equipment, and to define utility requirements. The startup time correlation parameter F also indicates the importance of this knowledge as shown in the startup section of this chapter. It is, therefore, important that the simulation models are reliable. Extensive validation of these models can be carried out by measurements of stream compositions, temperature, and pressure and comparing them with model predictions.

D: Training operating personnel of the future commercial plant

A pilot plant can also be used to train operating personnel of the future commercial plant. Nowadays this training is mostly carried out using dynamic simulation models, but additional training with a real pilot plant helps to get acquainted with, for instance, taking samples for analysis and seeing what the intermediate stream samples and product samples look like and how misoperation also shows up in the changing colors of those samples.

E: Improving the estimates of catalyst life times

Long-term pilot runs under actual process conditions will show the catalyst lifetime and how the process stream compositions change with deteriorating catalysts. This type of testing can also be done in a miniplant.

F: Testing construction materials for corrosion

By placing many material test samples in many pilot plant streams and running the pilot plant for a long time, accurate corrosion rate information can be provided, which can be used to improve the final materials selections for the commercial scale plant. Many problems in commercial scale plant operations are still connected to materials corrosion [53]. This corrosion testing can also be done in a miniplant; see the section above on miniplant.

G: Validating process control and operating procedures

In process designs with strong dynamic interactions between various process parts, process control may be difficult to achieve. Testing the chosen process control design and tuning the process control in the pilot plant can be very important. Operating procedures can also be tested by process operators.

This validation is particularly important for batch processes, who by nature have dynamic behaviors. The validation of process control design and operating procedures is a necessity for them.

3.10.4 Pilot plant engineering, procurement, and construction (EPC) company choice

The detailed design and construction of the pilot plant should be done with the same rigor as the commercial scale plant. The selection of the Engineering, Procurement, and Construction (EPC) company for the pilot plant is, therefore. important. A company dedicated to pilot plant EPC is strongly preferred. If an EPC company is chosen that handles both commercial scale projects and pilot plants, then it is likely that the pilot plant will get a far lower priority and lower qualified personnel, as the size of the pilot plant project in money terms is a fraction of a commercial scale project. This in turn can mean that the agreed timing is not adhered to and that the pilot plant will contain errors. For this reason, an EPC company only concerned with pilot plants is the strongly preferred choice.

3.10.4.1 Equipment scale-up effects determination

Commercial scale equipment will be larger than preceding test equipment in the development stage. In general, this means that the hydrodynamics in the equipment will be different. The following transport phenomena will change with the change in hydrodynamics and in turn can affect the performance of the equipment:

3.10.4.2 Phenomena in equipment sensitive to scale-up and influencing performance

- Residence time distribution
- Micromixing
- Mass transfer
- Dispersion of a second phase
- Heat transfer
- Momentum transfer

For most standard pieces of equipment these transport phenomena are known as a function of geometric sizes, fluid velocities, and fluid properties; often in the form of correlations. The relations between the phenomena and the performance are also known and often expressed in models.

A lot of conventional reactor equipment and gas-liquid contact equipment correlations for the transport phenomena are available in textbooks such as Perry's Chemical Engineers' Handbook [55].

A word of warning is needed here. Often the mass transfer correlations assume implicitly that the residence time distribution of the fluids is both plug flow, while deviations from plug flow occur. Often these mass transfer correlations have been obtained under the conditions of plug flow. At scale-up the deviations from plug flow can become larger and then the commercial scale performance is worse than predicted using the mass transfer correlation.

For tray type gas-liquid columns, often the residence time distribution of the liquid is assumed to be completely back-mixed, as well as for the gas phase. For very large diameter trays with one or a few down comers for the liquid phase this assumption is sometimes not valid. Technology providers of large distillation trays have this knowledge available for their clients.

If the relation between the equipment size and the phenomena are not known then that relation must be determined. There are two established methods for determining the relation:

Method A: The phenomena are determined in the same piece of equipment but used in a different process.

Method B: A large scale of that equipment is constructed and the relevant phenomena are studies in that piece of equipment. To keep the cost low often simple fluids such as air and water are used, rather than the actual fluids. Such a test facility is called a mockup or cold flow model.

3.10.4.3 Mockup design and testing for hydrodynamic scale-up effects

Nowadays computational fluid dynamics (CFD) flow models are often made for equipment with complex geometries and complex hydrodynamic flow behavior affecting performance upon scale-up. These models are then used to design and optimize the equipment geometries of the commercial scale. These models, therefore, need validation. This validation is carried out in so-called mockups where inert fluids such as water and nitrogen are used and mixing, residence time distribution, and mass transfer are measured and compared with the model predictions.

3.10.4.4 Development stage gate evaluation

The purpose of development stage gate evaluation is to decide whether to go to the detailed design stage or not. As the detailed design stage is very costly, this is an important decision. The economic evaluation of the commercial scale design for this stage gate is, therefore, very important.

Formal evaluation methods found in the evaluation chapters can be used as input for the gate decision. It is advised to consider all criteria for the decision, not just the economics. The gatekeepers for the decision should be experienced and have the authority to take the go/no-go decision.

3.11 Detailed design stage

3.11.1 Detailed product design

The product has already been designed in the development stage. However, product pricing adaptation to latest developments in the market, details of packing, logistics, and transport planning and if needed additional buildings, are now also to be defined.

3.11.2 Detailed process engineering

The detailed process engineering stage comprises detailed process design, procurement of all process equipment, and construction of the process. These three steps are in general executed by Engineering, Procurement, Construction (EPC) contractors. For a detailed description of the activities of this stage and how the manufacturing company cooperates with the EPC contractor, the reader is referred to other books such as Bakker's [23].

The starting point for the detailed design is the commercial scale concept design provided by the manufacturing company. If only a preliminary concept process design is delivered then the EPC contractor will make a concept design. In general first a process design is made with sufficient definition to allow an investment estimate of ± 30% with which the manufacturing customer can decide to continue the project or not and then a design allowing a ± 10% accurate estimate. These two levels of definition and accuracy required differ a little from company to company. Clarification between the customer and the EPC contractor is always needed. Also, a complete evaluation for all other criteria for now and future trends on safety, health, environment, social acceptance, and legal items of the local country (license to operate) is carried out as part of concept design. The customer then decides to go for detailed engineering, procurement of equipment, and construction, or not.

Nomenclature of process concept designs varies enormously between companies and clarification should always be asked for. Process concept design with 30% capital investment inaccuracy, is also called front-end loading 1 (FEL-1). The next concept design with 20% inaccuracy, is also called FEL-2, and the third process concept design with 10% inaccuracy is also called FEL-3 [23].

Detailed engineering is an enormous effort with many engineering specialists involved. Here only some main elements are mentioned.

(1) A complete piping and instrumentation diagram (P&ID) is made for every part of the process. It shows all piping connections to all equipment and all controls and instrumentation. This is an enormously important piece of information for process construction and safe and proper operation. To get a feeling for the importance, consider that, on average, piping concerns 20% of the total process investment cost. A large part of that cost is making specifications for each pipe on, construction material, tracing or not, sizing, and precise 3D location.
(2) A complete process description is made for each unit operation containing dimensions, capacities, temperatures, pressures, construction material, chemical and physical knowledge, and data.
(3) A 3D plant layout plan is made showing both all equipment and piping.
(4) A detailed commissioning and startup instruction manual is made.
(5) Equipment detailed specifications for purchasing are defined.
(6) Detailed drawings for special equipment, to be custom made, are sketched.

3.11.3 Choice of EPC contractor

The choice of the EPC contractor for process design and construction is very important. The selection should be made by the project leader of the manufacturing company. Any involvement by business leadership in the selection of the individual contractors is inappropriate. Any involvement of the lead sponsor's purchasing organization in the selection of contractors is likely to be catastrophic. Even more important is the selection of the right contractor team, a team that the manufacturing company team can work with creatively [33].

The EPC contractor should have experience in the relevant process industry branch. For the food industry branch for instance, that contractor will know how to prevent contamination, how to ensure complete cleaning of equipment, and how to avoid undesired micro-organism growth.

For a breakthrough novel process unit within the overall process, additional contractors to the main contractor may be involved, with specific knowledge of designing for the materials flowing through the process unit and with knowledge of designing for similar stream compositions. That contractor will, for instance, design in additional measures to prevent or removing local fouling, or select at critical locations an instrument that is robust to fouling.

3.11.4 Demonstration plant

For a radically novel process that in the end will be built for very large capacities and with a large investment, first a demonstration process is in most cases designed and operated. This demonstration process is in general 1–10% the capacity of the final commercial scale. It may be 10–30% of the full-scale investment cost.

The main reasons for having this demonstration plant are to reduce real and perceived risks. The risk can be technical and/or business-related (uncertain new market development).

The technical risk of a novel process (and product) is reduced simply by the lower investment cost. The higher investment cost per ton of product for this demonstration scale may also be earned back later by incorporating the learning points of the demonstration plant in the final full capacity plant. The savings through these learning points can be small, but also be as much as 40% of the final investment, depending on how much redundancy has been built in by the engineers in the demonstration plant [53]. The redundancy measures can be operating safety margins, oversized equipment, or even spare equipment.

The business risk due to new market development is uncertain. Sometimes this demonstration plant is needed, to convince investors with no experience in the process industry (inside or outside the company) that reliable processing is feasible and that promised return on investment can be made.

It is very important that up front it is made certain that the demonstration plant will make a profit. It is also important that the lower return on investment of the demonstration plant is reported clearly to the business department of the company. If either is not carried out the commercial scale plant will not become reality.

An example of a demonstration plant prior to the final full-size plant is the Shell GTL plant in Bintulu. It has a capacity of 14,700 bbl./d of liquid products. It started up around 1990. Its capacity is a factor of 100 smaller than the commercial scale PEARL plant in Qatar of 140,000 bbl./d of liquid products. The latter plant started up in 2011 [56].

3.12 Process startup and product launch

3.12.1 Introduction

A product market introduction and in particular a consumer product introduction requires an enormous amount of planning and preparation. Here are some main elements for consumer product introduction. A business to business product introduction will contain similar elements, but to a lesser extent.

3.12.2 Panel for product testing

The consumer product will be tested by a panel first consisting of company personnel and later by a panel of outside potential customers of the product. If it is a food product the panel will be asked to mention the look, the smell, mouth feel, and the taste. If it for instance is a laundry soap the potential users will be asked to read the instructions and use it in a washing machine. All reactions of the panel will be monitored and the product and instruction may undergo a final adaptation, based on the panel feedback.

Here is an example of such a panel test for a new, more concentrated laundry detergent. Because of the more concentrated detergent, far less is needed per laundry batch. The external panel members, however, could not believe that so little soap would work and still were tempted to add (nearly) the same amount as before.

3.12.3 Product launch planning

Detailed planning will be made by the producer for all elements of the product launch. Market intelligence will be used for the product launch planning. It will include how much, when, and where to supply the product for the first time. A prelaunch will nearly always be planned in which the product will be launched in a certain location (a town, or a region) only. The purpose of the prelaunch is to validate market expectations and to detect imperfections in the logistics. After the successful prelaunch, the full product introduction will start. Contract preparation and signing in time with all relevant stakeholders is of course an essential element in the planning.

3.12.4 Matching marketing and manufacturing timing

It is of utmost importance that the planning is made by intimate contact between marketing and manufacturing to fit the manufacturing production availability and capacity to the market planning.

In some cases, the pilot plant will be used to make sufficient product for panel testing or even the prelaunch. Up front, in the design of the pilot plant years earlier than the product prelaunch, special attention will also be paid to product packaging. This has all been considered by having intimate contact between marketing, development, and manufacturing.

For consumer product oriented companies, this regular exchange of information and matching marketing with development manufacturing is a common practice. For business to business manufacturing oriented companies that move to consumer products, this will be new, and structured connections between marketing, development, and manufacturing are often absent, causing risk of mismatching the product prelaunch and manufacturing.

3.12.5 Information to supply chain and customers

All relevant stakeholders in the supply chain (transport, storage, shop) and customers will be informed about the prelaunch and the launch of the new product by proven and novel media means. The product description for each stakeholder is an important aspect, and the fine-tuning of the product introduction price is an important element of the product introduction to the market.

3.12.6 Process implementation

In the implementation stage of a novel process for an existing business to business product, the startup should be viewed as an experiment that is not allowed to fail. If it fails then the client does not receive the product and in turn cannot produce its own product. That client will look for other suppliers, as security of supply is enormously important to stay in business.

This means that precaution measures must be taken. The measures are regarding the process design, construction, preparation, startup, and regular operation.

3.12.7 Recognition of new commercial implementations requires special preparation

The first step in any implementation is to recognize that something new is to be implemented. Most often failures are due to not recognizing that the process is new and perceiving that the process is the same as before. An analysis of projects containing revamps, expansions, and green fields, highly innovative and lowly innovative, revealed that the most successful projects in time and cost were innovative projects. However, the worst projects were also innovative, but in the latter cases the technical difficulty was grossly underestimated. The innovative projects were treated as standard technology and in many cases adequate testing of the process was absent as well as sound basic data packages [57].

Heuristic: A process is novel if it has not previously been in operation with that feed, that product, that catalyst, that piece of equipment, or those conditions at commercial scale

This heuristic rule is often violated. A case I have experienced was a crystallization process step applied for decades in the beverage industry that was introduced at a bulk chemical plant. Because the process step was considered commercially proven, smooth operation was expected. The process step was interrupted every few weeks and requiring cleaning before regular operation could be continued. After questioning it appeared that in the beverage industry cleaning was carried out every week as part of the normal operating procedure. The bulk chemical process, however, was expected to run continuously for four years.

Another case I have experienced is the use of a new batch of catalyst that is perceived as the same as before. The same procedure for startup is followed and then all kinds of problems appear, because in reality the catalyst behaves differently from the previous catalyst. What can happen if novelty is not recognized is dramatically illustrated in the Shell case hereafter.

3.12.8 Case of Shell Moerdijk where new catalyst charge leads to explosion

A dramatic explosion occurred at the Styrene process at Shell in Moerdijk, the Netherlands in 2014, because a new catalyst was charged to a reactor. Fortunately, nobody was injured, because nobody was present in the plant. The damage, however, was so large that it took one year and a half to rebuild the process.

The catalyst was perceived as an improved version of the previously used catalyst. Performance tests under production conditions had been carried out at lab scale. However, nobody had considered the consequences of the difference between chemical composition of the catalyst during preconditioning and startup and no formal new startup and operation risk assessment was carried out. The new catalyst contained 5% Chromate VI instead of the 0.2% in the old catalyst.

The preconditioning of the catalyst in the commercial reactor occurred with ethylbenzene, as always had been used before. Ethyl benzene was considered an inert flushing liquid just to heat up the catalyst bed. In reality, the chromate reacted exothermically with the ethylbenzene causing in the end a runaway reaction resulting in an explosion destroying large parts of the process. As always with a calamity like this there were additional causes resulting in the explosion. The spray installation for the trickle-bed reactor did not function optimally for distributing the ethylbenzene by which local hotspots in the catalyst bed occurred that could not be noticed by the operators. At these hotspots the copper oxide in the catalyst started to react with the ethylbenzene, causing more heat to be produced, which caused evaporation of the ethylbenzene and thereby an enormous pressure increase. In the design, this high pressure was not considered and a pressure relief valve was not part of the design. In addition, the operators of the process at that time had never carried out a startup of this reactor. They did not notice differences such as an unusual pressure profile, a strongly fluctuating ethylbenzene flow, and a controller alarm. Manual operation rather than automatic control operation also contributed to the explosion [58].

3.12.9 First commercial scale process startup

3.12.9.1 Startup preparation

From literature on industrial process startups, Harmsen extracted four critical success factors for the design stages and six for the startup stage [59].

The design success factors are:

- Project is identified as new.
- An integrated pilot plan, as downscaled version of the commercial scale process has been used
- R&D, detailed design and construction personal have meeting to ensure that all knowledge is integrated in the design and construction.
- Scale-up knowledge for unit operations is available.

The six startup preparation measures follow here:

- Potential Problem analysis
- Complete startup team
- Operators trained in startup and normal operation
- Precaution measures
- Startup plan
- Documentation

With these preparation measures, he analyzed ten Shell chemicals startup cases. It appeared that for cases where the process design success factors were fulfilled and where all preparation measures were taken the startup time was 5–10 times shorter than the startup time predicted with the startup time correlation [59].

3.12.9.2 Startup manual

An important element of successful implementation is having an operation manual and a startup manual. Process startups still fail often due to a lack of proper startup procedure and lack of operator training [53]. The startup manual includes the precommission period when all individual equipment is cleansed and tested using inert gas

Tab. 3.17: Startup time correlation derived from industrial cases [54].

		$t_{startup} = 3.3 + 3.7\,N - 3.2\,F + S$
Parameter	**Dimension**	**Description**
$t_{startup}$	Months	Startup time from moment start all feeds till design targets met
N	–	Number of new process steps
F	–	Fraction of mass and heat flows that are known $F = 0$, if mass and heat flows are not known $F = 1$, if mass and heat flows are all precisely known
S	Months	Solids processing parameter $S = 0$, if no solids are processed $S = 0.7$, if refined solids such as plastics are produced $S = 10.8$, if raw solids are fed to the process

and liquids. Also process controls are tested as much as possible for the limits of the process conditions.

Independent project analysis has made a statistical correlation for the startup time of a new process derived from their large database of startups of commercial scale processes. Most of them concerned solids processing units but also others [54]. The startup time correlation is shown in Table 3.17.

A process step is new, if it has not been applied before at commercial scale for that specific input or output. A process step is here defined as a section of the process such as a reaction section or a purification section with a defined input and output. A process step will include one or more unit operations.

This startup time correlation can be used to plan the startup and the related marketing and sales planning. It can also be used to justify a certain development effort, such as having a pilot plant to get the value of F from near 0 to near 1. The difference in startup time of those two conditions is on average 3.2 months. The amount of money lost by not having a specification product for that period in general easily outweighs the cost of a pilot plant.

Further reading on how to make innovation projects successful is provided by Savelsbergh [60].

References and further reading

[1] Goffin K, Mitchell R. Innovation management, strategy and implementation using the pentathlon framework, 2nd edn, Basingstroke: Palgrave Macmillan, 2010.
[2] Cooper RG. Winning at New Products: Creating Value Through Innovation, New York: Basic books, 2017.
[3] Arora A. Implications for energy innovation from the chemical industry, NBER, 2010.
[4] Valencia RC. The Future of the Chemical Industry by 2050, Hoboken: J. Wiley, 2013.
[5] Seider WD, et al. Product & Process Design Principles, 3rd edn, Hoboken: John Wiley & Sons, USA, 2010.

[6] Almeida-Rivera C, Bongers MM, Zondervan E. A Structured Approach for Product-Driven Process Synthesis in Foods Manufacture, Computer Aided Chemical Engineering, 39, 2017, 417–441.
[7] Little AD. Pathway to Innovation; A Global Benchmark study April, 2010. Available from: http://www.adlittle.com/downloads/tx_adlreports/ADL_InnoEx_Report_2010.pdf. Last resourced 15 November 2010.
[8] Harmsen GJ. Dutch Chemical Technology, Top of the world, lecture for AIChE, section BENELUX, the Hague: oral presentation, 2006.
[9] Jaber MY. Learning Curves: Theory, Models, and Applications, Michigan: CRC Press, 2017.
[10] Swanson RM. A vision for crystalline silicon photovoltaics. Progress in Photovoltaics, Research and Applications, 14, 2006, 443–453.
[11] WBCSD, CEO Guide to the Circular Economy, 2017. Available from: http://www.wbcsd.org/Overview/Resources. Last resourced 10 January 2018.
[12] Chakravorti B. What Businesses Need to Know About Sustainable Development Goals, HBR, 2015.
[13] Cheng B, et al. Corporate Social Responsibility and Access to Finance. Strategic Management Journal 2017. Last resourced 10 January 2018. Available from: https://dash.harvard.edu/handle/1/9887635.
[14] Thomson Reuters. ASSET4 ESG scores on credit views key ESG scores on over 3,200 global companies, 2017. Available from: https://customers.reuters.com/community/fixedincome/material/ASSET4ESGSCORES.pdf.
[15] Gorak A. Delft Skyline Debate, Chemical engineering and processing, Special Issue Vol. 51, 2012.
[16] Verloop J. Insight in Innovation, Amsterdam: Elsevier, 2004.
[17] Mehdi Miremadi M, et al. Chemical innovation for the ages, 2017. Available from: http://www.mckinsey.com/industries/chemicals/our-insights/chemical-innovation-an-investment-for-the-ages. Last resourced 10 January 2018.
[18] Palucka T. The Wizard of Octane: Eugene Houdry never trained as a chemist but he made the greatest advance in the history of petroleum chemistry, American Heritage of Invention & Technology 20(3), 2005, 36–45.
[19] Harmsen GJ. Reactive distillation: the frontrunner of industrial process intensification: a full review of commercial applications, research, scale-up, design and operation, Chemical Engineering & Processing, Process Intensification, 46(9), 2007, 774–780.
[20] Walpole H. Letter to Horace Mann, 28 January 1754. Available from: http://www.thefreedictionary.com/serendipity, Last resourced 17 September 2017.
[21] Drent E, et al. Polyketones, Encyclopedia of Polymer Science and Technology, 3, John Wiley and Sons, 2001.
[22] Synthon, 2017. Available from: http://www.synthon.com/Corporate. Last resourced 10 January 2018.
[23] Bakker HLM, et al. Management of Engineering Projects – People are Key, Nijkerk: NAP, 2014.
[24] Nagji B, Tuff G. Managing your innovation portfolio, HBR, 90(5), 2012, 67–74.
[25] Meesters G. DSM technology platforms, 2017. Available from: https://www.dsm.com/corporate/about.html. Last resourced 25 January 2017.
[26] Markham SK. Traversing the Valley of Death: A practical guide for corporate innovation leaders, 2014. Available from www.TraversingTheValleyOfDeath.com. Last resourced 10 January 2018.
[27] Cooper RG. Where are all the breakthrough new products?: Using portfolio management to boost innovation, Research-Technology Management, 56(5), 2013.
[28] Cooper RG. Stage-gate systems: A new tool for managing new products, Business Horizons, 33(3), 1990, 44–54.

[29] Cooper RG, Edgett SJ. Best practices in the idea-to-launch process and its governance, Research Technology Management, Vol. 55, 2012, 43–54.
[30] ConQuastor. Innovative Business Process Management, Utrecht: ConQuastor BV, 2013.
[31] Verganti R. Design-Driven Innovation: Changing the Rules of Competition by Radically Innovating What Things Mean, Boston, USA: Harvard Business Press, 2009.
[32] Kahneman D. Thinking, Fast and Slow, London: Penguin Books, 2012.
[33] Merrow ED. Industrial Megaprojects, Concepts, Strategies, and Practices for Success, Hoboken: John Wiley & Sons, 2011.
[34] Mankins JC. Technology Readiness Levels: A White Paper, NASA, Office of Space Access and Technology, Advanced Concepts Office, 1995. Available from: http://fellowships.teiemt.gr/wp-content/uploads/2016/01/trl.pdf. Last resourced 15 January 2018.
[35] ESA, Technology Readiness Level (TRL) The ESA Science Technology Development Route. European Space Agency, Future Missions Office, Technology Preparation Section. Technology readiness levels (TRL) (PDF). European Commission, G. Technology readiness levels (TRL), HORIZON 2020 – WORK PROGRAMME 2014–2015 General Annexes, Extract from Part 19 – Commission Decision C(2014)4995.
[36] DOE, Technology Readiness Assessment Guide (DOE G 413.3-4). United States Department of Energy, Office of Management, 2011.
[37] EC2014 Technology readiness levels (TRL) (PDF). European Commission, G. Technology readiness levels (TRL), HORIZON 2020 – WORK PROGRAMME 2014–2015 General Annexes, Extract from Part 19 – Commission Decision C(2014)4995.
[38] Graettinger D, Caroline P, et al. Using the Technology Readiness Levels Scale to Support Technology Management in the DOD's ATD/STO Environments: A Findings and Recommendations Report Conducted for Army CECOM (CMU/SEI-2002-SR-027). Carnegie Mellon Software Engineering Institute, 2002.
[39] EARTO, 2014, The TRL Scale as a Research & Innovation Policy Tool, EARTO Recommendations, 2014. Available from: http://www.earto.eu/fileadmin/content/03_Publications/The_TRL_Scale_as_a_R_I_Policy_Tool_-_EARTO_Recommendations_-_Final.pdf. Last resourced 15 January 2018.
[40] Harmsen J. Novel sustainable industrial processes: from idea to commercial scale implementation, Green Processing and Synthesis, 3(3), 2014, 189–193.
[41] Nijstad BA, Paulus PB. Group creativity: Innovation through collaboration, New York: Oxford University Press, 2003, 326–338.
[42] Herrmann D, Felfe J. Effects of Leadership Style, Creativity Technique and Personal Initiative on Employee Creativity, Brit J Manage, 25, 2014, 209–227.
[43] Rademaker E. 101 gouden tips voor de business case, Teteringen: Van Ierland Uitgeverij, 2007.
[44] Anderson JC, et al. Customer Value Propositions in Business Markets. Harvard Business Review, 84(3), 2006, 90–99.
[45] Vogel GH. Process Development From the Initial Idea to the Chemical Production Plant, Berlin: Wiley-VCH, 2005.
[46] Yole. Micro fluidic applications, 2015. Available from: http://www.yole.fr/iso_album/microfluidicapplications_emergingappli_yole__june2015.jpg. Last resourced 15 January 2018.
[47] FluidicMEMS, microfluidics lab on a chip, 2016. Available from: http://fluidicmems.com/list-of-microfluidics-lab-on-a-chip-and-biomems-companies. Last resourced 15 January 2018.
[48] Sekhon BS, et al. Microfluidics technology for drug discovery and development – an overview, Int. J. PharmTech Research, 2(1), 2010, 804–809.
[49] Schwalbe T. Microstructured reactor systems, CHIMIA, 56(11), 2002, 636–646.
[50] Boodhoo KVK, Harvey AP (eds). Process Intensification for Green Chemistry, Chichester: John Wiley & Sons, 2013.

[51] Günther A, Jensen KF. Multiphase microfluidics: from flow characteristics to chemical and materials synthesis, Lab Chip, 6, 2006, 1487–1503.
[52] Kenig EY, et al. Micro-separation of fluid systems: A state-of-the-art review, Separation and Purification Technology, 120, 2013, 245–264.
[53] Harmsen J. Industrial Process Scale-up: A practical innovation guide from idea to commercial implementation, Amsterdam: Elsevier, 2013.
[54] Merrow EW. Estimating start-up times for solids-processing plants, Chem. Eng., 24, 1988, 89–92.
[55] Perry RH. Perry's Chemical Engineers' Handbook, 8th edn, New York: McGraw-Hill, 2008.
[56] Helvoort T, et al. Gas To Liquids Historical development of GTL technology in Shell, Amsterdam: Shell Global Solutions Int, 2014.
[57] De Groen T, et al. 2 × 2 Your choice of projects, twice as cost effective, twice as fast, a guide for key decision makers in the process industry, Leidenschendam: napDACE, 2002.
[58] Biesboer F. Explosie Shell illustreert falend risicomanagement: onvoorzien scenario, de ingenieur, 2016, 50–53.
[59] Sekhon BS, et al. Microfluidics technology for drug discovery and development – an overview, Int. J. PharmTech Research, 2(1), 2010, 804–809.
[60] Savelsbergh C, Martijn Jong M, Storm P. Project success: creating excellent projects. How can I make my project successful? Maastricht, AMI 2011. Resourced from: http://ami-consultancy.com/wp-content/uploads/2016/04/Project-Success-creating-excellent-projects.pdf. Last resourced 15 January 2018.

4 Designing for innovation

4.1 Introduction

This chapter provides theory about design for innovation, an integrated method for breakthrough concurrent product-process design, planning for design and experimentation for innovation, setting criteria for design, and shows roles of modeling in design.

This chapter is meant for industrial product and process designers and PDEng students to generate novel product-process solutions. The provided thinking modes and methods facilitate breakthrough concurrent product-process designs.

For graduate students, step-by-step design methods for novel products and processes are described in Chapters 6 and 7. Modeling and optimization is treated in Chapters 8 and 9. Evaluating designs is treated in Chapters 10–12.

4.2 Design thinking

4.2.1 Design for innovation theory

Design as a verb, is a problem-solving thinking activity. It starts with assigning a specific problem and ends with presenting a solution to the client [1]. The design approach to a problem is to lay out possible solutions and try to improve upon them. The approach is mainly cycling between two thinking modes, a creativity mode and an analysis mode. These two steps are repeated until a satisfactory solution is obtained. The focus is thus on getting a specific solution to a specific problem.

Dorst [2] shows design thinking (and other types of thinking modes) in a scheme, which is summarized in the Table 4.1.

Abduction-1 and abduction-2 are the types of thinking carried out in the field of engineering design. Understanding these differences help companies to organize their innovation and prepare an organization environment, so that both designers and scientific researchers can cooperate. It also helps students, first educated in scientific

Tab. 4.1: Types of thinking to solve the unknown (based on [2]).

Type of thinking	What (thing)	How (working principle)	Result
Deduction	Known	Known	Unknown
Induction	Known	Unknown	Known
Abduction-1	Unknown	Known	Known
Abduction-2	Unknown	Unknown	Known

https://doi.org/10.1515/9783110467741-004

research and subsequently educated in engineering design to understand the large differences in the modes of thinking they must acquire.

In abduction-1, which may be called normal abduction, the engineering designer generates intermediate design solutions to the 'what', and simulates the result of the intermediate solution (by using knwowledge of 'how') and by adjusting the solutions of 'what' until the desired result is obtained. Simulation can be just a sketch and visualizing its working principle ('how'). There are many iterations between solution generation, simulation, and evaluation [2]. Conventional process design follows abduction-1 reasoning. The product is specified. Most feeds are specified. The chemical reaction stoichiometry for the conversion of the feed to the product is known. Working principles for reaction and separations (the 'how') can be selected from textbooks. Unit operations can be selected from textbooks, while sizing is determined using available relations. Chapters 6 and 7 deal systematically with conventional design for novel products and processes.

In abduction-2 both the what (the desired end solution) and the how (the working of the end solution are unknown. The what and the how are then obtained by cycling between what and how to simulate intermediate solutions. Often the intermediate solutions also involve reframing the solution space by thinking first which problem is hardest to solve, redefining the framework of that problem, and then continuing to find intermediate solutions. The reframing of the solution space may also be done several times. This type of design thinking is particularly fruitful for solving complex problems for which no off-the-shelf solutions exist, and which do not lend themselves to solutions generated by scientific research because they cannot be assigned to a specific discipline.

In concurrent product process design abduction-2 design thinking is applied, as both the product and its related manufacturing process is unknown at the start. Section 4.3 provides methods to help the designers in generating combined solutions.

4.2.1.1 Design and risks

An additional view on what characterizes design is from Petroski [3]. He views engineering design as a succession of hypotheses that such and such an arrangement of parts will perform a desired function without fail. Petroski stresses that a design result is always novel and thereby always will have a risk of failure. Petroski also remarks that implicitly the designer assumes that the design will not fail if it is used as intended.

The risks and how to reduce them associated with these two aspects of design are treated in this book explicitly by advocating the testing of novel designs and writing startup and operation manuals for commercial scale implementation.

4.2.1.2 Design links with society and nature

The International Council of Societies of Industrial Design (ICSID) defines design (includes services, processes, and systems) as such: "Design is a creative activity whose aim is to establish the multifaceted qualities of the objects, processes, services, and their systems in whole life cycles. Therefore, design is the central factor for cultural and economic exchange." [4].

The exchange between society and designers is treated extensively in this book in Chapter 2 and in Chapter 4 showing the relations between modal aspects of reality, including society and nature on the one hand and on setting design criteria derived from these aspects on the other hand.

4.2.1.3 Further reading on design

The field of design thinking is rapidly expanding. Good entries are Dorst [2], Verganti [4], Verkerk [5] Lawson [6], and Bont [7]. The latter treats designing all kinds of industrial products and services in complex contexts.

4.2.2 Design knowledge types

Knowledge of designing can be categorized in many ways. The authors found the categorization of Vincenti fitting with their view on the knowledge of designing. Vincenti distinguishes the following categories as presented by Verkerk [5] (relevant chapters in our book are mentioned for each category):

(a) Knowledge of fundamental design concepts; elements a design is made of.
This knowledge is explicitly treated in Section 7.2 on synthesizing product and process designs, and Chapters 8 and 9 on modeling.

(b) Knowledge of criteria. The kinds of demands a design should meet.
This subject is treated in Section 4.5 of this chapter, in Chapter 6, and in Chapters 10–12.

(c) Knowledge of quantitative data.
In our book, these are called basic design data, and are treated in Chapter 6.

(d) Knowledge of theoretical methods and models.
Chapters 8 and 9 deal with modeling for products and processes in detail.

(e) Knowledge of practical considerations.
This aspect is treated in Sections 4.6 and 4.7.

(f) Knowledge of design approaches.
This main topic is explicitly treated in Chapters 4–12.

Experienced designers have gained large amounts of knowledge in all these categories. Novice designers do not have the notion of these types of knowledge. Entry courses on process design treat the knowledge types (a), (b), (c), and (d) explicitly. Knowledge types (e) and (f) are obtained when designers execute design projects.

4.2.3 Differences between design thinking and scientific research

The main differences between design thinking and scientific research thinking are found in Table 4.2. The first four items have been obtained from Verganti [4]. The others have been obtained from Lawson [6].

Tab. 4.2: Differences between creative design and scientific research (based on [4] and [6]).

Parameter	Creative design	Scientific research
Number of ideas	Create numerous ideas	One vision or hypothesis
Peer values	Beginner approach to problems	Knowledgeable scholar approach
Focus	Variety and divergence	Convergence on one paradigm
Work culture	Neutral: if it works it is good	Personal vision
Goal	Specific solution	General theory
Stopping criterion	Good enough	Best
Method	Many (anything that helps)	Hypothesis and experiment
Reliability	Judgment and calculation	Experimental validation
Essence	Prescriptive	Descriptive

4.3 Designing for innovative products and related processes

4.3.1 Introduction

This section provides methods for designing both the novel product and the related novel process effectively and efficiently. It covers all innovation classes and for all innovation stages. It has been used by PDEng students at TU Delft and also in industry. It facilitates team work planning.

4.3.2 Product-process design method: Delft Design Map (DDM)

The presented design method was first defined and developed by Grievink and Swinkels in the period 2002–2014 [8]. In this period, it was also further refined and tested by case work inside a consumer product company by Grievink and Almeida [9]. It combines all important elements: the value chain (with subsystems supply chain and market), product, process, design cycle steps, and innovation stages.

To aid the designer, all these elements have been systematically placed in the Delft Design Map (DDM); shown in Table 4.3.

The first column in this table lists the design steps from the engineering design cycle [8] and [9]. These 'Design (Cycle) Steps' range from 1 Scope to 7 Report.

The top rows of this table lists 12 'Design Levels' shown horizontally in the DDM (3rd row, Table 4.3). The design levels are grouped into substages (framing, supply chain imbedding, process technology, process engineering, and final design):

Framing:
- 'Project Framing' (PF)

Supply chain imbedding:
- Customer Wants – Product Concept (CW-PC)
- Product Concept – Property Function (PC-PF)
- Input-Output structure (I-O)
- Subprocesses (SP)

Process technology:
- Task Network TN)
  Unit Network (UN)
- Process Integration (PI)

Process engineering:
- Equipment Design (ED)
- Operability Integration (OI)
- Flowsheet sensitivity and Optimization (FO)

Final:
- Final Design (Final)

The design levels have a clearly ordered sequence that is synchronized with the innovation stages (DDM, first row) defined in the Stage Gate™ framework [10].

Each cell in the resulting table stands for a set of focused design activities to be carried out. A cell is called a 'Design Activity Space (DAS)'. Each activity space is confined by the design level and design step. The required activities in each space are thereby relatively limited and straightforward. Advice for the activities to be executed in each design activity space is indicated with a section reference number of this book.

4.3.3 Explaining the Delft Design Map for product-process design

The DDM table's purpose is threefold. First the DDM table is used to quickly find information on what design activity to perform and how. The second purpose of the table is facilitating the planning of the work to be done. The third purpose of the table is to help categorize and check the design result, and whether all activities have been carried out.

In this paragraph, the engineering 'design (cycle) steps' and the 'design levels' are described in further detail. Recall that the design (cycle) steps are carried out for each

Tab. 4.3: Delft Design Map structure (based on [8] and [9]).

Stage gate™:			**Concept**			
Substages:		**Framing**	**Supply chain imbedding**			
Design levels ⇒ ⇓ **Design steps**		**(PF) Project Framing**	**(CW-PC) Consumer Wants – Product Concept** (product performance), quality function	**(PC-PF) Product Concept – (product) Property Function**	**(I-O) Input-Output structure** Products, feeds, wastes, utilities, process functions	**(SP) Subprocesses** Subprocesses, action blocks, technology, I/O structures
1	**Scope** (of design)	6 ⇓	6 ⇓	6 ⇓	6 ⇓	6 ⇓
2	**Knowledge** (about objects)	6.5 ⇓	6.5 ⇓	6.5 ⇓	6.5 ⇓	6.5 ⇓
3	**Synthesize** (alternatives, structure, scale)	7 ⇓	5.2, 5.4, 5.5, 7.2.2, 8.1–8.3 ⇓	5.2, 5.4, 5.5, 7.2.2, 8.1–8.3 ⇓	7.2.3 ⇓	7.2.3 ⇓
4	**Analyze** (physical behavior)	7.3 ⇓	7.3, 8.4–8.6 ⇓	7.3, 8.4–8.6 ⇓	7.3, 9 ⇓	7.3, 9 ⇓
5	**Evaluate** (performance vs specifications)	7.4 ⇓	4.5, 7.4, 10–12 ⇓	7.4, 8.7, 10–12 ⇓	7.4, 10–12 ⇓	7.4, 10–12 ⇓
6	**Select** (alternatives)	7.5 ⇓	7.5 ⇓	7.5–7.6 ⇓	7.5 ⇓	7.5–7.6 ⇓
7	**Report** (decisions & results)	13	13	13	13	13

design level, and that each combination of a design step and a design level forms a cell in the DDM, called a Design Activity Space (DAS). The numbers in each cell of the DDM table refers to a chapter or section in this book.

Tab. 4.3: (continued)

Feasibility			Development			
Process technology			**Process engineering**			**Final**
(TN) Task Network Allocation, targeting, mechanism, rates of task integration in compartments	**(UN) Unit Network** Purification, conversions, separations, assembly stabilization	**(PI) Process Integration** Energy, solvents, water, emissions, production time	**(ED) Equipment Design** Structure, size, materials, reliability	**(OI) Operability Integration** Safety, monitoring, control, flexibility, availability	**(FO) Flowsheet sensitivity and Optimization**	**(Final) Final design** Product & process (PFD)
6 ⇓	6 ⇓	6 ⇓	6 ⇓	6 ⇓	6 ⇓	
6.5 ⇓	6.5 ⇓	6.5 ⇓	6.5 ⇓	6.5 ⇓	6.5 ⇓	Design data, heuristics & models
7.2.3 ⇓	7.2.3 ⇓	5.6,5.7,5.8, 7.5 ⇓	7.2.3 ⇓	7 ⇓	9 ⇓	Flowsheet & control structure
7.3, 9 ⇓	7.3, 9 ⇓	5.8, 9 ⇓	9 ⇓	9 ⇓	9 ⇓	Mass & energy yield factors
7.4, 10–12 ⇓	7.4, 10–12 ⇓	7.4, 10–12 ⇓	7.4, 10–12 ⇓	7.4, 10–12 ⇓	7.4, 10–12 ⇓	SHEETS evaluation
7.5 ⇓	7.5 ⇓	7.5–7.6 ⇓	7.5 ⇓	7.5 ⇓	7.5–7.6 ⇓	Selected design(s), 3.12
13	13	13	13	13	13	Go/ No Go?

4.3.3.1 Design (cycle) steps description:

In the bullet list below the design (cycle) steps are described, with a few examples of what design aspects and activities are relevant for selected design levels. The design levels will be elaborated in more detail after the bullet list. Chapter 6 covers items (1) and (2), Chapter 7 covers items (3–6), and Chapter 13 covers item (7) reporting.

(1) Set the design scope, involving:
   - Design goal
   - Constraints
(2) Acquire and assess knowledge about potential building blocks (depending on design level):
   - Target market size and dynamics
   - Competitor activities/patents
   - Raw material supply specifications
   - Pure component properties (physical, safety, health data) for components, and combined streams
   - Thermodynamic equilibria, kinetics
(3) Synthesize alternative designs (networks/structures) to meet the design scope by:
   - Selecting suitable building blocks/objects at this design level
   - Varying order and connectivity of these building blocks

   It is important to aim for a quantity of solutions first before moving to step 4, Analyze, to avoid the too-early elimination of design options that are also valuable in combination with other options. Creativity methods and tools discussed by Tassoul [11], 'TRIZ' methodology [12], and synthesis heuristics (Section 4.7.1) are sources to help create a large number of design options.
(4) Analyze the physical-chemical behavior of generated alternative designs (networks):
   - Estimates, computations, and/or model simulations connecting resource flows to building blocks
   - Expert analysis by sketches and drawings of product and process block flow diagrams

   The aim of this step is to quantify the design options behavior at the design level at hand. Aim to arrive at quantitative data like yield estimates mass and heat balances, separation efficiency, utilities used, etc. as much as possible. These quantitative data are the input into the next design step, evaluate, where the quantitative data are translated into performance criteria.
(5) Evaluate the performance of alternative solutions against the SHEETS (see Section 4.5.3) sustainability goals and constraints:

   S: Safety
   H: Health
   E: Environment
   E: Economy
   T: Technology
   S: Social

   For the supply chain related design levels the estimates will be less accurate than for the estimates at the equipment design level; it is crucial to make every attempt to quantify the performance of the various design options for each design level, as it will make clear where design options perform well or not, and where the design focus should be drawn to in the subsequent design levels.

(6) Select a limited number of preferred solutions from the alternatives for the next design level.
   - Eliminate/weed out unfeasible options.
   - Keep promising parts/interesting aspects of eliminated options for implementation in other options.

(7) Report all design outcomes, drawing, mathematical models, analysis and evaluation results, key decisions, underlying rationales, and numerical data for transfer to the next design level.
   - Write, store, and share activity reports for key design activities at each design step (see Chapter 13 on communication).

4.3.3.2 Executing design activities in all twelve design levels

In this section design activities for the design levels in the framing, supply chain imbedding, process technology, process engineering and final substages of the Delft Design Map will be shown.

1. Design level PF: Project Framing (substage: framing)

Important design activities in the PF-Project Framing design level can be summarized as follows:

- Frame the product and/or process design task in the overall business innovation project ('BOSCARD': Background, Objectives, Scope, Constraints, Assumptions, Reporting, Deliverables):
  - **B**ackground of business context
    - Marketing and supply chain aspects and opportunities
      - 6P-model (proposition, promotion, position, packaging, price, product)
    - Design driver (company profile, market pull, technology push)
    - Design type (new product family and process platform, platform derivative, manufacturing improvement, platform development)
  - **O**bjectives
  - **S**cope
  - **C**onstraints
- Describe the foreseen design approach.
  - Design methodology
  - **A**ssumptions
- Project organization
  - Team members and roles
  - **R**eporting

  - Deliverables and milestones
  - Project and time planning
- Result: ***Project Brief***

2. Design level CW–PC: Consumer Wants – Product Concept (substage: supply chain imbedding)

- Identify direct product users.
- Identify stakeholders in product/process life cycle.
- List (qualitative) stakeholder/consumer needs.
- Translate (qualitative) needs into product performance attributes (quality function) and their quantitative metrics.
  - E.g., using Quality Function Deployment (QFD: House of Quality [HoQ]) methodology (see Section 8.3)
  - Rank product performance attributes.
  - Performance attributes (quantitative) specifications
    - Targets and acceptable ranges
  - Make relationships explicit between different product attributes (using HoQ method).
  - Identify the key (product) design challenges.
- Result: ***House of Quality (1)*: product performance (quality) attributes mapped onto consumer wants**

3. Design level PC–PF: Product Concept – (product) Property Function (substage: supply chain imbedding)

- Create/synthesize product concepts (with existing or novel product application processes, aiming to fulfill the product (quality) performance metrics.
  - Find thermodynamic phases and physical scales to compose different product structures.
  - Find chemical/biological compositions of these various phases.
  - Analyze the behavior of the structured products using mathematical modeling and/or experimentation.
  - Evaluate product design options against performance metrics.
- For each product design option also review:
  - Tolerances on variability of product structure and compositions per grade
  - Expected economic life span of product(s) and evolution of its price range
  - Overview of suitable alternative feeds to make the product(s)
  - Recovery of materials from spent product as secondary feed for manufacturing or in a circular economy
- Result: ***House of Quality (2)*: product property function**

  - Product composition, structure, physical properties – combined with the product application process, **mapped onto the product performance (quality) attributes**

4. Design level I-O: Input-Output structure (substage: 'supply chain imbedding')

- Imbedding of the process in a manufacturing supply chain and choice of future site
- Specification of products and options for cogeneration of chemical products and energy
- Selection of chemical product grades to be manufactured and available feeds
- Demarcate the process from its environment by setting boundaries ("battery limits").
- Specification of production modes (campaigns) and operation mode(s)
- Estimation of mass balances based on conversion yield factors
- Identification of possible waste streams
- Estimation of expected economic and physical life span of process plant
- Duties and SHEETS (Safety, Health, Environment, Economy, Technology, Social) specifications for process plant and external permits
- Result: ***Input-Output Diagram*** **(incl. feeds, (by)products, solid/liquid/vapor (S/L/V) waste streams, technologies**

5. Design level: subprocesses allocation (substage: supply chain imbedding)

- Decompose the main process into subprocesses (see **NOTE**) with high level manufacture process functions – for supply chain mapping.

NOTE: *these subprocesses should be envisaged as large supply chain blocks (not as a split of a single product plant) to be located at different locations, and/or run by separate legal entities, and/or using a single technology and/or selling their products to more customers. It is not meant as the split of a single production plant into reaction, separation, and polishing steps. Each subprocess will be further designed designing its own tasks network and unit network design spaces. For maximum process intensification, the subprocesses should be defined quite broad as is realistically possible.*

- Choose the use of key technologies per subprocessing function (licensing).
- Decide on the making, buying, or selling of intermediates or product components.
- Design the connectivity and recycle structure between subprocesses.
- Decide on (re-)use of standard processing equipment for short process life span.
- Or: design subprocesses with innovative processing operations for a long economic life.
- Result: ***Block Diagram*** **with selected subprocesses and processing actions (and different feeds, (by)products, S/L/V waste streams; including preliminary mass balances (yield factors) and evaluation regarding gross margin, transportation cost estimates, etc.**

6. Design level TN: Task Network (substage: process technology)

- Task network: allocation, targeting, and integration
  - Decomposition of subprocesses into a network of elementary tasks (*allocation*)

A task achieves a change in the state of matter from an inlet to a target outlet state, so demanding a processing duty (targeting). Per task, a (bio) chemical and physical mechanism (e.g., reactions, phase transfer) and operating conditions are selected.

- Balancing of rate processes and finding space-time requirements
- Identification of compatible tasks suitable for integration in spatial compartments
- Sizing of compartments, with due consideration of various production modes
- Results: **task network(s), represented in a *Task Block Diagram***

7. Design level UN: Unit Network (substage: process technology)

- Create spatial structures (e.g., cascades) and operating modes (e.g., fed-batch) for the (task integrated) compartments.
- Integrate compartments in operational units achieving the main processing functions (e.g., conversion, separation).
- Adjust connections and recycle structures between operational units.
- Result: **process unit network at mesolevel, represented in a *Unit Block Diagram***

8. Design level PI: process integration (substage: process technology)

- Efficiently match of internally consumed and generated physical resources by creating a network of exchanges between process units, involving mass (solvents), and various forms of energy (thermal, mechanical, power).
- Obtain any remaining unbalanced duties from the utility systems.
- Consider the operability aspects of the resource exchanges networks.
- Deal with ceilings on integrated emissions from the plants (e.g., SO_2, CO_2).
- Account for multiproduct integration aspects with respect to unit run times and storage.
- Set targets for the system's availability as built-up from availability of units.
- Result: ***Resource exchange networks* at the macrolevel of the entire subprocess.**

9. Design level ED: Equipment design (substage: process engineering)

- Select and size *existing* type of equipment, which can host process unit operations.
- Or: design better performing *new* equipment for task integrated operational units.
  - Spatial structures and space-time requirements fix the main equipment dimensions.

- Choose materials to safely accommodate processed materials and conditions.
- Set targets for 'RAM' (= Reliability, Availability, Maintainability) for equipment items.
- Fix multiplicity and overdesign of identical equipment (for size, safety & reliability).
- Identify (inequality) constraints confining feasible window of operation of equipment.
- Determine process variables in equipment to be controlled with sensors and actuators.
- Result: ***Equipment Specification sheets***.

10. Design level OI: Operability integration (substage: process engineering)

- Apply a safety analysis, evaluation and adaptations (e.g., DOW Fire & Explosion Index).
- Perform flexibility analysis and feasible windows of plant operation.
- Specify key features of an operator – process information interface with time scales.
- Select instrumentation for process performance & product quality monitoring.
- Develop models to convert measured plant data to process management information.
- Design a plant-wide stabilizing control structure per operating mode.
- Check on feasibility of mode switches, start-up & shut-down.
- Determine 'RAM' features and run-time per production mode.
- [Optional: design an optimizing control structure for product quality and throughput.]
- Result: ***Process Flow Diagram*** **optimized regarding process control.**

11. Design level FO: Flowsheet sensitivity and Optimization (substage: Process engineering)

- Select the key performance indicators (KPIs) of the process plant.
- Identify sensitivities of KPIs to main design variables of process units. Perform an uncertainty analysis of the design and make necessary adaptations.
- Do a Pareto trade-off optimization of some KP's for final designs.
- Result: **Optimized** ***Process Flow Diagram*** **(*PFD*)**

12. Design level Final: Final design (substage: final)

- Consolidate and compile:
  - All design data, heuristics and models
  - Product composition and structure
  - Process Flow Diagram, control structure, equipment design

  - Mass & energy balances, including yield factors
  - SHEETS Performance metrics (Safety, Health, Environment, Economy, Technology, Social)
  - SWOT (Strengths, Weaknesses, Opportunities, Threats) of the design and risk analysis and mitigation suggestions
- Evaluate:
  - Project results
  - Project design process
- Recommendations for next steps:
  - Go/No Go? To Manufacturing & Product Introduction Stage-Gate™ phases.
- Write and file Final Report.
- Close design project.

4.3.4 How to plan and execute design activities using the Delft Design Map

The general way of working with the DDM is based on the presumption that the design results in the preceding design level ($N - 1$) are available for transfer to design level (N). Here, the design is further expanded, for which the steps of the engineering design cycle are executed (again). The design results so obtained in design level (N) are passed on to design level ($N + 1$).

For a complete novel product and process design the team would indeed start at the first design level and carry out the design activities as indicated by the engineering design steps (see Figure 4.1). Then move to the next design level and repeat the cycle of design steps again. This can be repeated until all design levels have been covered. This is indicated with the arrows in Table 4.3. No design process is strictly sequential, so knowledge and understanding generated in the later design levels can be 'ploughed back' into earlier design levels to improve the design. The circular setup of the design steps around the design team aims to suggest that steps can be worked on simultaneously and iteratively by the team members, rather than in a strictly sequential manner. As long as the information is reported, classified and shared between the team members, the design results can be captured.

Another way of working with the DDM is to first assess the design case at hand, and then use the DDM to assess what design levels and design step combinations (Design Activity Spaces: cells in the DDM) are crucial at the start of the project. Execute the design activities, and use the results to plan further navigation through the DDM, especially when a complete novel product and process design is not required. The DDM can help the team making a navigation plan and allocate resources accordingly. Yet a different way is to use one of the predefined 'fingerprint DDM tables' relating to a design case's innovation class. This will be further explained in Section 4.3.5.

Almeida [9] describes the strengths and weaknesses of the strictly sequential, the simultaneous, and the 'anticipating sequential' approaches. The strictly sequential

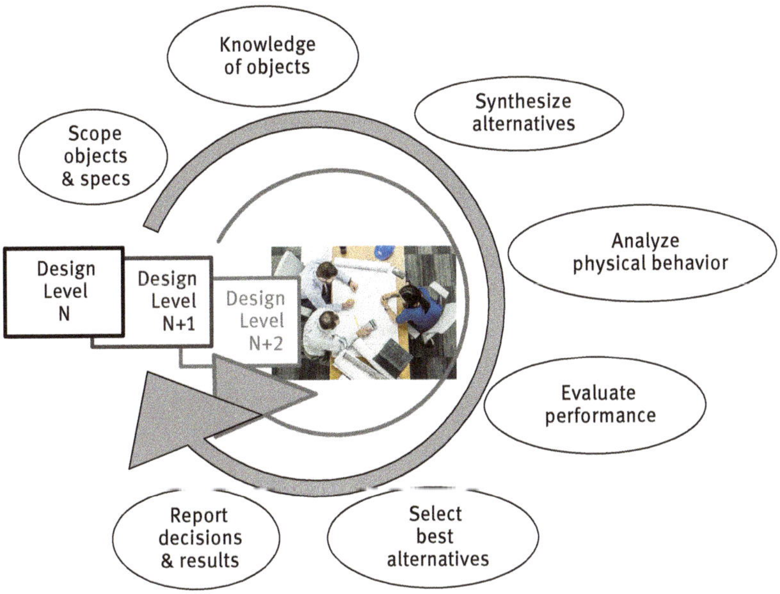

Fig. 4.1: Design steps of the generic engineering design cycle. Photo: Creative Commons.

approach is the most intuitive and, therefore, currently most adopted, but it also the one that needs multiple iteration cycles. Attempts are being made for simultaneous designs of products and processes, but this requires the capability of formulating mathematical descriptions of the product's physical and chemical functions in use and the manufacturing process to produce the required product structure/composition. For some cases, this modeling is feasible and Section 4.7 provides a method for this modeling. However, for most product-process cases, modeling solves only a part of the design problem.

As this field has only recently begun to further develop, the current best bet is the anticipating sequential approach that can use mathematical optimization at restricted design levels, but follows the sequential approach to a large extent.

The key aspect is to stress the anticipating sequential approach as a hybrid to concurrent design of product and process: you do not necessarily move through the design levels in a sequential way and without recycling. Multiple design levels can be started, especially when novel product/manufacturing processes are targeted and new knowledge and insights need to be acquired that potentially and most probably are affecting each other mutually.

The design levels are merely a demarcation of product and process aspects that need to be addressed, but the design team needs to maintain overview of the impacts the different design levels have. Moving back and forth between design levels for analysis, evaluation, and certainly also synthesis steps is key for the overall best result, and the Delft Design Map caters very well to this [9].

4.3.4.1 Practical benefits of working with the Delft Design Map

The practical benefits of using the DDM are:

- Sure-footed and faster design processes due to having a clear design structure
- More inventive power by greater awareness of the design levels, design steps, and design options
- Better options for supply chain integration and process intensification at multiple scales in the process
- Comprehensive coverage of process integration of all shared resources
- Applicability to processes for structured products
- Better integration of operability aspects into design
- Proper documentation of design decisions and rationales
- Easier reuse of a design for retrofitting and revamps using such documentation

4.3.5 Design planning with Delft Design Map for various innovation classes

Chapter 3 provides a table with innovation classes for products and processes with their markets and supply chains, including several examples. In this paragraph, we are linking each innovation class to the design focus that is required. For several innovation classes, so-called fingerprint tables are defined using the DDM. The design activities are classified as 'key design focus' (K), 'optional design focus' (o), 'default design activity' (D), and 'check implications of design decisions' (C). With these fingerprint tables the designer can quickly see where the main design efforts should be focused.

The fingerprint tables have been derived from observing about 200 PDEng product/process innovation design projects carried out for industrial partners in the past decade by Delft University of Technology (TU Delft).

4.3.5.1 Fingerprints of Delft Design Map for design task planning

Tab. 4.4: Legend belonging to fingerprint Tables 4.5–4.10.

Legend	Design levels	
K: Key design focus	PF: Project Framing	UN: Unit Network
o: Optional design focus	PC/CW: Product Concept/ Consumer Wants	PI: Process Integration
D: Default design activity	PC/PF: Product Concept/ Property Function	ED: Equipment Design
C: Check implications of design changes	I-O: Input-Output structure	OI: Operability Integration
_: No activity	SP: Subprocesses	FO: Flowsheet sensitivity and Optimization
	TN: Task Network	Final: Final design

Tab. 4.5: Fingerprint for innovation class 6: radically new
(New market, completely new product, new supply chain, and new process).

Stage gate™:		Concept				Feasibility			Development			
K: Key design focus o: optional design focus D: Default design activity C: Check implications design changes _: No activity	**Framing**	**Supply chain imbedding**				**Process technology**			**Process engineering**			**Final**
Design levels ⇒ **⇓ Design steps/ activities**	**(PF)**	**(CW-PC)**	**(PC-PF)**	**(I-O)**	**(SP)**	**(TN)**	**(UN)**	**(PI)**	**(ED)**	**(OI)**	**(FO)**	**(Final)**
1 **Scope**	K	D	D	D	D	D	D	D	D	D	D	D
2 **Knowledge**	D	K	K	K	K	K	K	D	D	D	D	D
3 **Synthesize**	K	K	K	K	K	K	K	K	K	K	o	K
4 **Analyze**	K	K	K	K	K	K	K	K	K	K	o	K
5 **Evaluate**	K	K	K	K	K	K	K	K	K	K	o	K
6 **Select**	K	K	K	K	K	K	K	K	K	K	o	K
7 **Report**	D	D	D	D	D	D	D	D	D	D	D	D

Tab. 4.6: Fingerprint for innovation class 5: radically new
(Existing market, new product [new platform], new supply chain and new process).

Stage gate™:		Concept				Feasibility			Development			
K: Key design focus o: optional design focus D: Default design activity C: Check implications design changes _: No activity	**Framing**	**Supply chain imbedding**				**Process technology**			**Process engineering**			**Final**
Design levels ⇒ **⇓ Design steps/ activities**	**(PF)**	**(CW-PC)**	**(PC-PF)**	**(I-O)**	**(SP)**	**(TN)**	**(UN)**	**(PI)**	**(ED)**	**(OI)**	**(FO)**	**(Final)**
1 **Scope**	K	D	D	D	D	D	D	D	D	D	D	D
2 **Knowledge**	D	K	K	K	K	K	K	D	D	D	D	D
3 **Synthesize**	K	K	K	K	K	K	K	K	K	K	o	K
4 **Analyze**	K	K	K	K	K	K	K	K	K	K	o	K
5 **Evaluate**	K	K	K	K	K	K	K	K	K	K	o	K
6 **Select**	K	K	K	K	K	K	K	K	K	K	o	K
7 **Report**	D	D	D	D	D	D	D	D	D	D	D	D

Tab. 4.7: Fingerprint for innovation class 4: new
(Existing market, modified product [platform derivative], modified supply chain, and new process).

Stage gate™:		**Concept**				**Feasibility**			**Development**			
K: Key design focus o: optional design focus D: Default design activity C: Check implications design changes _: No activity	**Framing**	**Supply chain imbedding**				**Process technology**			**Process engineering**			**Final**
Design levels ⇒ **⇓ Design steps/ activities**	**(PF)**	**(CW-PC)**	**(PC-PF)**	**(I-O)**	**(SP)**	**(TN)**	**(UN)**	**(PI)**	**(ED)**	**(OI)**	**(FO)**	**(Final)**
1 **Scope**	K	D	D	D	D	D	D	D	D	D	D	D
2 **Knowledge**	D	D	D	K	K	K	K	D	D	D	D	D
3 **Synthesize**	K	_	_	K	K	K	K	K	K	K	o	K
4 **Analyze**	K	C	C	K	K	K	K	K	K	K	o	K
5 **Evaluate**	K	C	C	K	K	K	K	K	K	K	o	K
6 **Select**	K	_	_	K	K	K	K	K	K	K	o	K
7 **Report**	D	D	D	D	D	D	D	D	D	D	D	D

Tab. 4.8: Fingerprint for innovation class 3: new
(Existing market, existing product, existing supply chain, and new process).

Stage gate™:		**Concept**				**Feasibility**			**Development**			
K: Key design focus o: optional design focus D: Default design activity C: Check implications design changes _: No activity	**Framing**	**Supply chain imbedding**				**Process technology**			**Process engineering**			**Final**
Design levels ⇒ **⇓ Design steps/ activities**	**(PF)**	**(CW-PC)**	**(PC-PF)**	**(I-O)**	**(SP)**	**(TN)**	**(UN)**	**(PI)**	**(ED)**	**(OI)**	**(FO)**	**(Final)**
1 **Scope**	K	D	D	D	D	D	D	D	D	D	D	D
2 **Knowledge**	D	D	D	D	D	K	K	D	D	D	D	D
3 **Synthesize**	K	_	_	C	C	K	K	C	K	K	o	K
4 **Analyze**	K	C	C	C	C	K	K	C	K	K	o	K
5 **Evaluate**	K	C	C	C	C	K	K	C	K	K	o	K
6 **Select**	K	_	_	C	C	K	K	C	K	K	o	K
7 **Report**	D	D	D	D	D	D	D	D	D	D	D	D

Tab. 4.9: Fingerprint for innovation class 2: novel
(Existing market, existing product, existing supply chain, and modified process).

Stage gate™:		**Concept**				**Feasibility**			**Development**			
K: Key design focus o: optional design focus D: Default design activity C: Check implications design changes _: No activity	**Framing**	**Supply chain imbedding**				**Process technology**			**Process engineering**			**Final**
Design levels ⇒ **⇓ Design steps/ activities**	**(PF)**	**(CW-PC)**	**(PC-PF)**	**(I-O)**	**(SP)**	**(TN)**	**(UN)**	**(PI)**	**(ED)**	**(OI)**	**(FO)**	**(Final)**
1 **Scope**	**K**	D	D	D	D	D	D	D	D	D	D	D
2 **Knowledge**	D	D	D	D	D	D	D	D	D	D	**K**	D
3 **Synthesize**	**K**	_	_	_	_	_	_	_	_	_	**K**	**K**
4 **Analyze**	**K**	C	C	C	C	C	C	C	C	C	**K**	**K**
5 **Evaluate**	**K**	C	C	C	C	C	C	C	C	C	**K**	**K**
6 **Select**	**K**	_	_	_	_	_	_	_	_	_	**K**	**K**
7 **Report**	D	D	D	D	D	D	D	D	D	D	D	D

Tab. 4.10: Fingerprint for innovation class 1: conventional
(Existing market, existing product, existing supply chain, and capacity increase process).

Stage gate™:		**Concept**				**Feasibility**			**Development**			
K: Key design focus o: optional design focus D: Default design activity C: Check implications design changes _: No activity	**Framing**	**Supply chain imbedding**				**Process technology**			**Process engineering**			**Final**
Design levels ⇒ **⇓ Design steps/ activities**	**(PF)**	**(CW-PC)**	**(PC-PF)**	**(I-O)**	**(SP)**	**(TN)**	**(UN)**	**(PI)**	**(ED)**	**(OI)**	**(FO)**	**(Final)**
1 **Scope**	**K**	D	D	D	D	D	D	D	D	D	D	D
2 **Knowledge**	D	D	D	**K**	**K**	D	D	D	D	D	D	D
3 **Synthesize**	**K**	_	_	**K**	**K**	_	_	_	_	_	_	**K**
4 **Analyze**	**K**	C	C	**K**	**K**	C	C	C	C	C	C	**K**
5 **Evaluate**	**K**	C	C	**K**	**K**	C	C	C	C	C	C	**K**
6 **Select**	**K**	_	_	**K**	**K**	_	_	_	_	_	_	**K**
7 **Report**	D	D	D	D	D	D	D	D	D	D	D	D

This Delft Design Map has been developed over decades with strong interaction between developments in the global field of product and process design, developments in industry, and developments in teaching at Delft University. Its complete historic development is found in Appendix A4.

4.4 Planning for design and experimentation in innovation

Design and experimentation both play important roles in innovation. The major methods for planning both activities are provided in Chapter 3. The Delft Design Map, however, can also be used for planning both design and experimentation. In all cases where knowledge cannot be obtained from written sources, and where analysis cannot be done by calculations, experimentation will be needed. Validating designs will also be done by using experimentation.

There is a general efficient approach to planning design and experimentation, namely by making a design first and listing all assumptions needed for that design. For all assumptions that are critical to success and for which no written data can be obtained, an experimental plan is made. If the execution of that experimental plan is expensive and takes a long time, then the design may be altered in such a way that the assumption is not needed any more, and the related experimental program is then avoided. In this way only the most essential experiments are carried out.

This working method is particularly fruitful if the concept designer and the experimentalist maintain strong interactions. As soon as the experimentalist foresees expensive long lasting experiments the designer can modify the design and ask again an estimate of time and effort for the major assumptions to be tested.

4.4.1 Decision sequence ranking method

Information and knowledge from many different knowledge fields are affecting the design. For strong interactions between many fields of information, a decision ranking method can be of value. This was proven for an industrial case with eight different fields of information by Korevaar in 2004 [13]. Each field is judged by its influence on the performance of the design by the team members with a simple score: strong, weak, or no. Each field is also scored on active, or passive related to the other fields. With these scores, the fields are classified as:

(I) High performance influence and highly active
(II) High performance passive
(III) Low performance active
(IV) Low performance passive.

The field I experiments are performed first. If their result is negative the whole design is stopped and a new design is made.

4.5 Embedding design by criteria and context setting

4.5.1 Introduction to the purpose of criteria and context setting

In designing a product, a process, or any artefact, the designer tries to take all relevant aspects into account, and in such a way that the new artefact fits well in reality. Only then the new artefact will be successful. At the start of the design, this can be done by deriving design criteria from a description of reality that is relevant to the design. This subject is treated in Section 4.5.2.

4.5.2 Comprehensive list of modal aspects for defining criteria from modal aspects of reality

The most comprehensive description of reality – in terms that are easy to understand and that are relevant to design problems – is provided by Verkerk [5]. He provides a list with 15 modal aspects of reality, shown in Table 4.11.

This table also contains suggestions of what design criteria type can be defined for each modal aspect for a design (2nd column). Guidelines for this procedure are provided for each modal aspect hereafter.

Tab. 4.11: Modal aspects and design criteria. Modal aspects from Verkerk [5].

Modal aspect	Criteria type	Examples
Belief	Agreement with belief	Halal, Kosher
Moral	Moral values	Safe, health, sustainable
Juridical	Legal	Within the laws
Esthetical	Esthetic	Attractive appearance
Economic	Economic	Profitable, efficient
Social	Social	Socially accepted
Linguistic	Communication	Clear reporting
Shaping	Shape	Correct 3D shape
Logical	Logical	Correct model calculations
Sensitive	Sense	Tasty product
Biotic	Biological	No patogen contamination
Physical/Chemical	Properties and principles	Technically feasible
Kinematical	Flow rates	Heat and mass flows correct
Spatial	Geometry	Within plot area
Numerical	Numerical	Mass balances closed

Regarding the modal aspect 'belief,' the generation of design criteria needs elaborate consideration. The governing beliefs in the countries in which the product and the process will be present will have to be considered for acceptance. The obvious criterion for food products and processes for Muslim belief for instance is whether it is halal. However, for nonfood products where the feed materials for the manufacturing process are from animal origins, this criterion also holds. Even the choice of location of the process can be subject to belief aspects. A process for Liquefied Natural Gas (LNG) was planned for a location in the north part of Australia. Late in the project, when its location was publicized, it appeared that the location was considered holy ground by Aboriginal inhabitants. The location had to be changed due to their protests. As a consequence, the design also had to be modified, due to the different conditions at the new location.

Moral values play a strong role in safety, health, environment, social, and sustainable development criteria. Most companies have developed specific design criteria for these aspects, and report on these in their annual sustainability reports.

Social acceptance of new products and processes are of course of enormous importance. If a product is not accepted, there will be no market for it. Social acceptance is very difficult to predict. In most cases it needs testing.

The linguistic model aspect is about human communication. A design in itself is information to be communicated to others. Therefore, the linguistic model aspect is very important. Clear reporting, presenting and obtaining feedback is vital for any design.

The shaping aspect of design is about the three-dimensional shape of product and process. The shape of a product is important for its use, transport, and storage. The shape and outline of the process must fit within the given plot limits.

The logical aspect is about reasoning, modeling, and calculations regarding the design. Mass and heat balances for instance must be correct.

The sense aspect is particularly important for product design. In most design cases for each sense, at least one criterion will need to be defined.

The biotic aspect is particularly important for biological and food products and processes. Contamination by pathogenic micro-organisms for instance must be avoided. Also for process design biotic criteria are important, for example, in preventing micro-organism induced corrosion.

Physical and chemical aspects concern all product and process designs. The technical feasibility of a product or a process will always depend on the physical and chemical properties of the materials selected.

The kinematic aspect is about movement. It is particularly important for flows inside the process. If the velocity becomes higher than the speed of sound, unwanted effects may occur.

The spatial aspect is the space taken up by the product and process, and by the material transport related to the product and process design.

The numerical aspect is involved in all computations made for the design. The calculations should be based on correct mathematical equations.

4.5.3 SHEETS criteria list

For students to learn to scope a design a limited set of criteria should be used. Often such a limited set is also used inside companies. This limited set can be seen as the minimum criteria set for each innovation project and its design. In this book the set is called SHEETS. It concerns the following criteria types:
- Safety
- Health
- Environmental
- Economic
- Technical
- Social

This list also incorporates the triple P (3P) definition of sustainable development of People (Social), Planet (Environmental), and Prosperity (Economics). Chapters 10–12 provide detailed descriptions of these SHEETS criteria and how to use these for evaluating designs.

4.6 Role of modeling and simulation in concurrent design

Models and simulations play various roles in design. In the discovery and concept stage it is a fast way of determining the feasibility of an idea or concept. The famous inventor Tesla could imagine or simulate in his mind the working of a novel concept, like the dynamo. Only when he could envisage it working, would he construct a prototype. Some product developers work in a similar way. They say things like, 'I envisage this to be working.' For novel process concepts, experienced engineers can also imagine the working of a complex process. To the authors' knowledge, a few guidelines can be given to facilitate this type of imagination and simulation. One guideline is to consult an experienced developer about your idea and discuss. A second one is to take time to visualize the working of your idea using drawings, schematics, or mockups as presented next.

A more common way of modeling a concept is to produce a drawing, construct a mathematical model, and (mathematically) simulate the working of the product or process. In this way problems with the concept surface soon, and modifications can be made to overcome the problems. This is an important role in the analysis of preliminary design solutions.

In the radical concurrent product-process design, modeling can also play this role. As the synthesis of a product and its process, concepts are carried out in a concurrent manner, the mathematical modeling role is more complex. A workable design flow diagram involving modeling and simulation of product and process has been generated by Grievink, Swinkels, and Almeida. A simplified version is shown in Figure 4.2 below (see also Section 8.5).

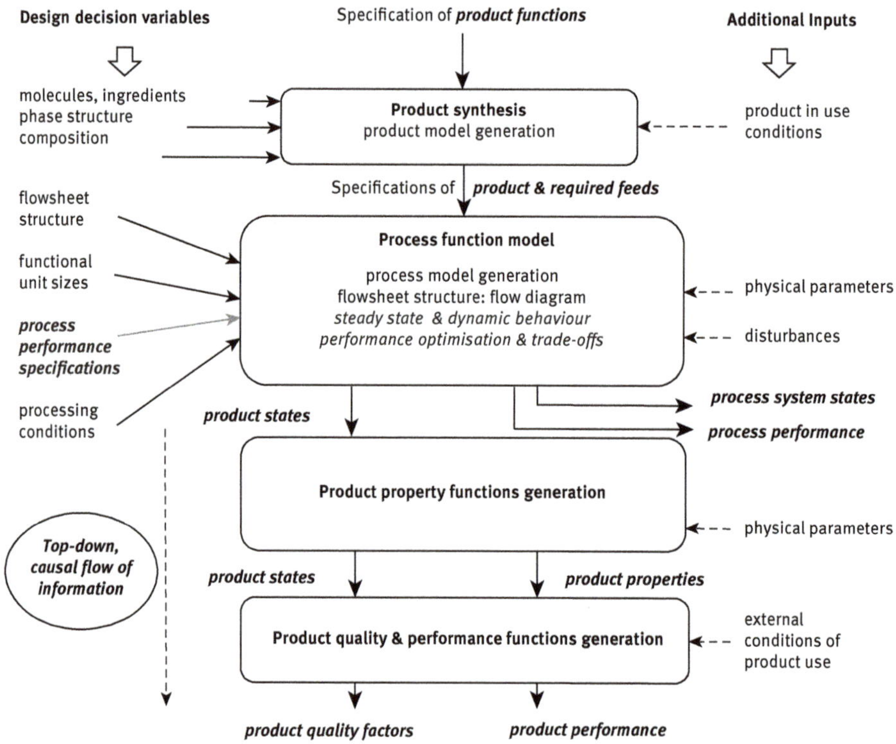

Fig. 4.2: Concurrent product-process-concept design flow diagram.

Specifications of product functions are input into the product synthesis step. A product model is built and its function is simulated based on external use conditions and design decisions. The design decision variables are molecule identities, which phase structure and composition are identified and manipulated. By model simulations the behavior of the product concept in its use/application is predicted. By trial and error, the product synthesis and model analysis steps are carried out until a satisfactory product concept is obtained.

The resulting product composition is then taken as the specification for the product the process concept design should deliver. A process concept design, based on various process functions is then synthesized and analyzed for its behavior. Again, this is done by simulating with a functional model: a process model. After several iterations of process design modifications and simulation an acceptable process design is obtained. The design results are the process system states, physical conditions of the process functions, flows between the function blocks, the process performance, and the product state.

In the third block (product property function), all (physical and chemical) product properties are predicted by a model that contains all relevant interactions between the constituting materials and the product structure.

In the fourth block (product quality and performance functions), the product quality and product performance in use is predicted. The product state, (physical) product properties, and the external conditions for using the product are the inputs for predicting the product quality factors and product performance.

Chapter 8 deals with product modeling, simulation, and optimization in detail. Chapter 9 deals with process modeling. The connectivity between product and process modeling and simulation will be covered in Section 8.5.

4.7 Exploiting experience in design (design heuristics)

4.7.1 Strength of design heuristics

Designing using heuristics is enormously fast. Particularly in the concept stage, this facilitates making several designs quickly, and selecting the best. The resulting concept design will also be more reliable than those made by using theory only.

Experience can be exploited for design in the following ways:

(a) By using rules of thumb (design heuristics)
(b) By asking experienced people for advice and feedback
(c) By going back in your memory on previous cases and deriving knowledge to use for the case at hand

(a): Design heuristics is coded knowledge obtained from many experienced designers. Many heuristics for process concept design are provided by Seider [14] and Douglas [15]. For process design in all innovation stages, Koolen is also a very good resource [16].

Heuristics for product design are scarce. Most of this type of knowledge is available within product innovation companies, and strongly context bound: to product design in foods, cosmetics, pharmaceuticals, etc. Martin provides heuristics for each of these product types [17].

(b): Obtain advice and feedback from experienced people. They have a lot of knowledge from experience and insight, but this is not easily transferred to others. This knowledge is called tacit knowledge. By stating the design solution, or a design problem, and asking them to go back in their own memory, give them time (can be minutes, can be days) on similar cases and probe them for stories during conversations or reflection on decisions. Through this process, a lot of useful information can be obtained.

An example from Harmsen's own experience is presented here. I had made a trickle-flow fixed bed reactor design and showed it to an experienced engineer. He said: "What is the bulk crushing strength of the catalyst and is it below the maximum pressure drop that could occur when the bed is in operation, and when dust fouling has happened on the bed?" I looked up the bulk crushing strength of the catalyst particles, and it appeared to be even below the pressure drop of the empty bed. I needed to redesign the reactor and had to select a larger catalyst particle size to further reduce the pressure drop.

(c): Consult your own experiences. By asking yourself the same question as you would ask experienced people, you will find that you have more knowledge then you imagined you had.

If while designing using heuristics, a conflict of rules occurs, then rethink the whole design in such a way that the conflict disappears. Philosophically this is the same as solving the conflict of a thesis and an antithesis, by moving the whole problem to a different level, or different context by which synthesis is possible.

4.7.2 Weaknesses in using design heuristics

Heuristics are based on truths related to a historically fixed context (sometimes called orthodoxies). Some heuristics for instance are based on the context that electricity is an expensive energy form. If in the future electricity becomes very cheap (albeit for certain periods during the day), caused by excess wind energy produced, solar cells becoming very cheap, etc., then these heuristics have become invalid.

4.7.3 Tapping into experience

Experiences shape our intuition. The intuition is used for practical purposes including design purposes. It is a very fast and a very effective form of thinking for all kinds of design activities. This is the reason you see so many experienced employees in design departments.

The intuitive form of thinking has one disadvantage: it blocks the individual pieces of memory about specific historic cases. The individual learning points of historic cases, however, can also be of interest to the design at hand. Kahneman [18] provides a simple method of drilling through the intuition barrier. His method is: Ask the experienced person (may be yourself), to go back into his own past and "envision" the cases that were a success, and those that were failures. The person will fall silent for a minute or more, as he goes back in his memory. Then ask the question, what is the learning point relevant to the design case at hand. With this method, drilling through the intuition barrier is easy. Kahneman calls this the outside view, as it gives us views beyond the barrier of intuition.

Heuristics for breakthrough products and their processes are not publicly available. Heuristics for product design and for process design are provided in Chapter 7.

4.8 Industrial example of design driving innovation

The prime example of a design driven breakthrough process innovation is the countercurrent extractive reactive distillation for the production of methyl acetate by Eastman Chemical. Agreda [19] describes this innovation in detail with all steps from concept generation to commercial scale implementation.

The implemented design is a one-column process containing two countercurrent extractions, a reactive distillation, and distillation sections at the top and bottom. It is dynamically controlled allowing for stoichiometric feeds of the two chemicals. At that time there was no equivalent concept of a complex combination of phenomena in a single column. Even simple reactive distillation was just being researched at the time of conception.

The following steps were taken in this design driven research, development, and implementation.

Step 1: A conventional design with classic unit operations would consist of two reactors and eight distillation columns. This design was tested at pilot plant scale. Because of its complexity, it was deemed desirable to search for a more economical process, requiring far less unit operations, hence; a challenging design goal was assigned.

Step 2: Many ideas were considered and computer simulations were used to test the ideas. After close examination of the problems of azeotropic distillations and the chemical equilibrium reaction, the concept was generated to use countercurrent extractions to avoid azeotropic distillations and to use reactive distillation to avoid reaction equilibrium limitations and to carry this all out in a single column. The top and bottom part of the column were designed as classic distillations to purify the product streams. An economic analysis revealed that the process was sufficiently economically attractive to pursue and laboratory testing was recommended.

Step 3: Laboratory experiments were carried out to generate kinetics and physical properties. The feasibility of the concept was also tested in a laboratory setup. These tests also confirmed the simulation calculations. The liquid holdup for reaction appeared to be a very critical factor for success. New economic analysis showed favorable results and large scale testing was recommended.

Step 4: A bench scale test glass column was built with a diameter of 0.1 m and a height of 9 m. With this column in operation, corrosion tests were carried out as well as detailed measurements of component concentration profiles over the height of the column.

Step 5: A commercial scale design was made using the data from the bench-scale tests. Steady state and dynamic computer simulation programs were developed to generate data for the design of the column. Again an economic analysis was made. The results reinforced the expectation that it could be a viable commercial process. A down-scaled pilot plant was then designed, constructed, and tested. Its dimensions were a diameter of 0.2 m and a height of 30 m. Special precaution was taken to ensure adiabatic operation. A dynamic control was also installed to allow stoichiometric feed ratio as intended for the commercial scale. Again corrosion tests were also carried out. The results proved that steady state production of methyl acetate of the required purity could be made over a significant period of time. Again an economic analysis was made of the commercial scale process.

Step 6: The decision was taken to make a detailed design, to construct and to operate the commercial scale process. The commercial scale process contained an additional methanol recovery column to allow for a lower temperature bottom flow reducing potential corrosion and to minimize chemical losses during upsets of the operation. Additional measures were taking concerning sufficient liquid holdup on the reaction trays.

Step 7: Operation procedures were written with special attention to startup. Plant commissioning started with inspection and water checkout. Many problems were identified and corrected at this stage.

The plant started up in May 1983. Many problems including froth related problems had to be solved in the startup phase. However, no production losses were incurred. Agreda attributes this to the pilot plant testing. The plant now operates routinely at rates above 100% of the design. The design production rate is 22,700 kg/h. It is about five times lower in investment cost and energy cost compared to the conventional process setup with 11 unit operations [19].

It is clear from his description that at the beginning it was not known how the process would work, nor what the process concept would be. Suitable simulation tools were not even available, but were developed during the research and development stages. Also a novel control method had to be developed. This innovation would not have happened if a design approach had not been followed from the start and through all subsequent stages to commercial implementation. It illustrates the power of design driven design. The case also shows the power of the stage-gate approach and having the commercial scale solution always in mind, and from there, designing whatever research and development is needed.

This design driven success led Siirola, the principal design engineer of Eastman Chemical, to develop a process synthesis method based on tasks (functions) rather than classic unit operations [20]. It is a nice illustration where first, design solutions are generated and then through reflection, a general design method is derived. Or as Henderson put it: “Science owes more to the steam engine, than the steam engine owes to science” [21].

Abbreviations

6P	Proposition, Promotion, Position, Packaging, Price, Product
BOSCARD	Background, Objectives, Scope, Constraints, Assumptions, Reporting, Deliverables
C	Check implications of design changes
D	Default design focus
DDM	Delft Design Map
ED	Equipment Design
FO	Flowsheet sensitivity & Optimization
HoQ	House of Quality
I-O	Input–Output
ICSID	International Council of Societies of Industrial Design
K	Key design focus
LNG	Liquid Natural Gas
o	optional design focus
OI	Operability Integration
PC-CW	Product Concept – Customer Wants
PC-PF	Product Concept – (Product) Property Function
PI	Process Integration
PFD	Process Flow Diagram
RAM	Reliability, Availability, Maintainability
QFD	Quality Function Deployment
SHEETS	Safety, Health, Environment, Economy, Technology, Social (aspect)
S/L/V	Solids/Liquids/Vapors
SP	Sub Processes
SWOT	Strengths, Weaknesses, Opportunities, Threats
TN	Task Network
Triple P	People, Planet, Profit
TRIZ	Theory of Inventive Problem Solving – (Russian acronym)
UN	Unit Network

References and further reading

[1] Dorst K. Understanding Design, 175 reflections on being a designer, Amsterdam, Netherlands: BIS Publishers, 2006.
[2] Dorst K. Frame Innovation: Create new thinking by design, Cambridge MA, USA: MIT Press, 2015.
[3] Petroski H. To engineer is human: The role of failure in successful design, New York NY, USA: Vintage books, 1992.
[4] Verganti R. Design-Driven Innovation: Changing the Rules of Competition by Radically Innovating what Things Mean, Boston MA, USA: Harvard Business Press, 2009, 4.
[5] Verkerk MJ, et al. Philosophy of Technology, New York, USA: Routledge, 2016.
[6] Lawson B. How designers think – The design process demystified, 4th edn, Oxford USA: Elsevier, 2006.
[7] De Bont C. Advanced design methods for successful innovation, Den Haag, Netherlands: Design United, 2013.
[8] Grievink J, Swinkels PLJ, Van Ommen JR. Basics of Process Design. In: De Jong W, Van Ommen JR (eds). Biomass as a Sustainable Energy Source for the Future: Fundamentals of Conversion Processes. 1st edn, Hoboken NJ, USA, Wiley & Sons Inc., 2014, 184–229.
[9] Almeida-Rivera C, Bongers P, Zondervan E. A structured approach for product-driven process synthesis in foods manufacture. In: Martin M, Eden MR, Chemmangattuvalappil NG (eds). Computer Aided Chemical Engineering Vol 39: Tools for chemical product design – from consumer products to biomedicine. Amsterdam, Netherlands: Elsevier BV, 2017, 417–441.
[10] Seider WD, et al. Product & Process Design Principles, 3rd edn, Hoboken NJ, USA: John Wiley & Sons Inc., 2010.
[11] Tassoul M. Creative facilitation. Delft, Netherlands: VSSD, 2009.
[12] TRIZ – a powerful methodology for creative problem solving, 2017. Resourced from: http://www.mindtools.com, https://www.mindtools.com/pages/article/newCT_92.htm.
[13] Korevaar G. Sustainable Chemical Processes and Products: New Design Methodology and Design Tools, Delft, Netherlands: Eburon, 2004.
[14] Seider WD, et al. Product & Process Design Principles, 3rd edn, Hoboken NJ, USA: John Wiley & Sons Inc., 2010.
[15] Douglas JM. Conceptual design of chemical processes. New York NY, USA: McGraw-Hill, 1989.
[16] Koolen J. Design of Simple and Robust Process Plants, Weinheim, Germany: Wiley-VCH, 2001.
[17] Martin M, Eden MR, Chemmangattuvalappil NG, eds. Computer Aided Chemical Engineering Vol 39: Tools for chemical product design – from consumer products to biomedicine. Amsterdam, Netherlands: Elsevier BV, 2017.
[18] Kahneman D. Thinking, Fast and Slow, London, UK: Penguin Books, 2012.
[19] Agreda VH, Partin LR, Heise WH. High-purity methyl acetate via reactive distillation, Chemical Engineering Progress, 1990, 86(2), 40–46.
[20] Siirola JJ. Industrial applications of chemical process synthesis, In: Anderson JL, Bischoff KB (eds). Advances in Chemical Engineering Vol 23: Process Synthesis. Academic Press, 1996, 1–92.
[21] Henderson LJ. Quote in Moore, WJ. Physical Chemistry, 4th edn, London, UK: Longmans Green and Co., 1962.

Part B: **Design generation**

5 Special design approaches

5.1 Introduction to special design approaches

This chapter is about special design approaches. Each approach has a specific lens through which the problem and its solution are looked at. The purpose of the chapter is to enhance the creativity of the designer by this special view on the problem and to improve the quality of the design solution. Each approach has its specific solution methods.

Some of the approaches, such as design for biomimicry, design for industrial ecology, and design for renewable energy sources are very suited for the discovery and concept stage. Other approaches such as design for energy efficiency, design for human factors, and design for six sigma are very suitable for the feasibility and detailed design stage.

5.2 Design for biomimicry

The term 'biomimicry' appeared for the first time in 1981 and was popularized by scientist and author Janine Benyus in her book *Biomimicry: Innovation Inspired by Nature* [1]. She defines biomimicry as a "new science that studies nature's models and then imitates or takes inspiration from these designs and processes to solve human problems." Since then an array of articles and books have appeared, in which the idea is further developed. Here are the basic ways to apply biomimicry.

The simplest way is to take a natural material with a very special property and use its specific property such as strength, hardness, or chemical reactivity and design a product with the same material for an application. The industrial process for making the material may also be copied from the way nature makes it.

A second way is to take a physical principle from nature and use that principle to make a novel product. An example is the use of a nanostructure that reflects sunlight in such a way that the material shows a specific color. The same nanostructure, but with different materials can then be designed and made to provide color on the surface of the material. In this way the need to paint is avoided.

A third way to apply biomimicry is to take a natural system's method and use the same method for an industrial system. An example is the natural system for closing the material cycles for carbon, oxygen, and water. Through photosynthesis, carbohydrates are formed from carbon dioxide and water, which in turn are used by other natural species as an energy source by oxidizing them back to carbon dioxide and water. This third way has been developed into industrial ecology and its subsystem, industrial symbiosis. This way is treated separately in this book in the section on industrial ecology.

https://doi.org/10.1515/9783110467741-005

A fourth way to apply biomimicry is to be inspired by nature in the way it has solved problems. Perhaps this is the most fruitful method. It has the advantage that it is not rigidly held to the specific natural method. Such rigidity can lock in the designers to solving the problem in a fixed natural way. A classic example is the idea that a flying machine should be flying by moving wings such as birds do. This led to many failures. Only by leaving this idea and separating the propulsion from the airlift did flying machines become a success.

The inspiration to improve the airplane wings for energy efficiency by looking at the design of bird wings, however, is still applied by wing designers.

For an elaborate description of biomimicry for product and process design and other usages, the reader is directed to references[1–3].

5.3 Design for industrial ecology

Industrial ecology is the study of the flows of materials and energy in industrial and consumer activities, of the effects of these flows on the environment, and of the influences of economic, political, regulatory, and social factors on the flow, use, and transformation of resources. Core elements are [4]:
- The biological analogy
- The use of system perspectives
- The role of technology change
- The role of companies
- Dematerialization and ecoefficiency
- Forward looking research and practice

Unfortunately, this handbook provides little information of use for designers. To my knowledge there is also no other book on designing for industrial ecology available.

5.4 Design for circular economy

The widely accepted definition of a circular economy is the one by the Ellen MacArthur Foundation: “a circular economy is one that is restorative and regenerative by design, and which aims to keep products, components and materials at their highest utility and value at all times, distinguishing between technical and biological cycles.” [6] Figure 2.3 of Chapter 2 shows the two closed mass cycles and their mostly applied process steps. This figure can be used by designers for inspiration to design for closed mass cycles for the entire life cycles of products and processes.

Cramer has developed a practical set of ten guidelines. She calls it the 10 R Ladder of Circularity [8]:

Refuse:	Do not use native material, but use refused material.
Reduce:	Use less material.
Renew:	Use waste material in a new way.
Reuse:	Use waste material again.
Repair:	Repair the device or material.
Refurbish:	Revive the material.
Remanufacture:	Make a new product from used materials.
Repurpose	Reuse product for other function.
Recycle:	Recycle material for use.
Recover	Recover some value from used material such as energy.

The ladder should be started at the top. If a certain step has been tested and cannot be further applied, a lower step should be tested.

As the used material often has a zero or even negative price, while the new product has a higher market price, profits can often be foreseen. In general, startup companies exploit this potential earning power, as shown in her webinar [8]. Cramer estimates that in the Netherlands alone, 7 billion €/year can be made by circular economies, just considering water, energy, and materials [7].

5.5 Design for cradle-to-cradle

Braungart's cradle-to-cradle design method for ecoeffectiveness is based on three essential elements [10, 11]:

(A) Design for recycling all product components staying within the technology cycle (see Section 5.4).

(B) Design for end-of-cycle products obtained from biological origins and ending in the biosphere, such that all decomposed components end up in the environment after their use, acting as nutrients for plants to grow.

(C) Use solar as the sole primary energy source.

His method is used for novel consumer product design and focuses on protecting the environment.

5.6 Design for industrial symbiosis

A subset of industrial ecology is industrial symbiosis. It is concerned with transferring the biology analogy of the ecosystem to the industrial system. Chertow investigated 22 industrial symbiosis cases and concluded that environmentally and economically desirable symbiotic exchanges are all around us and now we must shift our gaze to find and foster them [12].

Table 5.1 provides an overview of the main elements of industrial symbiosis [13].

Tab. 5.1: Industrial symbiosis elements derived from nature.

Nature ecology system elements	Derived industrial symbiosis elements
Food chain and closed material recycles	Closed material cycles
Many different species connected in network	Industry and society parts connected in network
Primary energy from sun	Solar energy based
Energy stored and transported via food	Energy in intermediate materials

The designer can use the Dow Chemical method finding and implement industrial symbiosis synergies [14]. It contains the following steps:

Step 1: Perform a high-level feasibility study of potential cases with environmental, social, and economic benefits and select cases with positive outcomes based on these three criteria.

Step 2: Research material properties and potential applications for those cases.

Step 3: Locate potential suppliers and users of streams and define potential connections.

Step 4: Obtain interest from partners in the project both from technical and economic points of view.

Step 5: Finalize project.

Step 6: Monitor and improve project.

Magan describes an actual industrial symbiosis case (which he calls a byproduct synergy case) and from which he derives following steps [14]

Step 1: Organize facilitated meetings between potential partners of an industrial site and explain the concept and the potential benefits. The meeting should be chaired by an independent person to avoid perceived bias for a certain partner.

Step 2: Form working groups of engineers from potential partners to analyze options and prioritize best cases.

Step 3: Overcome barriers, such as legal, technical, economic, communication, competing priorities, geographic, and perception challenges by maintaining communication and cooperation within the working groups.

Step 4: Make action plans.

In Magan's experience, steps 1–4 can be carried out within a year. After that first year quarterly meetings should be held to implement action plans and bring in new companies [15].

Synonyms used for industrial symbiosis are:

- Byproduct synergy
- Eco-industrial development
- Eco-industrial parks
- Industrial ecology
- Industrial ecosystems

5.7 Design for renewable energy sources

5.7.1 Present bulk chemicals production from fossil fuel resources

Present industrial processes are still mainly driven by fossil sources. This is mainly performed by burning fossil energy resources such as natural gas in furnaces and using the heat in the form of steam or a hot medium to provide the energy for reaction, separation, extrusion, and mass movement.

For the formation of hydrocarbon products and their derivatives, the reaction energy is provided as follows. First the hydrocarbons are transformed into olefins by the steam cracker process. This is a highly endothermic process where the required reaction energy is provided by burning fossil hydrocarbons in a furnace. By placing the reactor tubes in that furnace, the reaction heat is directly provided to the tubes.

The resulting olefins are then used in subsequent reactions to produced plastics, and virtually all other chemicals. All these latter reactions are exothermic, so they then produce the product, and heat. That heat stems from the energy stored in the olefins. The advantage of exothermic reactions is that the reactions run to complete conversion at moderate temperatures as the main way to obtain a negative free enthalpy is the negative reaction enthalpy.

5.7.2 Design for renewable energy based on product-process chain methods

When renewable energy sources are used to produce chemicals and other products, ways must also be found to deliver energy for driving the reactions and for driving the other steps in the process.

Renewable energy is provided from the sun, directly or indirectly. Solar radiation is abundantly available but at a low energy density. It is so low that it must be concentrated first to be applied in industrial scale processes. A proven way is to use photovoltaic cells to convert solar radiation into electricity that can be transported and concentrated.

Another way is using Concentrated Solar Power (CSP) technology. Using mirrors, solar radiation is concentrated and can then be used to generate a very high temperature that can be transferred to a medium such as molten salt, which then can be used to drive a chemical reaction or converted subsequently to electricity.

Wind mills are also used to harvest solar radiation and convert it into electricity.

Yet another way is to grow biomass, which then in turn can be used as a renewable energy source. It can also be directly used as feedstock for chemical production.

In all cases the renewable energy should be used efficiently and at low investment cost for the production of chemicals and other products. This requires new innovative solutions. An example of such a solution is the production of ethylene glycol from carbon dioxide, water, and solar radiation described in Dautzenberg's lecture [16].

The first technology is electricity from Concentrated Solar Power (CSP). Already 99 of these are in commercial operation and 28 were under construction in 2017.

The second technology is high-pressure water electrolysis. This technology produces hydrogen and oxygen from water at low cost, as no compressor is needed. Several Japanese patents have been published.

The third technology is the electrochemical catalysis of carbon dioxide and hydrogen to ethylene glycol. The company Liquid Light has developed this technology [17].

The combination of these three technologies means that from the inputs carbon dioxide, water, and solar radiation, ethylene glycol can be made. As the primary feedstocks have little to no cost it is likely that this ethylene glycol production could be competitive with fossil based production.

It is clear from this example that renewables based products and processes can be designed and developed cost effectively, but they require a lot of considerations, such as the basic elements of solar radiation, electrolysis, and catalysis. The latter two may be combined.

Murzin provides process design methods for products from renewables [18].

5.8 Design for energy efficiency

'Energy efficiency' is a generic qualification that describes the use of a minimum of external energy resources to obtain a desired service or an amount of product. For chemical products and processes the common energy resource is power, any form of fuel or utilities that is derived from these two main resources. Their use is generally related to the amount of product produced or feed processed to get the 'process specific energy' (in J/kg), as shown in Figure 5.1. Aiming for energy efficiency, a low process specific energy minimizes both the environmental footprint (use of resources and CO_2-emission) and the operating cost to obtain the required external energy resources. For products, the energy efficiency of the intended service of the product may also be relevant, like fuel savings by low rolling resistance tires [19]. This section focuses on processes and it gives a brief introduction how to assess their energy efficiency and how to improve it.

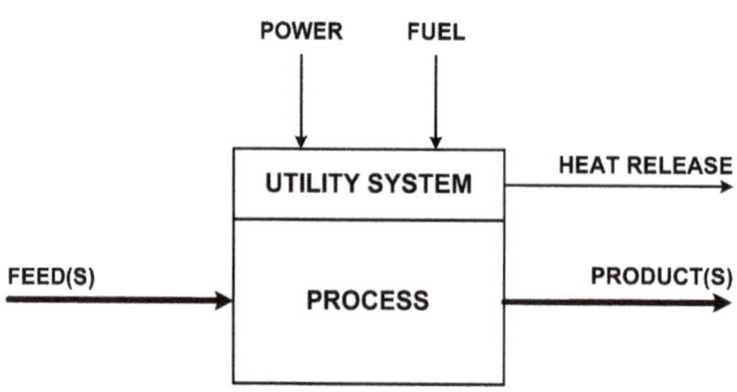

Fig. 5.1: Generic process input-output scheme.

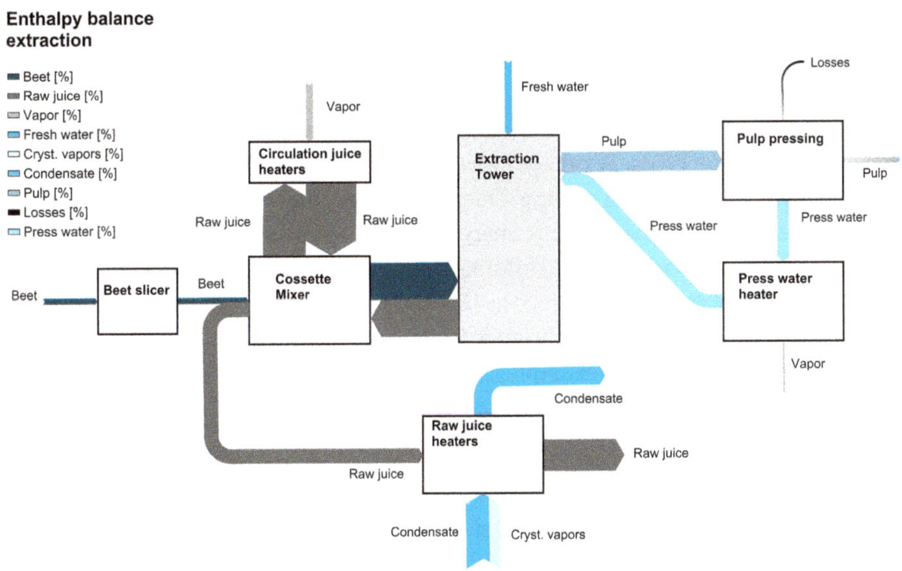

Fig. 5.2: Example Sankey diagram: sugar extraction (Courtesy Cosun R&D).

Energy efficiency can be achieved in any form of energy that plays a role in the process to produce the desired product. Changes in composition and the physical state of material streams are achieved by chemical and physical processes. All of these cause changes in energy states and as we deal with practical processes, they have their specific irreversibilities. The most energy efficient process is the reversible process. Thermodynamic approaches are available to determine the minimum energy resources required to perform the desired process in a reversible way. The ultimate low process specific energy found this way is the ultimate target to aim for in energy efficient design. This is the basis for 'exergy analysis' and design approaches to minimize the exergy loss, also referred to as lost work [20] [21]. Exergy losses in a process can nicely be visualized in a Sankey diagram (Figure 5.2), which shows the exergy supply to the process, the losses in each operation within the process, and the exergy sinks in products, waste streams, and losses to the environment [22].

The exergy analysis gives useful insight in the 'irreversibilities' that are important in a process that is available. This may guide changes to the process to make it more reversible and improve the energy efficiency. It relates, however, energy efficiency to an ideal and unreachable target. For designers it is generally more relevant to get guidance about what is practically feasible. One source of practical guidance is the current state of the art: existing operating plants have a proven process specific energy. Values for known processes are available in the literature. They may also be available from governmental and consultancy sources that monitor the efficiency of specific processes and industrial sectors [23]. Typical process specific energy values can be used as alternative targets for new facilities, but these values should be treated as the *maxima that can be allowed* for a new process, as it will have to compete with the existing ones.

Traditional (petro) chemical bulk industries have been keen on energy efficiency especially since the 1970s. For these processes the main energy transformation is related to physical heating and cooling. This has resulted in a practical tool kit to analyze such processes, determine practical energy efficiency targets, and design the heat exchanger network, the system that performs the required physical heating and cooling. This tool kit is known as Pinch Analysis [24] and is the main commonly used analysis and design method for energy efficiency in the industrial practice. We will illustrate the approach using a simple example case.

Example 5.8.1 (Example of reactor heating and cooling). Consider the scheme of Figure 5.3, with the stream data given in Table 5.1. Cold feed S1 is heated up and fed to a reactor. The reactor product S2 is cooled down. Stream S1 we refer to as a 'cold stream' as it needs to be heated up. Stream S2 we refer to as a 'hot stream' as it needs to be cooled down. Exchanger HX1 has a duty of 5 kg/s × 2 kJ/kg/°C × (100 °C − 50 °C) = 500 kW, HX2 a duty of 5 kg/s × 1.6 kJ/kg/°C × (105 °C − 45 °C) = 520 kW. Both the heating and cooling duty may be supplied by an external source: steam to heat S1 and cooling water to cool S2. The process specific energy can now be defined as:

$$\text{process specific energy} = \frac{\text{steam duty supplied}}{\text{product flow rate}} = \frac{500\,\text{kW}}{5\,\text{kg/s}} = \frac{100\,\text{kJ}}{\text{kg product}} \tag{5.1}$$

Obviously, it should be possible to make this process more energy efficient and lower the process specific energy. If we look at the overall heat balance we find a heating requirement of 500 kW and a cooling requirement of 520 kW. The net result suggests a net cooling requirement of 20 kW. No heating would be required. Clearly, this approach is too simple and only based on the first law of thermodynamics (heat conservation), while it ignores the second law of thermodynamics. This second law re-

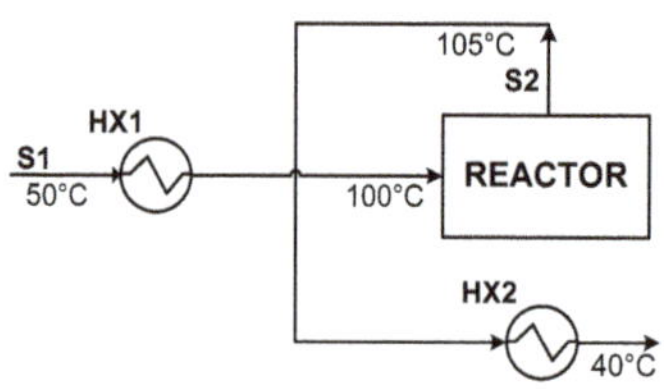

Fig. 5.3: Simple reactor system with preheat and aftercooling.

Tab. 5.2: Example 5.8.1 stream data.

Stream	Source °C	Target °C	Mass flow kg/s	Heat capacity kJ/kg/°C
S1	50	100	5	2
S2	105	40	5	1.6

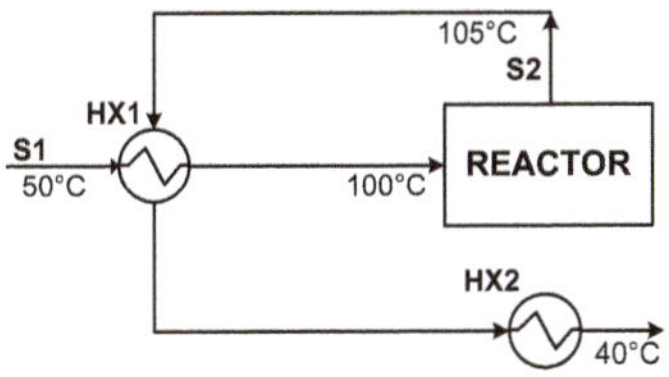

Fig. 5.4: Simple reactor system with feed-effluent heat exchange.

quires that for any feasible physical heat transfer the supply heat has to have equal or higher temperature than the demand heat. For a reversible heat transfer the supply (hot) and demand (cold) temperatures are the same, but in practice the exchanger area has to be limited in size which produces the requirement of a minimum temperature difference (well) above zero. This becomes clear from the generic heat transfer relation: $A = Q/U/\mathrm{DT}$, in which A is the heat transfer area (m^2), Q the duty (kW), U is the overall heat transfer coefficient ($kW/m^2/°C$) and DT is the effective temperature difference (°C) that is equal to the logarithmic mean temperature difference for a pure countercurrent heat exchange [25].

The reactor effluent is a little hotter than the reactor feed and can be used to heat this feed. If we try to supply all the heat (500 kW) required in Stream S1, as illustrated in Figure 5.4, it becomes clear that Stream S2 is inadequate to do that. After cooling 500 kW, S2 is cooled to 42.5 °C, which is below the supply temperature of the cold feed S1. Apparently, there is a limit to the heat transfer from the hot product to the cold feed.

To find this limit we draw the hot supply and cold demand in an enthalpy – temperature plot, as illustrated in Figure 5.5. In this plot we can graphically construct the feasible overlap where the hot (supply) curve is hotter (higher temperature) than the cold (demand) curve. When we allow the curves to touch, we allow the thermodynamic minimum approach of 0 °C and we get the ultimate heat transfer that is feasible. The resulting graph shows the following:
- It is possible to transfer 440 kW from the hot product to the cold feed;
- We will need 60 kW additional heating for the cold feed to reach the target reactor inlet temperature;
- We will need 80 kW additional cooling for the hot product to reach the target product temperature.

This is the best thermodynamic solution for the given system but as the temperature approach is 0 °C it will require an infinite transfer area. For any practical design it may be used as a reference. It will be necessary anyway to have three exchangers as shown in the scheme depicted in Figure 5.6.

For a practical design the temperature approach should be selected to allow the application of real heat exchangers. Choose for example a minimum temperature difference of 10 °C. Note that a realistic value depends on the exchanger type that can be applied; for the more common types the minimum difference is 10 to 20 °C, but more

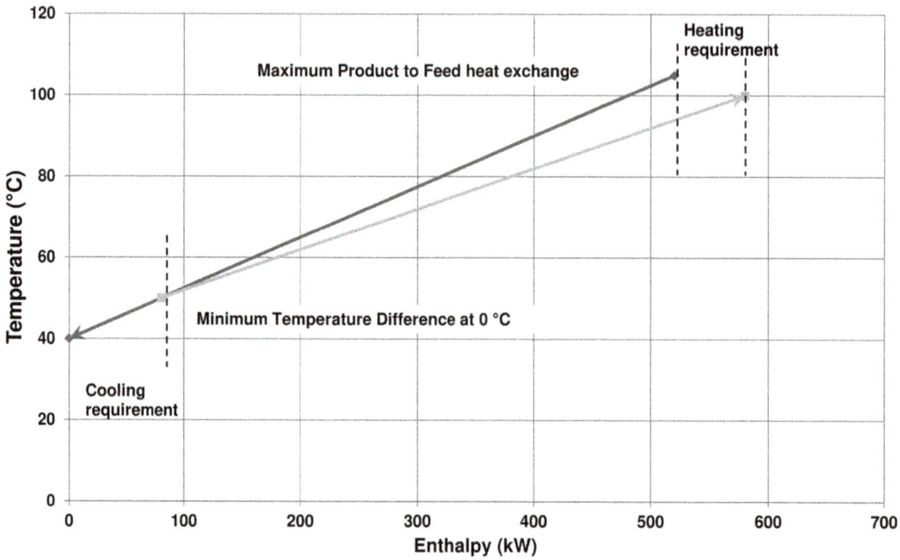

Fig. 5.5: Enthalpy–temperature plot Example 5.8.1 with minimum temperature difference of 0 °C.

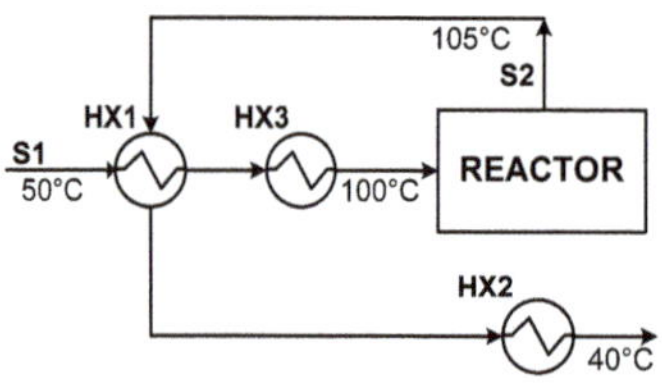

Fig. 5.6: Heat-integrated reactor system.

advanced exchangers allow a minimum difference as low as 1 °C. With this selected value of 10 °C we can do a similar graphical construction, see Figure 5.7, to determine the heat that can be transferred in the enthalpy – temperature plot. The resulting hot to cold process stream transfer is 360 kW, the heating requirement 140 kW and cooling requirement 160 kW.

Obviously, we can do this exercise for any other minimum temperature difference. If we increase the temperature difference we will reduce the process-to-process heat transfer and increase the need for additional heating and cooling. This will make the process less energy efficient. Additionally, we will need less transfer area for process-to-process heat exchange (S2 to S1) as the duty goes down and the temperature differences go up. Meanwhile, the transfer area for the additional heating and cooling (by utilities) will increase as these duties go up. The net effect is case dependent.

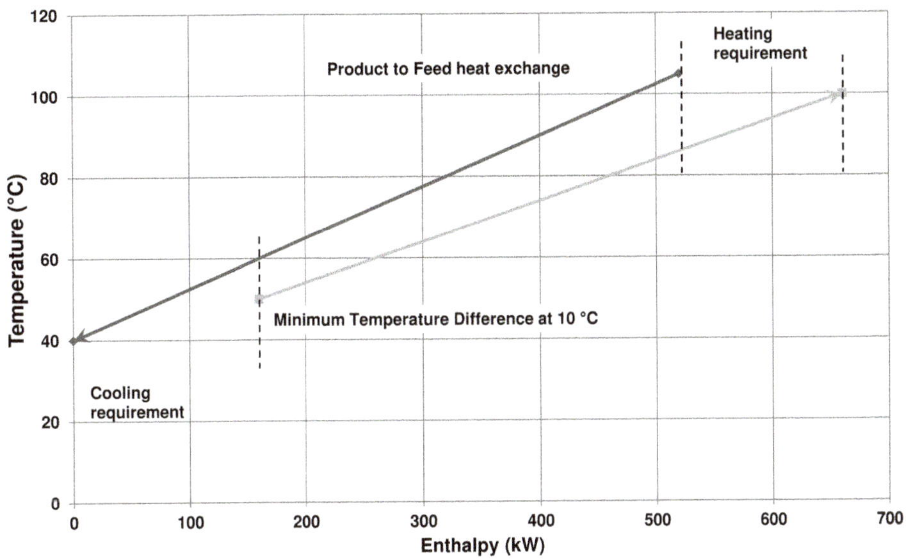

Fig. 5.7: Enthalpy–temperature plot with minimum temperature difference of 10 °C Example 5.8.1.

Example 5.8.1 shows that a heat curve plot can be used to determine realistic targets for process-to-process heat exchange and the need for external heating and cooling resources. Pinch Analysis is an extension of this simple analysis for processes with any number of hot and cold streams. To make this pinch analysis all cold process streams are lumped in a cold composite stream. Plotted in a temperature – enthalpy plot this gives the 'cold composite curve,' Figure 5.8, which shows the heat that is required in the process at each temperature. In the same way all hot streams are lumped to the 'hot composite curve.' Graphical construction, similar to the heat curves of the example, as illustrated by Figure 5.9, allows the determination of the maximum overlap, depending on the minimum temperature difference that we allow between the hot and cold composite curves. The point at which the composite curves are closest together to the selected minimum temperature difference is called the 'pinch,' which is considered the key area in designing an energy efficient heat exchanger network.

Pinch Analysis is the thermal analysis of a process to determine realistic targets for process-to-process heat transfer, heating and cooling by external energy resources similar to the analysis for one hot and one cold stream in the example. Commercial software like Supertarget [26], Aspen Energy Analyzer [27], and various process simulation packages support such analysis and provide easy interfaces to generate these curves from a process model. The composite curves are a powerful visualization to show what heat recovery (process-to-process heat exchange) is possible and what quality of external energy resources or related utilities are necessary. It also gives an indication how difficult it will be to get the desired heat transfer between the hot and cold process streams to and from the utilities. Figure 5.10 illustrates that if curves are

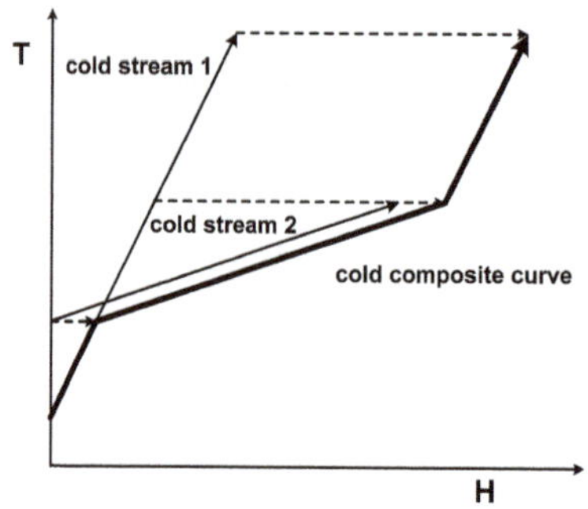

Fig. 5.8: Cold composite curve.

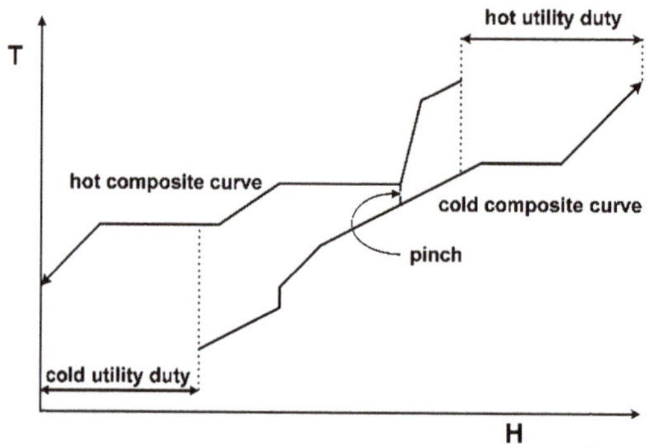

Fig. 5.9: Typical composite curves showing pinch, process-to-process heat exchanger, heat demand, and release to external sources.

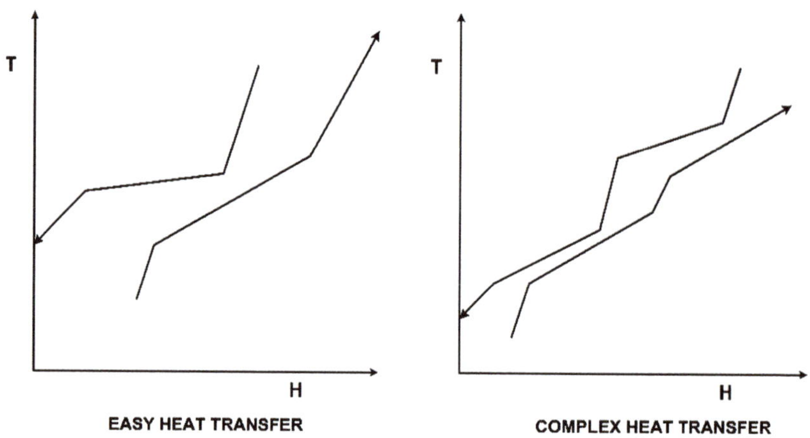

Fig. 5.10: Composite curves for easy and complex heat transfer.

far apart and the pinch is a narrow region (left) the heat transfer will be easier than when the curves are close together and have a large parallel region (right). Based on this insight it is possible to design an efficient utility system that converts the primary external energy resources to the utilities required in the process. Also it is possible to derive from the composite curves process modifications that will ease or enhance the heat transfer and finally allow for a more energy efficient design.

Once the process is fixed and the utilities selected, the composite curves give guidance in designing the heat exchanger network to heat balance the process following the Pinch Design Method. Most pinch related literature focuses on this last design stage, but in practice this network design is only a small part of the process design work. Pinch Analysis is useful in the entire process design work and it is recommended to apply it early when the first process concepts are evaluated to identify the scope in which to improve the energy efficiency of the process, especially when physical heat transfer is important in this process. Excellent explanations of Pinch Analysis and the Pinch Design Method can be found in Smith [24] and Kemp [28].

5.9 Design for human factors

Processes and products are sometimes two sides of the same coin. A process facility is in essence a large machine that is operated and interacted with by personnel and can thus be regarded as a sociotechnical system. Several large scale incidents with fatalities have occurred in the industry where human factor aspects such as poor human-machine interface, lack of understanding of human error, lack of situational awareness, etc. were the cause. Human factors are all those things that enhance or improve human performance in the workplace. Human factors engineering (HFE) focuses on the application of human factors knowledge to the design and construction of sociotechnical systems. The objective is to ensure systems are designed in a way that optimizes the human contribution to production and minimizes potential for design-induced risks to health, personal or process safety, or environmental performance [29].

Good HFE will increase HSE (health, safety, and environment) performance and reduce the likelihood of error, which in turn helps to improve the operational cost of the facility. There are a variety of HFE studies that can be performed depending on the project phase. OGP [29] has produced a guidance in which the most common ones are described per project phase. Studies that can be performed during the basic engineering phase are for example:

- Working environment health risk assessment (WEHRA);
- Valve criticality analysis (VCA);
- Vendor package screening;
- Task requirements analysis (TRA);
- Human machine interface (HRI) requirements analysis;
- Control room requirements analysis;

- Control system and alarm management analysis;
- Safety critical task inventory;
- Critical task analysis;
- Human error ALARP demonstration.

One HFE study discussed here as an example and typically performed during basic engineering and detailed engineering studies is called the "working environment health risk assessment" (WEHRA).

The goal of the WEHRA is to identify work tasks that may be associated with potential risks to health. The purpose is to uncover these risks, investigate if they can be avoided with proper design and suggest measures to reduce the risk to an acceptable level. The first step is the mapping and creation of an activity chart, where the nature of the activity, area of work, duration, and frequency of the work tasks are described; see Table 5.3.

Tab. 5.3: Example WEHRA activity chart for a single item.

No.	Area	Equipment	Activity	Duration	Frequency	No. Persons
1	**Cellar deck**	Choke	Replacement of actuator	8 hours	2 times per lifetime for each choke	1 on board the installation

To identify the events/tasks that may lead to any potential health hazard, a ranking table is used. In this table, all activities involving personnel are listed and evaluated per working environment factor used. The scoring and evaluation is done in a meeting, where several disciplines can provide their input. The working environment aspects that are covered are:
- Noise;
- Vibration (whole body, hand-arm);
- Ergonomics;
- Indoor climate;
- Outdoor operations/climate;
- Lighting;
- Radiation;
- Chemical substances and preparations;
- Biological agents;
- Psychosocial and organizational conditions.

A qualitative risk ranking is performed for each working environment factor; see Table 5.4. Evaluations are noted down and actions formulated for follow-up in the design.

Tab. 5.4: Example row of a ranking outcome that is created for each activity in the WEHRA activity chart.

Noise	Vibration	Ergonomics/ access	Climate	Lighting	Radiation	Chemical	Biological	Psych./ org.
2	1	5	2	1	1	1	1	1

5.10 Design for six sigma

The term 'design for six sigma' is used in two ways. The first is just about making the design such that the six sigma (sigma stands for standard deviation) value of the product output is within the specifications that the customers agreed to. This means that the chance that at any time the product is outside the specification limits is extremely small. This method is applied in the detailed design stage by using all statistical information on input parameters and choosing the design parameters such that this input variation is so limited that the product variation defined by six sigma is within the product specification.

The second way is a method to improve the quality of the generic business processes related to design and innovation used inside a company. It is derived from the quality improvement method six sigma. The latter is focused on improving existing business processes using statistical data and information.

For further reading the reader is referred to textbooks by Cudney [31] and Yang [32].

References and further reading

[1] Benyus, J. Biomimicry: Innovation Inspired by Nature, New York, William Morrow Paperbacks, 2002.

[2] D. Baumeister, Biomimicry Resource Handbook: A Seed Bank of Best Practices Createspace, New York, Independent Publishing Platform, 2014.

[3] Martin-Palma RJ. Engineered Biomimicry, New York, Elsevier, 2011.

[4] Ayres RU, Ayres LW. A handbook of Industrial ecology, Cheltenham USA, Edward Elgar Publ. Ltd, 2002.

[5] Graedel TE, Allenby BR. Industrial ecology, 2nd edn, Upper Saddle River, N.J., Pearsen Education Inc. 2003.

[6] Ellen MacArthur Foundation. Circular economy scheme. Sourced from: https://www.ellenmacarthurfoundation.org/circular-economy/interactive-diagram. Last resourced: 15 January 2018.

[7] Ellen Macarthur foundation. Sourced from: https://www.ellenmacarthurfoundation.org/. Last resourced: 15 January 2018.

[8] Cramer J. Circular cities, plenary lecture, 10th World Congress Chemical Engineering, WCCE10, Barcelona, 2017.

[9] Cramer J. Circular cities webinar. Sourced from: http://www.nzwc.ca/events/circular-cities-amsterdam/Documents/Webinar-JacquelineCramer-Mar14-17.pdf.

[10] Braungart M. Eco-effectiveness Cradle-to-cradle design method, plenary lecture, Watervisie Congres te Rotterdam op 18 februari 2016.
[11] Braungart M, McDonough W. Cradle to cradle – Remaking the way we make things, New York: North Point Press, 2002.
[12] Chertow, Marian R. Uncovering industrial symbiosis. Journal of Industrial Ecology 11(1), 2007, 11–30.
[13] Harmsen J. Sustainable Development and the Process Industries, NPT, 2013, 18–19.
[14] Wu Q. Industrial Ecosystem Principles in Industrial Symbiosis: By-Product Synergy. Chapter 14, Harmsen J, Powell JB (eds.), Sustainable Development in the Process Industries: Cases and Impact, Hoboken NJ: John Wiley & Sons, 2010.
[15] Magan A, Olivetti E. By-product Synergy Networks: Driving Innovation Through Waste Reduction and Carbon Mitigation, Chapter 6, Harmsen J, Powell JB (eds.), Sustainable Development in the Process Industries: Cases and Impact, Hoboken NJ: John Wiley & Sons 2010.
[16] Dautzenberg F. A journey towards Green(era) technologies, Kivi lecture, Technical University Eindhoven, 2017.
[17] Liquid Light. Sourced from: http://cen.acs.org/articles/93/i43/Liquid-Light.html. Last resourced: 15 January 2018.
[18] Murzin DY (ed). Chemical Engineering for Renewables Conversion, Advances in Chemical Engineering, 42, 2013, 2–371.
[19] Pike E. Opportunities to improve tire energy efficiency, The International Council on Clean Transportation, White paper 13, 2011.
[20] Dincer I, Rosen MA. EXERGY: Energy, environment and sustainable development, 2nd edn, Amsterdam: Elsevier, 2012.
[21] Smith JM, et.al. Introduction to Chemical Engineering Thermodynamics, 8th edn, New York, McGrawHill, 2018.
[22] Sankey diagram. Sourced from http://www.sankey-diagrams.com. Last resourced: 15 January 2018.
[23] Johansson TB, et al. GEA, *Global Energy Assessment – Toward a Sustainable Future*, Cambridge: Cambridge University Press, 2012.
[24] Smith R. Chemical process design and integration, Chichester: John Wiley & Sons, 2005.
[25] Thulukkanam K. Heat exchanger design handbook, 2nd edn, Boca Roca: CRC Press, 2013.
[26] Sourced from: http://www.kbcat.com/energy-utilities-software/supertarget. Last resourced: 15 January 2018.
[27] Sourced from: http://home.aspentech.com/en/products/pages/aspen-energy-analyzer. Last resourced: 15 January 2018.
[28] Kemp IC. Pinch Analysis and Process Integration – A User Guide on Process Integration for the Efficient Use of Energy, 2nd edn, Oxford: Butterworth-Heinemann, 2007.
[29] OGP. Human Factors Engineering in Projects, Report 454, 2011. Sourced from: https://humanfactors101.files.wordpress.com/2016/02/human-factors-engineering-in-projects.pdf. Last resourced: 15 January 2018.
[30] Edmonds J. Human Factors in the Chemical and Process Industries: Making it Work in Practice, 1st edn, New York: Elsevier, 2016.
[31] Cudney EA, Agustiady TK. Design for Six Sigma: A Practical Approach through Innovation, Cleveland Ohio: CRC Press, 2016.
[32] Yang K, El-Haik BS. Design for Six Sigma: A Roadmap for Product Development, 2nd edn, New York: McGraw-Hill 2009.

6 Scoping the design

6.1 Introduction to scoping the design

Scoping the design involves defining the design goal, its constraints, its scope, and its context. This set of actions is carried out in discussion with the client of the designer as described in general in Chapter 4. That chapter is well suited for experienced designers. This chapter, with its specific guidelines, is particularly suited for students learning to design. The focus of this chapter is on the concept and feasibility innovation stages but advice for the other stages will also be provided.

6.2 Defining design goal and name

6.2.1 Purpose design goals and names

A design goal is a clear short statement on what should be achieved by the design. In a business environment this will be obtained after several rounds of discussion with the client. In design education it will, in a simple form, be provided by the teacher. The students should then redefine it so that each design team member and the teacher agree on the design goal. The design goal statement is crucial in communicating with anyone what the design is about. A vague design goal will lead to a mediocre design. A clear design goal is a good start to a superior quality design. A good name will further enhance the communication inside the team and with others outside the team.

After the design goal is defined the design should be given a name. The discussion in the design team to generate and decide on the design name will also help to create a good team spirit.

6.2.1.1 Design goal setting

The design goal should be a very clear short statement that the internal client (business case owner) and all participants of the innovation group understand and agree upon. It is the shortest summary of what the design is about. For product concept design the goal can be formulated in terms of what functions the product has to have. An example is: design a chewing gum that does not foul the streets. The process design will state how much of what product has to be produced.

For starting a breakthrough innovation it is very important to define a challenging design goal statement. It should state the aim of the project in very concrete terms. It takes time to formulate this problem statement but you only can work efficiently if you know what your goals are. If it is not concrete, sooner or later conflicts will arise between people working in the concept stage who cannot decide what the better solution is [1].

https://doi.org/10.1515/9783110467741-006

If the goal statement is not different enough from the goal statement that gave rise to the current commercial process then the end result of the whole innovation process is that you merely rediscover what was good about the original process. All innovation work has then been in vain [1].

The goal statement should therefore contain a quantitative strong reason for change. The best way of achieving this is by explicitly referring to the best commercial process or product and stating in the goal and also in the constraints how much better the novel solution should be. This means that a good reference case needs to be defined. Section 6.4 discusses this subject in more detail. The main purpose of this challenging design goal is to stimulate design team members to come up with new solutions.

The design goal stated in the concept stage may be modified at the start of subsequent stages. For product designs this modification will often be needed, due to added insight from the interaction between the market desires and the product composition. For process designs the goal often stays the same or is only slightly modified.

6.3 Defining the design scope (system levels, boundaries, and context)

A major choice when scoping the design is the choice of system level in which the design will take place. All system levels relevant to design are described in Chapter 2 and a conscious choice of the system level for the design should be made. This choice will involve a discussion with the client. Unconsciously, a system level may already have been selected by the client. The designer can then show the other system levels to the client and suggest advantages for including other system levels in the design scope. In a design course the teacher may first choose a system level in which the students have to make a design and then choose a higher system level and again ask the students to make a design. In this way students become aware of the importance of choosing the system level.

Even within a system level the boundaries for design still have to be chosen. The boundaries define which elements are outside the design objective and are treated as context (a given or a constraint). These boundaries will also be defined in discussion with the client. A simple example of a process design is whether to include storage of feedstock and product in the design scope or not.

After the system level has been chosen for which the design will be made, it is useful to define the other system levels that form the context of the design. If, for example, it is decided to take the chemical reactor and the catalyst level as the design scope of the process, then the other higher system levels will not structurally be changed by the new design. However, the size of mass and energy flow to and from the surrounding system levels may change, due to the new reactor and/or catalyst design, if the amount of byproduct relative to the main product is reduced.

6.4 Defining constraints (specifications)

6.4.1 Design constraints

A design constraint in general consists of a type, a parameter, and its upper and lower limits. These constraints may also be called design specifications. A good starting point for defining constraints is to use the generic criteria check list provided in Chapter 4 and decide for each criterion what type of constraint and what quantified limits should be included. For a concept design by students the reduced constraint list SHEETS of Chapter 4 can be used.

In addition to these there will be constraints imposed by the company strategy. If the company has an explicitly written down innovation strategy then defining the strategy related constraints is easy. This may however not be the case. The written strategy may not be available, or the strategy may not be sufficiently defined to cover the specific innovation design and project. Once the design is made and presented, a conflict with the implicit strategy may be revealed. Then the strategy may be further refined and the design constraints adapted to this new situation.

Here is an example from the author's experience. A team had to design a sustainable process for ethylene production. The design solution by the team was to take biomass as a feedstock. Via fermentation and dehydration ethylene was produced. When the design result was presented to top management, the reply was, that an existing output stream from a crude oil should be used as feedstock and that the sole use of biomass as feedstock was outside the company strategy. That latter statement was until that time not written down. Shortly afterward, it was included in the company strategy.

This example shows that a strategy can also be underdefined. A stage-gate approach in which concept design is presented to management in the discovery stage will quickly reveal this underdefinition, resulting in a more defined strategy and thereby avoiding large research and development efforts on projects that do not fit the business strategy.

In the discovery stage the constraints will be limited to that which fits the company strategy and perhaps that for which potential value for the company can be created. For the concept stage the minimum set of constraints that should be used is SHEETS:

- Safety
- Health
- Environment
- Economics
- Technically feasible
- Socially accepted

In addition it is becoming more and more common to include a constraint on the contribution to sustainable development. A full set of constraints inside a company will be longer. The full set provided in Chapter 4 may be used as a checklist to make sure that no constraint type is forgotten.

It is good to consider however that at the start of the concept design the problem will still be underdefined and ill-defined. In some cases the problem is first defined by the client of the designer and during some meetings, due to the knowledge provided by the designer, the problem is better defined. Some preliminary solutions may be presented to the client to test whether the problem is properly defined. If so, the problem is specified further into a goal and constraints (specifications). This subtle game between designer and client has been described by an experienced designer as such: “Our job is to give the client, on time and on cost, not what he wants, but what he never dreamed he wanted- and when he gets it he recognizes it as something he wanted all the time.” [1]. Read also in Chapter 4 more about this subtle game. For students it is very important to learn that design problem definition is not just accepting what the teacher provides as the design problem to work on, but redefining it. In education a dedicated effort has to be spent on this aspect.

For the feasibility stage a complete set of constraints should be defined and discussed with internal and external clients, as now the product and the process for commercial scale production has to be defined. The same holds for the development and detailed design stages. The table on all aspects relevant to a design as presented in Chapter 4 can be used to generated a complete set of criteria and can also be used as a checklist to determine whether an item was forgotten.

6.4.2 Product specifications

Product specifications are of tremendous importance in both product design and process design. In product design the specifications will be developed through all innovation stages by interaction with marketing and customers. It is crucial that the developed specifications reflect the customer needs, differentiate the product from competing products, and are technically as well as economically realizable. In the initial project stages the specifications represent the hopes and aspirations of the team, often without being constrained by what can be technologically achieved. During the course of the project the specifications are further refined to include actual technological constraints and expected production costs by making difficult trade-offs between desirable product characteristics.

For consumer products in particular it is important that the product to be developed will be appealing to all senses. Here is the list of senses to ensure that for each sense, product specifications are made:
- Sight
- Feel
- Hearing
- Smell
- Taste

6.4.3 Process specifications

Process specifications will also have to be made. These will start with product output specifications, which are mostly obtained from the business part (marketing and sales) of the organization. If the product is radically new then not all characteristics of the new product are captured in quantified specifications. Some product specifications are then likely to be defined/provided later on, when the product prototype has been tested by the client, resulting in additional specifications. The amount of product to be made will also be part of the output specification.

For radically new products with functional concept specifications, the process designer is advised to make multiple process designs based on assumptions on the product composition and to discuss the process designs and their potential consequences for product performance properties with the product developer.

Input process specifications also start from the business part of the company, which provides which types of inputs are allowed and desired and also which part of the supply chain is now considered inside the design scope. These pieces of information will be used to define the input specifications. In addition, the process concept designer will further define and refine the input specifications using the relations with the process output specifications based on yield figures. As a starting point a 100% theoretical molar yield of product on feedstock may be used, or empirically obtained molar yield data from scouting experiments. The designer should be aware that often the precise yield dimension (molar or mass based) is not reported in the experimental results. A check on the precise yield definition will be needed. How the output and input specifications have been obtained will also be part of the design report. Chapter 13 contains more details of what and how process outputs and inputs should be reported.

6.4.3.1 Identifying reference cases to improve design constraints

Identifying a reference case and defining constraints for the new design with respect to this reference case is a very powerful method to make sure that the final design solution is far better than the reference case. The reference case is the best existing commercially available product and/or process, upon which the design solution should deliver a significant improvement.

For a radically new product design, defining a meaningful reference case can be complicated. Often the novel product has no single commercial reference product. The reference cases may be obtained by observing the functionality of the novel product and then finding existing products fulfilling the same functionality as group, so that a combination of existing products is selected as the reference case.

For a novel process design the reference case selection is relatively easy. It should be the best process in commercial operation for the same product. If that process is in in the company where the innovative design takes place then all information will be easily available. If the process is operated by a competitor then information is less easy to obtain. In some cases the information can be gathered from environmental reports on that process that had to be made public.

For a novel product or process concept design it is important to set a few hard-to-meet constraints related to the reference case. An example is setting the constraint that the novel design should be a factor of 4 times better than the reference case. The reason for this requirement is the same as for defining the product goal, namely setting only challenging constraints for the initial new concept design will ensure that the final design in the detailed stage will still be superior to the reference case.

A final remark on constraints: it is not very important whether the statement on the requirement to improve significantly relative to the reference case is put under the design goal or in the constraint section, or in both. The only guideline is: the design goal should be short. The number of constraints should be many.

6.5 Generating basic design data

For a design, input information to facilitate the design will be needed. This type of input information is called basic design data. These data need to be generated. This is done by searching in the literature and patent sources, by making best guesses, by making calculations, and by carrying out experiments. Not all information will be generated prior to the design start. Some will be generated during the design, when it appears to be needed to complete the design. For each subsequent innovation stage, more of these basic design data with an increasing level of detail will be generated. The ultimate challenge here is to generate just the minimum amount of data required to make reliable decisions in each stage, as generating more will result in unnecessary project expenses for data that may never be used when a project is not continued. This chapter provides guidelines on generating these data. The focus is on the first innovation stages up to and including development.

Basic design data includes not only data, but also information, knowledge, understanding, and wisdom as defined by Ackhoff [2]. Here is a short elaboration on the elements covered by basic design data:

- 'Single discipline' design data is data that can be easily understood by people skilled in that discipline, such as physical property data, safety data, and economic data. Single discipline data in general consist of a parameter (symbol), an amount (figure), a dimension, and may contain a reference. An example is: the density of water is 1000 kg/m^3 (at 277 K). A further distinction is between scale independent data, such as physical properties, and scale dependent data, such as investment cost, which depend on the scale (production capacity) of the design.

- 'Information' means relations between data. An example is reaction kinetics; providing information on the relation between reaction rates, concentrations, temperature, and catalyst type.
- 'Knowledge' means applying the data and information to the design. For example, it could mean selecting a best residence time distribution concept, such as plug flow, and using reaction kinetics information for reactor concept design.
- 'Understanding' means appreciating why the design choices have been made. It is cognitive and analytical. It is about explicit reasoning on why the whole or a major design is made as it is. It involves for instance the design choices made on a particular separation and reaction combination using safety and economic criteria.
- 'Wisdom' means evaluated understanding. It calls upon data, information, knowledge, and understanding and in particular on moral and ethical codes. Wisdom is in general very seldom described in basic design data, but when it is practiced, it is of enormous importance.

An example from the author's experience on the role of wisdom is a design for shale mining and oil production in Morocco in the eighties by the company I worked for. One of the design questions was, what are the environmental constraints? It appeared that the Moroccan government had a general set of environmental emission constraints for business activities. Experienced process designers in the company felt insecure about this set, and initiated an investigation into the local environment for the mining and the conversion process. The investigators reported that a migration bird sanctuary was near the spot of the mining and that it could be disturbed by the mining and processing activities. The company then decided to make sure that this would not happen and also decided to apply European environmental standards for the design. The reasoning was that causing environmental problems in Morocco could damage the company's global image of being a responsible business.

It is very important that generated data are reported, discussed, stored, and made retrievable to all project members. Usually such a database can be best established in the concept stage. Chapter 7 deals with the subject of reporting, storage, and retrieval in detail. Table 6.1 provides an overview of potential sources to generate basic data relevant for chemical product and process design. In the following sections, guidelines for basic design data generation in each innovation stage are provided.

6.5.1 Ideation stage

The main objective in the ideation stage is to identify which data are available and need to be obtained when deciding to pursue an idea further in the concept phase. Data generation in the ideation stage is therefore limited to those that can be easily obtained from literature sources or simply and quickly derived using simple equations and correlations or best guesses made by an experienced researcher or process development engineer. Physical properties, safety, health, and environmental data will be obtained from literature sources. If not available, educated guesses will have to be made. Consulting experienced employees of the company is often the quickest way of obtaining the required data and information. Economic data such as earning power can be derived by substracting feedstock cost from sales prices; for more information, see Chapter 10 on economics.

6.5.2 Concept stage

The main challenge in the concept phase is to just generate the required data to be able to decide on the most promising design concept(s). There is a huge risk in wanting to generate lots of nice to know data, which are in the end discarded because a different concept is chosen or the project is abandoned as a whole. Because in the concept stage quick decisions have to be taken and the budget is limited, the emphasis should be on rapidly finding these data, even at the expense of accuracy and reliability.

Some physical-chemistry data are always needed in the concept stage. Crucially lacking physical and physical-chemical data may be generated by specific experiments. These experiments can be done in-house but specific experiments such as physical basic data can also be obtained from specialized outside laboratories. Alternatively modern flowsheet computer packages that have a library of physical property models can be used to quickly generate physical properties. Care should be taken in using these models. They need a specialist to choose the best model for a particular property. It may also involve additional experiments to validate the calculation results. More detailed physical-chemistry data measurement and modeling studies are usually performed in the feasibility stage.

Generation of chemical kinetics data in the concept stage for process concept design is in general restricted to:

- The chemical stoichiometric equations of the main reactions involved
- Order of magnitude reaction times required for specified temperature and concentration windows
- Experimental results of proof of principle experiments showing products made and the yields of product on feedstock for a particular catalyst, some temperatures, and reaction times.

Economic data in the concept stage is in general limited to the purchase prices of feedstock and sales prices for products. For new products these prices will not be available. Pricing new products in the concept stage is an important step for innovative product companies. The reader is referred to the classic article by Dean [3] and to a McKinsey article [4] to get some insight into the art of new product pricing. Detailed methods for economic evaluations of novel processes and products are found in Chapter 10. Here we simply state the following guidelines:

(1) Estimate the new unit product value for the customer and do not base it on manufacturing cost.
(2) Determine the feedstock cost per unit of product.
(3) Estimate the Earning Power EP_0.

Safety, Health and Environmental (SHE) data are also very important for first a rapid analysis of preliminary concept designs and for evaluating concept designs when finalized. The product or a process that is inherently unsafe will not pass the concept stage gate. A product or process with no SHE information will also not pass the concept stage gate, or if it does then it will not pass the feasibility stage gate.

6.5.3 Data generation feasibility stage

Data generation in the feasibility stage is focused on detailing the required information for the selected design to such a reliability level that it can be used to establish a feasible commercial scale product and/or process design and when needed also a pilot and/or demo plant design. Alternative technical options for each process step will be contemplated. Selection criteria will be generated, and the best options selected. Equipment will be sized and stream compositions calculated. All these data need to be documented, stored, and made retrievable. For a plant design inside an existing processing site it is important to have all local plant and site related boundary conditions documented.

For reliable experimental physical property generation and modeling, large enterprises will have their own specialist group. Smaller companies will turn to specialized research institutes to obtain key pieces of information. Similarly much more detailed and reliable chemical kinetics data need to be obtained as the best reactor concept with its connected separation concept plus recycling can only be designed with sound kinetics. Here is a checklist of required chemical kinetic data:

- (Bio)chemical reactions
- Possible reactants/pathways
- Stoichiometry
- Conditions (pressure, temperature)
- Yields, selectivity vs conversion
- Product distribution
- Side reactions
- Reaction rates
- Equilibrium
- Phases (gas, liquid, multi)
- Exothermic/endothermic
- Residence time, space velocity

Tab. 6.1: Sources for basic design data.

Data Type	Reference books	Web based sources
Physical-Chemistry data	– *The properties of gases and liquids*, McGraw-Hill 2004 – *Materials Science and Engineering Handbook*, 4th Ed, CRC, 2004 – *CRC Handbook of Chemistry and Physics*, 97th Edition, 2016 – Dechema Chemistry Data Series, Dechema (2004)	– Properties of organic compounds – Chemexpert Chemical Directory – NIST Chemistry Webbook – Beilstein online database – Knowledge Centre (Bio) Chemical Engineering

Technology and scientific information	– Ullman's *Encyclopedia of Industrial Chemistry*
	– Kirk Othmer's *Encylopedia of Chemical Technology*
	– *Encyclopedia of Chemical Processing and Design*
	– Riegel's *Handbook of Industrial Chemistry*
		– www.sciencedirect.com
		– www.scopus.com
		– SciFinder
		– Wikipedia
		– www.google.com
		– Patents
		  – http://nl.espacenet.com/
		  – www.patsnap.com
		– www.nexant.com
		– www.ihs.com
		– www.intratec.us
		– www.cheresources.com

SHE data	– BP process safety series – Handbook of chemical health and safety – Handbook of chemical compound data for process safety – Chemical safety handbook – Property estimation methods for chemicals – Handbook of environmental health – Handbook of reactive chemical hazards	– KCCE: databases – ChemWatch (product safety) – INCHEM (chemical safety and environmental health) – International chemical safety card (by CAS number) – MSDS – http://chemiekaarten.sdu.nl/chkonline/ – http://hazard.com/msds – http://www.ilpi.com/msds – EU Directives and Law (for industry) – http://eur-lex.europa.eu/ – http://ec.europa.eu/enterprise/ – www.msdssearch.com
Economic data (market, pricing)		– www.icis.com/chemicals – www.nexant.com – www.ihs.com – www.orbichem.com – www.intratec.us – www.business.com/directory/chemicals – www.alibaba.com

Tab. 6.1: (continued)

Data Type	Reference books	Web based sources
Equipment cost	– *Product and Process Design Principles*, 3rd Edition, John Wiley & Sons (2010) Chapter 22 – *Plant Design and Economics for Chemical Engineers*, 5th Edition, McGraw-Hill (2004) Chapters 12–15 – Dutch Association of Cost Engineers	– www.matche.com/equipcost/ – www.mhhe.com/engcs/chemical/peters/data/ce.html – www.equipnet.com – www.ipps.com – www.alibaba.com

Also market and economic information will be detailed further to establish the business case of the project and assess its economic viability. Important aspects to be included are:
- Local society and cultural dimensions of market
- Market share, competitive position
- Desired production rate
- Desired product purity
- Product sale price versus purity specifications
- Feedstock materials compositions and their prices
- Location
- Climate
- Depreciation agreement or policy
- Feedstock prices
- Product sales prices
- Investment cost estimate
- Utilities availability, specifications and price
- Process cost structure
- Raw materials availability and composition

A minimum set of safety, health, and environmental data for the feasibility phase comprises:
- Safe operating window for the process parts
- Toxicity of all components applied for product and process
- Life cycle assessment of environmental impact
- Waste disposal options
  Waste streams/emissions
- Environmental legislation

Chapter 11 provides comprehensive information on safety and health data and Chapter 12 discusses the sustainability data needed for evaluation.

6.5.4 Development stage

During the development stage a large number of data for many design object aspects have to be generated on top of the already generated data in previous stages. These data concern:
- The commercial scale product and process and their related product prototype and pilot and/or demo plant
- Data generated by prototype and pilot plant testing to validate the feasibility of the novel product and process
- The marketability of the novel product
- Intellectual property protection

Table 6.2 provides a checklist of items for which basic design data have to be generated in the development stage.

Process performance data

Phenomena that cannot be predicted from modeling are: corrosion, fouling, foaming, catalyst decay solvent decay, and others. These potential problems should be part of the reasoning for having the pilot plant; see also Chapter 3 on reasons for having a pilot plant. The pilot plant and the testing program should be such that observations and measurements on corrosion, fouling, foaming, and catalyst decay can be made.

Tab. 6.2: Basic design data items of the development stage.

Item	Obtained from
Safety performance: product and process Chapter 11	Literature and experiments
Health performance product and process Chapter 11	Literature and experiments
Environmental performance LCA data Chapter 12	Literature
Sustainable development performance Chapter 12	Literature
Economy performance: see Chapter 10	Calculations
Product technology data	
Process technology data	
Technology performance process	Pilot plant test results
Technology performance product	Prototype product test results
Product market data	Market surveys
Product sales data	Market and client surveys
Intellectual property	Design rights, trademark, patents, copyright

References and further reading

[1] Dorst K. Understanding Design, 175 reflections on being a designer, Amsterdam: BIS Publishers, 2006.
[2] Ackoff RL. From Data to Wisdom, J. Appl. Sys. Anal., 16, 1989, 3–9.
[3] Dean J. Pricing policies for new products.,Harvard Business Review, 54(6), 1976, 141–153.
[4] Marn MV, et.al. Pricing new products, McKinsey Quarterly August 2003. Sourced from: https://www.mckinsey.com/business-functions/marketing-and-sales/our-insights/pricing-new-products. Last resourced: 6 October 2017.

7 Executing designs

7.1 Introduction to executing designs

This chapter describes methods and guidelines for executing novel product and novel process designs for innovation classes 4 and lower, as defined in Chapter 3. Its focus is on designs for discovery and concept stages, and is particularly useful for MSc students and for young professionals.

Figure 7.1 shows all major steps in executing any design in any stage. The first step; scoping is treated in Chapter 6. This chapter provides general methods and guidelines for the steps: synthesize, analyze, evaluate, and report.

The design steps synthesis and analysis, have strong interactions. An initial synthesis result is analyzed by how well it meets or fits the design scope. The difference between the analysis and evaluation steps is, that in the analysis step, the synthesized design options are often modeled and qualitative and/or quantitative data generated (e.g., mass and heat balances) and simulation outcomes are compared with the design performance requirements (like atomic efficiency, economic margin, etc.). These synthesizing and analyzing steps will be repeated to the satisfaction of the design team until a design is achieved that meets the scope goal and criteria. Then the result will be evaluated by people from outside the design team and the result will be reported. Methods and guidelines for all these steps are provided in Sections 7.2–7.5 of this chapter.

In the discovery and concept stages it is recommended to create several design alternatives and then select the best (or the top two or three), to proceed with. The other alternatives are reported and noted for optional use later. Therefore, also the subject of ranking and selecting best designs is added in Section 7.6. Special attention is also

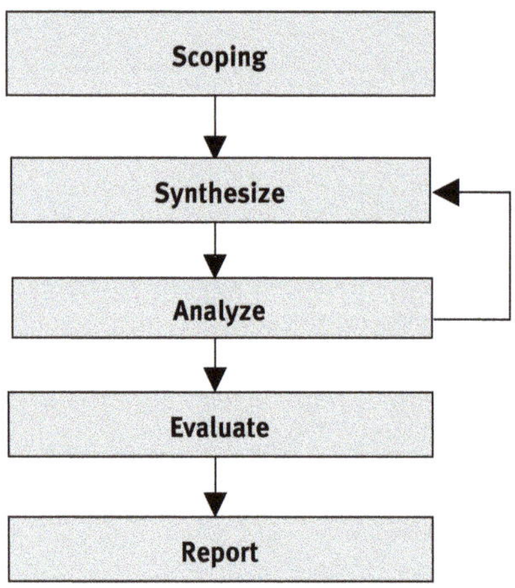

Fig. 7.1: Design steps.

https://doi.org/10.1515/9783110467741-007

paid to modify the design such that it is robust to uncertainties and its intellectual property is protected. Guidelines on these subjects are also in Section 7.6

How to analyze and optimize a selected design by making models and simulating a design performance is covered in Chapter 8 and 9. Specific guidelines on how to evaluate a design for the important criteria of safety, health, environment, economics, technical feasibility, and social acceptance are found in Chapter 10–13. Guidelines on how to report are found in Chapter 14.

7.2 Synthesizing preliminary solutions

Generating preliminary solutions by design is called (design) synthesis. This synthesizing can be done by using heuristics and by following specific steps and guidelines. For products and processes synthesis in general heuristics information is provided in Section 7.2.1. Guidelines for synthesizing products and structured products are provided in Section 7.2.2. Guidelines for synthesizing processes are provided in Section 7.2.3.

7.2.1 Synthesizing solutions in general using heuristics

Synthesizing design options in each design level is about generating preliminary solutions. There are many ways to generate preliminary solutions. Here are some practical, but very general heuristics to generate (synthesize!) preliminary design solutions:

(1) Look at the design scope generated by the methods of Chapter 6. Read the design goal. Then scribble down any idea that comes to mind that may partly meet the design goal. The scribbling may be a simple sketch.
(2) Ask yourself: what does the solution have to do to (partly) meet the design goal? Write down what the solution must do. Verkerk [1] calls this the qualifying function. Do not bother determining how the solution would or if it could work. If you think you need more than one function to meet the design goal, then write down those functions.
(3) Look for information on existing solutions (literature, patents, interviews of people) that may partly fulfill the qualifying function. Write down these partial solutions.
(4) Deconstruct existing solutions into constituting functions. So, make an abstract description of functions.
(5) Fill in some columns on the left-hand side of the Delft Design Matrix of Chapter 4 with keywords.
(6) Put the design aside and do something completely different.

(7) Anytime an idea comes to mind, note it down directly in your smartphone, voice message or a piece of paper at hand.
(8) Organize brainstorming, brainwriting sessions (or any other creativity method as proposed by Tassoul [2] to generate solutions.
(9) Use any of the following words: substitute, combine, adapt, modify, put to other use, eliminate, rearrange part-solutions to improve the solution (SCAMPER) [3].
(10) Generate families of (part-)solutions with each family having the same 'ancestor.' An ancestor is the common element of one family of solutions [3].
(11) Apply the 'TRIZ' creative problem solving methodology [4].

7.2.2 Synthesizing products

7.2.2.1 Product synthesizing design using simple heuristics

For novel product design of innovation class 4 all synthesis methods of Section 7.2.1 can be applied. As the new design is a modification of an existing product a lot can be learned from deconstructing the existing product, hence; method numbers 3 and 4 are in general fruitful.

Here is an example from a group of students that for the first time in their life had to make a product design. The design goal was to design a chewing gum that would not foul the streets. They looked for information through literature searches, patent searches, and by asking a chewing gum manufacturing company to provide publicly available information about chewing gum. They deconstructed the conventional chewing gum, into the gum that has an elastic plastic chewing function and taste additions. They concluded that the plastic gum of polypropylene is the part that fouls the street. It stays on the street for 28 years on average, if it is not removed.

They then looked for a polymer material that would depolymerize into water-soluble components and found a polyvinyl alcohol with the desired hydrolyzation rate. They also made the chewing gum very attractive by putting the taste components in microglobules. These depolymerized in the mouth, by which the taste components were slowly released into the mouth [5].

7.2.2.2 Synthesizing physically structured products in concept stage

First the difference between physically structured products and other products is explained. Two types of products can be distinguished in our book:
- Simple homogeneous products down to the molecular scale
- Structured products homogeneous down to a few microns and then heterogeneous and structured products

(Heterogeneous products such as cars and devices, which are heterogeneous on every scale, are not treated in this chapter.)

Simple homogeneous products are gases, liquids, and solids, such as polymers and ceramics. Particulate matter such as crystalline powders also qualify for this type.

Physically structured products are for instance emulsions, foams, and gels (see Chapter 8, Table 8.5 for more examples). They have one or more thermodynamic phases, with at least one of the phases having distributional features at the tiny and microscales. For instance, some chemical species in the product may have a distributed molecular structure (e.g., polymeric chain length distribution). In addition, the product may be made up of some coexisting thermodynamic phases with a geometrical distribution on a microscale, such as particle size- and/or shape distributions. 'Structured' products have properties that are related to the ratios of the thermodynamic phases, the chemical composition of these phases, as well as their internal spatial distributional structure (e.g., droplet size distributions in emulsions). Designing physically structured chemical products and a more precise characterization of the inner structure are presented in the following section.

7.2.2.3 From functional product specification to product structure

The complexity of the internal product structure arises from the presence of multiple constituents and from the distributive properties of its constituents. Examples are:

(1) Supramolecular structures such as polymers with molecular chain length distribution
(2) A thermodynamic phase present as particles or droplets of varying size
(3) Dispersion of multiple phases

The joint effect of all distributive and lumped properties of the internal structure determines the practical performance features of the product. These performance features and product costs determine the competitiveness of the product on the market in meeting customers' expectations. An example of practical performance requirements for a structured product such as a crystalline material, is given by Bermingham [6]:

- For downstream handling: ability to filter, to wash, to dry, and to let flow, freedom from dust, mechanical strength;
- For customer applications: no caking in storage, dissolution rate, mechanical strength, porosity, freedom from dust, aesthetic appearance;
- For food, some practical performance requirements [7, 8] are, in order of importance: safety > nutritional value > taste > smell > mouth-feel.

Also, Cussler and Muggeridge [9] and Wesselingh [10] give a number of diverse examples of (prioritized) practical performance requirements.

When developing and designing a new product and its associated manufacturing process, a top-down approach is followed from market analysis to process design, as indicated by downward arrows in Figure 7.2.

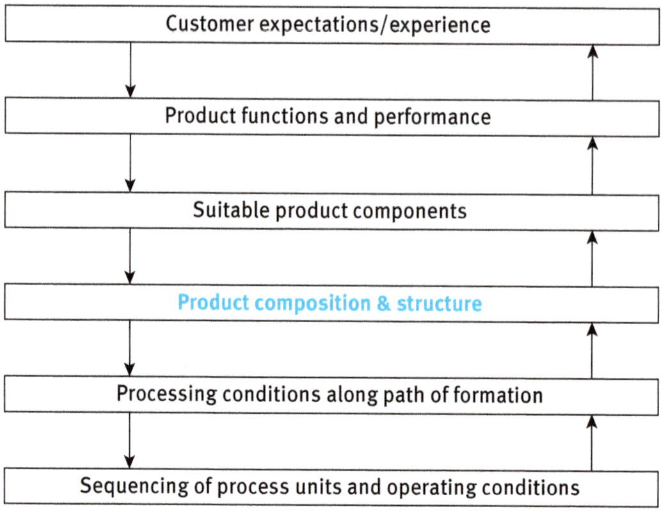

Fig. 7.2: Product and process design information downflow and physical causal effects upflow.

The product functions and performance requirements must be derived from the future wishes and expectations of customers. To support such translations, a systematic procedure has been developed for product design in general. This procedure is called 'House of Quality' [11], also known as 'Quality Function Deployment.' It allows one to crosscheck to what extent customer wishes are covered by the generated set of functional specifications of the product and if the specifications in the set are more or less in harmony (free of gross contradictions). A similar translation can be made from product functions to the selected components that will comprise the product. These components can be of a physical, chemical, biological, mechanical, or electronic nature. During product design and development, the product functions are materialized into a physical structure for the product. The use of the House of Quality approach is also discussed in further detail in Section 8.3.1.

The bottom-up arrows in Figure 7.2 show the physical cause and effects. The selection of process units with associated operating conditions and equipment design and their sequencing into a process flowsheet determine the path of processing conditions inside the process units when going from feed to product. The processing conditions involve the distributions of thermodynamic phases, pressure, temperature, concentrations, residence times, et cetera. The processing conditions along the path from feed to product influence the formation of the internal structure of the product. This internal structure involves chemical composition per thermodynamic phase, the spatial dispersion of the phases, and any internal distributive properties of the thermodynamic phases of which the product is made up (e.g., droplet size distribution).

The interactions between the 'down' flow of design information and the 'up' flow of physical cause and effect indicate that having proper information about the physical and chemical structure of a chemical product is essential to be able to synthesize the processing plant. The internal product structure is truly the conceptual linking pin between product design and process design. Knowing such a structure enhances the effectiveness of the design process as the physical features of the product serve as a functional specification for the process design.

What are the important generic physical features? Cussler and Moggridge [9, p. 130] list three central physical features of a chemical product:
- Structural attributes, involving physical and mechanical properties based on internal product structure
- Changes in thermodynamic equilibrium, induced by external variations (T, pH changes)
- Key rate processes, which can occur (reactions, mass and heat transfer, fluid flow)

They capture 'internal product structure' by means of four main items (pages 129–130):
- Chemical composition
- Physical geometry
- Chemical reactivity
- Product thermodynamics

When putting a product to its application, it is necessary to trigger the product activity by means of external controls, such as:
- Solvents
- Temperature change
- Chemical reactions
- Physical changes, like pressure, electrical field, and surface tension

E.g., the acidic environment of a stomach triggers the decomposition of the coating of a drug and the release of its chemically active substances.

The reader is invited to also use the Delft Design Matrix of Chapter 4 to further synthesize the product and its process, as it is noticed that in virtually all product design books, the outline of the internal product structure is rather short and verbally informal. These outlines are not sufficient for conceptual product and process design purposes. It is especially not so when process simulations are used, requiring a quantitative representation of the product structure. In Section 8.4.3 a mathematical modeling approach for multilevel structured products is given to analyze the product concepts.

7.2.3 Synthesizing processes

7.2.3.1 Radically novel process synthesis

Innovative process design follows the abduction-2 thinking described in Chapter 4. The desired output (the product) is known but the process type is unknown. Radically innovative processes are in general obtained by methods of process intensification [12].

The industrially most successful method of process intensification is that of function identification and integration into a single unit, such as in reactive distillation.

The following steps can be helpful in obtaining radically novel process designs.

(1) Identify chemical conversion routes to transform input molecules into product molecules.
(2) Define functional blocks to transform inputs into outputs.
(3) Connect blocks by mass flows represented by arrows from inputs to outputs.
(4) Check by simple atom balances that all input atoms are present in the output.
(5) Look for options to combine functions and use of streams.

Ad Rule 1: Perform literature searches to obtain options for the desired chemical conversions.

Ad Rule 2: Functional blocks can be selected from the following list and placed in a figure; see Figure 7.3:
- Transport
- Mixing
- Reaction
- Heat exchange
- Separation
- Forming

Ad Rule 4: Atom balances can be quickly made as they are not made or destroyed. Often atom balances reveal missing conversion and separation blocks.

Ad Rule 5: Often functions can be combined. Examples are mass movement and mixing (a pump is also a mixer), reaction and separation (reactive distillation, hybrid combination of membrane, and distillation).

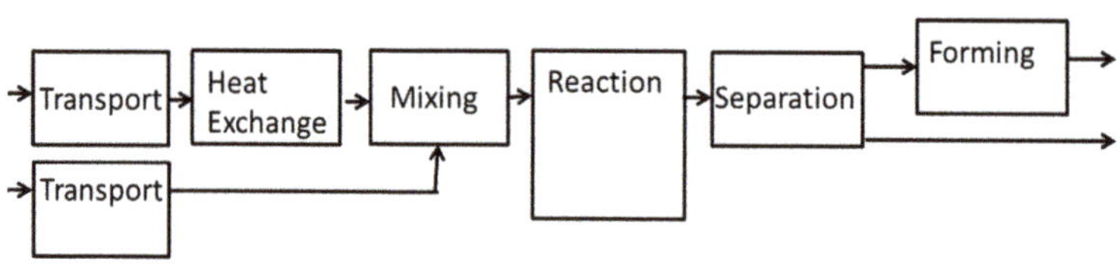

Fig. 7.3: Most frequently used process functions in process design.

7.2.3.2 Modified process synthesis

The first four steps for radically novel design are also useful for modified process synthesis. The difference starts at step 5 where for each functional block a conventional unit operation is chosen to execute the function.

7.2.3.3 Media choice

The choice of media (solvents, absorption liquid, etc.) is often not considered in process synthesis. Often the media chosen in the scouting experiments, such as a solvent, a crystallization medium, an absorption medium, or a stripping agent are taken as fixed for the process design. But using these media has all kinds of adverse effects on cost, safety, health and environment. Additional separation to recover the media will be needed, as well as additional storage. Diffusive emissions to the atmosphere will increase due to the additional equipment with their flanges and pumps.

It is far better to consider using available process streams such as feed streams, product streams, and process streams inside the process for a solvent, absorption liquid, crystallization medium. In bulk chemicals production, where processing costs are very important, only available media inside the process are used. So, it can be done. Here are some heuristics:

Media heuristic 1: need a solvent: consider all process streams including feedstock and product streams.

Media heuristic 2: want to strip a component from a stream: Consider using distillation or an available process gas stream for stripping.

Media heuristic 3: want to wash a component from stream with water: Consider using distillation, or extract component for an available process stream.

7.2.3.4 Unit operation choice

After the required process functions have been defined, unit operations can be selected, fitted to the functions. In general, the most important unit operations of a process are reaction and separation. These two units define the product yield and thereby the amount of feedstock needed for the required amount of product. Heuristics are found in the next section. Selection of other unit operations for mass flow, heat exchange, and product shaping can best be carried out by experts inside and outside the company. For the concept design stage, confection types such as centrifugal pumps, shell-and-tube heat exchangers, and single screw extruders can be chosen from the *Handbook of Chemical Engineers* [13]. In the detailed design stage, dedicated equipment selection can be made by experienced design engineers, involving equipment providers.

7.2.3.5 Heuristics for reaction plus separation synthesis

Defining and choosing the reaction section in combination with the separation section is very important for obtaining high product yields on inputs. For the reaction section itself the best residence time distribution must be chosen as well as the degree of conversion in combination with the separation and recycling of unconverted materials. Here are useful and easy ways to apply heuristics obtained from Levespiel [14] and Douglas [15]:

HR1: Single irreversible reaction: plug flow concept reactor

HR2: Single autocatalytic reaction such as in fermentation: back mixed concept reactor

HR3: Single reaction and adiabatic exothermic: back mixed reactor followed by plug flow reactor

HR4: If unwanted reaction is high order in B then distribute feed B along plug flow reactor.

HR5: Consecutive reaction: plug flow reactor with small conversion followed by separation and recycle unconverted component back to reactor.

HR6: Product inhibits reaction: plug flow reactor with small conversion followed by separation and recycle unconverted component back to reactor.

7.2.3.6 Heuristics for reactive distillation synthesis

Reactive distillation is now also a conventional unit operation for which heuristic rules for selection are available. Reactive distillation is a promising option if [17]:

HRD1: The reaction time is less than 30 minutes and the distillation temperature profile is compatible with the reaction temperature. By changing the distillation pressure, the distillation temperature may be changed to required reaction temperature.

All reactive distillation hydrogenations applied in refineries at very large scale obey this rule.

If furthermore either of the following conditions apply:

HRD2: There is a consecutive reaction forming an unwanted byproduct.

HRD3: The reaction is an equilibrium reaction.

HRD4: The distillation has an azeotrope.

7.2.3.7 Heuristics for separation synthesis

For selecting promising separation concepts, Table 7.1, based on heuristics obtained from PDC can be applied [16]. The top technique has the lowest cost. In descending order, the techniques become more expensive. So, by checking from top to bottom which technique can be applied, the most promising option is selected.

Tab. 7.1: Separation techniques selection hierarchy.

Separation techniques descending order more expensive	Practical solution examples
Phase separation	Decantation
Phase generation single step	Flash, condensation, crystallization
Phase generation multiple steps	Distillation, membrane
Separation with recycling	L/L Extraction, extractive distillation, adsorption

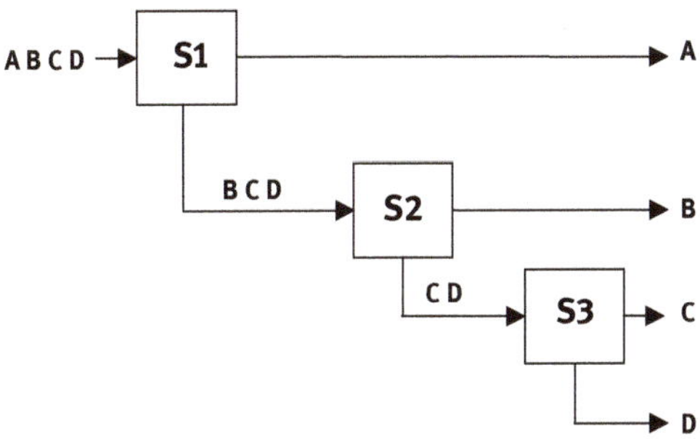

Fig. 7.4: Functional binary separations block flowsheet for four products from single feed.

7.2.3.8 Process separation synthesis from complex input

Process separation synthesis in the discovery and concept stages is facilitated by first making a simple block-flow diagram, connecting input streams to output streams. Each block can be defined as a binary separation. Here is an example for one input stream to be separated in four product streams. Figure 7.4 shows one block flowsheet for this case. Table 7.2 shows all 15 conceivable alternative block flowsheets and all 25 alternative separation block functions.

In brainstorm sessions with industrial experts for each separation task, several potential separation solutions are to be generated for each block. Typically about five to eight separation techniques for each block can be defined, resulting in typically 100 different flowsheet separation options.

7.2.3.9 Choice of batch processing versus continuous processing

The choice of batch-wise processing versus continuous processing is important as it determines most of the subsequent design choices. The orthodoxy is that regarding economics, large scale processing with production rates > 5 kt/y is always continuous and that small-scale processing < 0.5 kt/y is batch [15].

The basic reason why batch is often chosen for small scale production rates is that most or all process steps are carried out in the same equipment. So, the amount of equipment is reduced. For large scale processing the size of the equipment for batch processing becomes much larger than for continuous processing as the idle time for

Tab. 7.2: All alternative flowsheets and all binary separation functions for four product streams from one input stream.

Flowsheet label	1st Separation and label		2nd Separation and label		3rd Separation and label	
FS1	A/BCD	S1	B/CD	S2	C/D	S3
FS2	A/BCD	S1	BC/D	S4	B/C	S5
FS3	A/BCD	S1	BD/C	S6	B/D	S7
FS4	AB/CD	S8	A/B	S9	C/D	S3
FS5	ABC/D	S10	AB/C	S11	A/B	S9
FS6	ABC/D	S10	AC/B	S12	A/C	S13
FS7	ABC/D	S10	BC/A	S23	B/C	S5
FS8	B/ACD	S14	AC/D	S15	A/C	S13
FS9	B/ACD	S14	A/CD	S16	C/D	S3
FS10	B/ACD	S14	AD/C	S24	A/D	S20
FS11	C/ABD	S17	AB/D	S18	A/B	S9
FS12	C/ABD	S17	AD/B	S19	A/D	S20
FS13	C/ABD	S17	BD/A	S25	B/D	S7
FS14	AC/BD	S21	A/C	S13	B/D	S7
FS15	AD/BC	S22	A/D	S20	B/C	S5

switching from one operation mode to the next and emptying and filling take up time, which is not the case for continuous processing.

The advantage of continuous over batch for large scale capacities is, therefore, clear and holds in general.

The advantage of batch versus continuous for small production rate thus holds for processes with several process steps.

For single process steps such as a reaction step the picture is often far more complicated and the single choice of batch versus continuous processing is a whole series of choices with dependencies.

Batch reaction processes are often carried out in mechanically stirred reactors in which cooling is carried out by the reactor wall and by evaporating a solvent. It is then cooled in an overhead condenser and then returned to the reaction liquid. Often no catalyst is used, or a homogeneous catalyst is used. This is often an acid which subsequently is neutralized by a base and then discarded as a salt. Also, the reactor must often be cleaned in between batches.

During batch processing, the reactor is closed. If a runaway reaction occurs the pressure build up can be high and rapid, causing an explosion. Batch processes have shown far more explosions than continuous processes. As for the latter the reactor is open and pressure build up will, therefore, be lower. DSM and GSK have made choice for continuous flow microreactors for economic and safety reasons [18].

Continuous reactors are often operated in vessels with no mechanical stirrers. Often the reaction is catalyzed by a heterogeneous catalyst, placed as a fixed bed in the reactor. In most cases no solvents are used.

This means that the choice of batch versus continuous also means the choice of solvent or no solvent, the choice of catalyst type, the choice of process design, and the choice of equipment type. For the latter two choices the concepts of process intensification by combining functions in a vessel and the choice of microreactors is relevant.

The combined effect of process design choices means that even for small production capacities of less than 1 ton/year, continuous processing in combination with the choice for no solvent and no chemicals consumption has lower overall cost. To illustrate this statement, Schwalbe made a comparison for an industrial case where an existing batch mechanically stirred tank reactor for a production of 500 kg/y of a fine chemical was compared with a new continuously operated microreactor. The latter has a payback time of 0.4 years and overall cost reduction of 2.3 M €/y, so a cost reduction of 4600 €/kg [19].

For cases outside the described conditions, such as 0.5 ton/y < production rate < 5 ton/y the choice between batch and continuous allows for other considerations to also be considered. A consideration can be product quality variation. In batch processing, product quality varies more than for continuous operation, as process control under inherently dynamic batch conditions is less tight than under continuous steady state conditions.

Still other choices can be made. A choice can be made to operate a certain process step batch-wise and the other process steps continuously. A hold-up vessel after the batch step is then needed to ensure continuous feeding to the continuous section.

The fed-batch variant in which part of the feed is batch-wise and the other part is fed (semi) continuously is considered here the same as fully batch. The choice heuristics still hold.

If batch is chosen then the fed-batch option can be chosen subsequently. This is because, for instance, from a reaction selectivity analysis, distributed feed of one component provides a higher selectivity.

For more detailed descriptions of batch design and modeling, the reader is referred to [20–22].

7.3 Analyze interim solutions

7.3.1 Introduction to analysis

This section describes rapid analysis methods for interim design solutions. This is particularly useful in the discovery and concept stages where many potential solutions for products and processes are generated. A rapid analysis is then needed so that only the best solution is further developed.

Analyzing differs from evaluating. Evaluating in this book is reserved for a finished design (for a stage). The evaluation is then used to check whether the final design meets all design criteria of the design scope. Evaluating is often done in cooperation with people outside the design team, for instance by the members of the stage-gate panel.

7.3.2 Qualitative analysis of interim solutions

In the discovery stage large numbers of ideas should be rapidly analyzed with little information on each idea. This rapid analysis is best carried out with a group of experienced people. First, they should generate a set of criteria to facilitate the rapid analysis. Here are three steps to generate the criteria and perform the rapid analysis.

Step 1: Ask the experts to go back in their memory to similar situations in the discovery stage and then to remember which ideas were successfully implemented, and then what features of the ideas appeared to be the key success factors. From these success factors some criteria can then be directly defined.
These criteria are then specific to the analysis problem at hand and specific to the company where the innovation takes place. The criteria often include capabilities of the company to develop and implement certain novel products and processes and not others.

Step 2: Show the experts the list of generic criteria for analysis of Table 4.8 in Chapter 4. Also show them the SHEETS criteria list of Chapter 4. In addition, other criteria such as strategic fit and intellectual property protection should be added, from the business perspective.
The total set of criteria is then made from the results of steps 1 and 2.

Step 3: The experts attribute to each idea for each criterion a value of 1–5. For each idea the total score is obtained by multiplication, rather than adding up. By multiplication, ideas that score on all criteria reasonably well obtain a higher final score then ideas that score high on a few criteria and low on others. So, using multiplication reveals ideas that score well over the range of criteria and thereby these ideas are more likely to have a higher chance of success in the end.

This method of developing criteria by experts and in particular, step 1 of drilling into the source of past experiences, is based on empirical findings on experts by Kahneman, which he calls: “the outside view.” It appears that even experts do not naturally use all their experiences of the past, unless this experience is specifically called upon, by these series of question to generate the outside view. [23, p. 245–247].

The results of this rapid analysis can then be used to go back and synthesizing additional solutions or generating modifications to the ideas so that they score higher on criteria that so far have a low score.

7.3.3 Quantitative analysis of interim concept solutions

In the concept stage ideas are designed in some detail. Here is an important question regarding these interim concept solutions: is the concept sufficiently better than the reference case for all criteria? Below some simple analysis methods for each criterion type are provided.

The analysis result can be that the design should be further developed to facilitate the analysis. The analysis result can be that the interim design does not meet certain criteria. In that case the interim solution should be improved for the aspects of concern. The analysis can be that the design seems to be good enough for the next design step a full evaluation.

Product and process modeling play an important role in determining whether the concept meets certain criteria, as with modeling detailed information on product compositions, mass and energy flows are provided and the performance effects of changing design parameters can be easily determined. Chapter 4 discusses the roles of modeling and Chapter 8 and 9 provide methods for modeling, simulation, and optimization of products and processes.

7.3.3.1 Safety and health

A rapid analysis of safety and health issues of the novel design can be carried out by a simple comparison of the novel design with the reference case. The novel design should be safer than the reference case. Often this can be achieved by eliminating the most unsafe parts of the reference case. For products safety is mostly concerned with toxicity. Using the toxicity index for each ingredient in the product will be sufficient to draw a conclusion. Chapter 11 provides the toxicity index method.

7.3.3.2 Environmental analysis

For a rapid analysis on intermediate results and for an assessment of concept design results for the concept stage, a semiquantitative method based on large improvements on a single impact type using the rapid LCA method of Jonker can be used [24].

The method consists of a few steps.

Step 1: Define the functional unit of the design as the same as for the reference case.
Step 2: Select the most critical environmental impact type.
Step 3: Compare the environmental impact of the design with the reference case.
Step 4: Conclude whether the design is sufficiently better or not.
Step 5: Improve the interim design until its meets the environmental criterion.

Chapter 12 provides more information on evaluating for environmental performance.

7.3.3.3 Sustainable development analysis

A rapid analysis on contributions to sustainable development can be done by asking whether the design provides a novel solution for the needs of people, whether it would be socially accepted, if it would be environmentally friendly and if it would be economically attractive. If some answers are negative, then the design should be improved. If some answers cannot be given then more work on the design should be done. Chapter 12 provides more information on sustainable development.

7.3.3.4 Economics rapid analysis in the concept stage

A rapid economic analysis in the concept stage can be simply based on adding the cost of all feeds for the product and comparing that with the anticipated market price. If the total feeds cost price is higher than the market price the design should improve on this aspect. Chapter 10 provides several other methods for analyzing interim design solutions.

For all other criteria no formal rapid analysis methods are available. Asking the opinion of an experienced expert for an interim solution regarding marketing or other criteria is a rapid method.

7.3.3.5 Role of modeling and analysis

By making a model and then simulating the design behavior, an analysis can be made of its technical feasibility. The model also generates other output parameter values, useful for many other analysis items. Chapters 8 and 9 provide detailed information on how to model, simulate, and analyze product and process designs.

7.4 Best selection from alternative concept solutions

In the discovery and concept stages very large numbers of interim solutions are generated from which the most promising options should be selected with little information at hand. For efficiency reasons these ideas should first be ranked. Only the top ideas are then further worked out.

It is recommended that ranking includes criteria development for the discovery stage that are generated by experienced R&D people mainly from inside the company. In particular, the criterion for technical feasibility should be defined such that it is meaningful in the particular company's context, because bringing ideas to successful implementation strongly depends on the capabilities of that company. The intuition of those innovation experts can be considered as reliable, because it has been built up in a "regular environment" [23]; namely the R&D area of their company.

However, it is also important to have an expert from outside the company in the evaluation panel. Preferably it is someone with experience with innovation in more than one company. This "outside view" as defined by Kahneman brings additional knowledge to the table based on a broader experience [23, p. 245].

An important element of activating the intuition and memory of experts is to prompt their memory. This is done as follows. The leader of the evaluation session asks the experts to go back in their memory to which projects were successful and what characteristics in the early discovery or concept stage appeared to be key success factors [23, p. 245–246]. The author has several times executed this memory prompting by asking the experts to go back in time in their memory and think of projects that in the end were successful. Every time the experts fell silent for minutes and then came up with stories and appropriate criteria that even surprised themselves. They did not know that they knew them.

First, criteria for the ranking should be generated. The generic table for problem definition and evaluation can be used as a starting point. It is not very practical to have many criteria. Six in total is often the maximum to keep the ranking manageable.

Safety, health, and environment may be combined for one criterion. Social acceptance, market opportunity, strategic fit, and economic attractiveness may also be combined in a criterion. Technical feasibility to develop the idea to commercial realization is also likely to be a criterion.

They should also carry out the description of the scaling of the criteria. A scale of 1–5 is in general sufficient for each criterion. A meaningful short statement for each score value helps later to speed up the scoring for each criterion. According to Kahneman, the use of weight factors will not improve the overall quality of the judgment, because the human mind unconsciously also makes a judgment of the outcome of the whole and will adjust the score using weight factors. In fact making the criteria list together and then directly asking the experts to give the total score will result in very similar conclusions to the result that adds up the scoring of individual criteria scoring [23].

The valuation of each idea or concept for each criterion is in general best carried out by scoring the experts individually for each idea. In this way the errors of the experts are independent, so that the averaged final score value is more reliable than when the scores are obtained by agreement. In this way, one avoids the halo effect of people tending to form a consensus [23].

However, for complex process alternatives and for complex product ideas this individual scoring may be inferior to a plenary session in which first a discussion is started on how the idea would perform. In this discussion each member's imagination is stimulated to envisage how the idea would perform and then plenary scoring is, in the opinion of the author, of more value.

The overall score is best obtained by the multiplication of the score values per criterion. In this way, alternatives that score a reasonable value for each criterion get a higher overall score than alternatives that score highly on one or two criteria and low on all others. This means that the top alternatives are more robust toward future uncertainties since they score well on all criteria and not on merely one or two. This is a better method than adding up the individual scores as advocated by Kepner and Tregoe [25]. The latter method compensates low scores with high scores and alternatives often obtain the same total sum score, although their score distribution is very different.

The use of weight factors for criteria is not recommended for this intuitive scoring of alternatives, as the human mind unconsciously "corrects" the score for the weight factors [23].

More sophisticated methods such as Analytical Hierarchy Process (AHP) and Data Envelopment Analysis (DEA) for ranking alternatives with multiple criteria take far more effort to obtain a ranking result. Moreover the ranking result obtained is harder to communicate to management, who must provide the budget to further develop the top alternative.

7.4.1 Selecting best process concept option

In the discovery and sometimes also in the concept stage, many process options are (and should be) generated. The design is often only a simple block-flow diagram. Inputs and outputs are known to some degree from scouting experiments. Process concept alternatives will be available from literature searches. For efficiency reasons these concepts should be ranked and only the best selected for further development. Ranking with a panel is, however, not always immediately available. For the individual designer the following simple method has been developed to still select the best option.

The simple selection method is carried out as follows. A large number of process alternatives are obtained from literature searches. Each alternative solution is then first tested whether for any of the safety, health, environment and social acceptance aspects there is a major problem foreseen. If so the alternative is filed and for the time being not further investigated. For the remaining alternative process alternatives, block-flow diagrams are made and the yield of product on feedstock is researched in literature. Then an economic ranking is made based on this design information, as provided in the next paragraphs.

7.4.2 Economic ranking of process concepts

For quickly economically evaluating process routes based on different conversion types such as thermal, heterogeneous catalysis, homogeneous catalysis, enzymatic catalysis or microbial a semiquantitative method has been developed [18]. It is based on historic statistics of process cost element distributions as found in Perry's chemical engineers' handbook [13].

These statistics show that the major process cost element is the feedstock cost. If the process routes are all from the same feedstock then the product yield on feedstock cost is the only parameter value needed as indicator for feedstock cost for the different process routes. If the feedstock differ then the feedstock cost price should be looked up. The feedstock cost per ton of product is then obtained from

$$\text{Feedstock cost/ton product} = \text{Feedstock price (\$/ton)/yield}$$

The other process costs such as investment cost charge, maintenance, utilities, and operations are all strongly related to the major pieces of equipment [29] and these are strongly related to the number of process steps for conversion and separation. The number of process steps is derived from the number of conversion steps and the minimum number of separation steps involved. The latter is directly derived from the stoichiometric equations showing what byproducts will minimally be formed, and hence what minimum separations are needed. If a solvent or a homogeneous catalyst is used then a separation step for their recovery should be added [26].

If the reactions form an equilibrium between product and feedstock then also an additional separation will be needed. Whether an equilibrium is present can be established from Equation (7.1) for the equilibrium constant K [28]:

$$K = \mathrm{e}^{-\frac{\Delta G}{RT}} \tag{7.1}$$

The value of the free enthalpy value, ΔG, can be looked up and K determined. If K is much larger than 100, then single pass deep conversion is feasible. If K is around 1 then single pass conversion is not feasible and an additional separation step has to be added to the total separation steps.

The selection method is then as shown in Figure 7.5: for each alternative feedstock the yield figure is noted down. If there are feedstock type differences between the alternatives then the feedstock cost per ton of product is determined from the feedstock price and the product yield on that feedstock. The total number of process steps is also determined from the stoichiometry equation based on the reaction, the equilibrium value K, and whether separation is needed for homogeneous catalysis or solvent recycle.

The best process route has the lowest feedstock cost per ton of product and the lowest number of process steps. If a best selection cannot be made, because the lowest feedstock and the lowest number of process steps belong to different alternatives then the lowest feedstock cost option should be chosen. In all other cases the top two or three options should be further investigated in detail, by more detailed process design and cost analysis, using information from Chapter 10.

An example of a process concept design with the essential information needed for the ranking method is shown in Figure 7.6 [26, 27].

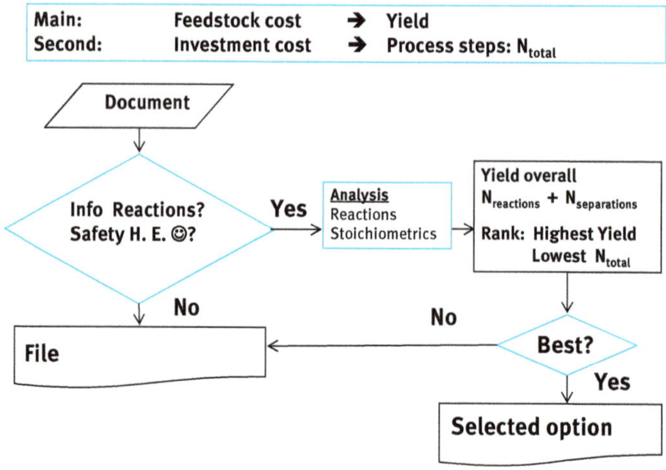

Fig. 7.5: Industrial process route selection method.

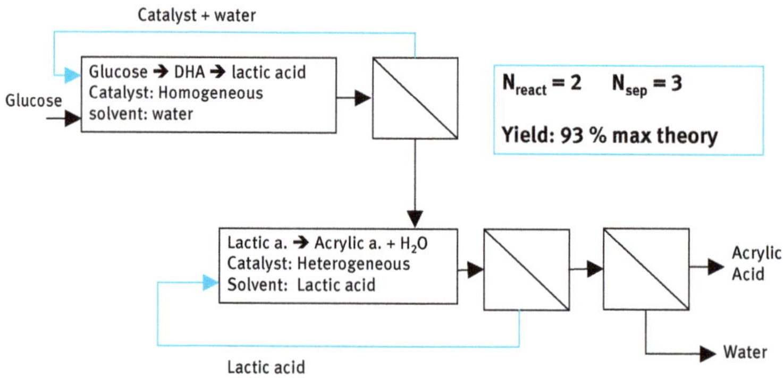

Fig. 7.6: Block flow diagram to acrylic acid: Route C. Reference: C.B. Rasendra, Platform chemicals from biomass, PhD thesis, U. Groningen, 2012.

7.5 Finalizing a design

7.5.1 Introduction

This section is about modifying the design such that:

(a) All criteria must be met.

(b) The design is optimized using the desired criteria.

(c) The design is modified such that the risks are evenly spread over the criteria, by which the overall risk of failure is reduced and robustness is obtained.

This design action is called a balancing design, or trade-off design. This design action is mainly situated in the feasibility stage, where for the first time a commercial scale design is made that meets all criteria, defined as a feasible design. This feasibility is strongly related to minimizing the risk of failure; hence the focus on minimizing the overall risk.

The subsequent sections deal with criteria types, optimization over all criteria, and increasing robustness toward future uncertainties by using scenarios.

The design criteria set in the scoping step of the design are often further defined as the criteria that must be met, otherwise the design will not be accepted, and desired criteria for which the design should be optimized.

7.5.2 Balancing designs

When a design nears its completion, a lot of knowledge will have been generated revealing the relations between design decisions and design performance. Then it is time to balance the design such that it becomes robust to uncertainty in the value of certain parameters such as feedstock prices. This robustness can be obtained to a certain extent by choosing design parameters such that the design output or performance is away from the most critical constraint boundaries and that the optimum is a smooth optimum and not peaky. Polar graphs are very useful in communicating the balanced design. For showing the effects of uncertainty ranges in design parameters on the design performance and showing that even for the uncertainty ranges the design stays within its constraints, Tornado diagrams can be useful.

7.5.3 Increasing robustness toward future uncertainties using scenarios

Because the design will be implemented in the future and may last many years it is important not only to take the present status of the evaluation criteria into account but also their future development. If this is not done then the product and process can rapidly become out of fashion and may need to be ended. This in turn then leads to loss of capital and loss of potential revenues.

One way of looking at the design in view of future uncertainties and by doing so making it more robust to these uncertainties, is by using a set of opposing scenarios. An example of such a set is provided in Figure 7.7, obtained from IBM. It shows four scenarios generated from using two axes with opposing society trends. The horizontal axis is about possible nature and industry trends, either being fragmented or consolidated and the vertical axis is about customer preferences either being individualistic or homogeneous. With these axes, four scenarios are built. Each scenario is internally consistent and plausible and has elements that are relevant to the design.

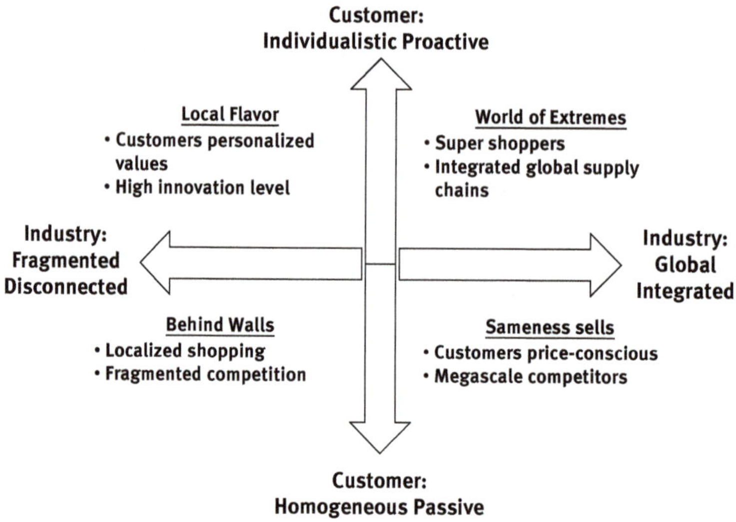

Fig. 7.7: Scenario set example.

The design is placed in each scenario and conclusions are drawn under each scenario how well the design would behave. If the design behaves well in each scenario it is likely that the design is robust toward future uncertainties. If the design behaves badly under at least one scenario then the design is not robust. If that is the case the design team should consider changing the design, or making it adaptable, in such a way that it will be successful under all scenarios, and hence is made more robust toward future uncertainties.

Scenario sets may be obtained from literature sources such as the United Nations or big enterprises such as IBM or Shell. Companies or design teams can build their own scenarios using the method of opposing potential trends and generating plausible worlds for each quadrant.

7.5.4 Intellectual property (IP) creation and protection

Intellectual property is created by design when the novel design is drawn and written down and the content is protected from use by others. Designs of a company can be protected in various ways. It can be done by patents, copyright, or design rights. Chapter 3, Section 3.6.5 provides more information. Most important is to have IP protection before the design is made public in any way.

7.6 Evaluating and reporting designs

7.6.1 Introduction

Evaluating a design means that the design result is checked whether it meets all design goals and design criteria. If several designs have been made than each design is individually checked in this way. This checking will be done by people outside the design team. The information for which this checking is carried out, however, is provided by the design team. So the quality of the evaluation is the responsibility of the designers and of the evaluators.

The purpose of this section is then to provide an overview of evaluation items for which the design team must provide information. A very important part of evaluating any design is comparing it with the reference case, being the existing commercial technology for the same or similar purpose (function). Therefore, first guidelines for selecting and defining the reference case are provided in Section 7.6.2. Then guidelines for evaluations are provided in Section 7.6.3.

7.6.2 Reference case for evaluations

A reference case is the present commercial product with the same function as the novel product. If the novel product fulfills a function that presently is fulfilled by more than one product then all those products combined are taken as the reference case. The reference process is the best commercial scale process for the same product or function.

Having such a reference case to compare with the novel product and process is a powerful communication means to the evaluation panel. For each criterion, the novel product and process is compared with that present commercial product and process, and the novel design must compete. All advantages and disadvantages of the novel product and process are directly clear.

Finding all relevant information about the reference case may take considerable effort. For reference products it may take a full chemical analysis of the competitors' products as well as performance tests. For a reference process it may involve reading public environmental impact reports of the competitors' process. In some cases it may even involve making a design of the competitors' process from limited public information available. However, this effort will pay off knowing the present performance of the best competing product and process so that improvements by the novel product and process can be determined for all evaluation criteria.

7.6.3 Evaluating designs

Evaluating designs in a commercial organization will in general be done at the gate from one innovation stage to the other. Criteria for the evaluation in general will concern:

- Business aspects, such as strategy fit and economics (foreseen benefits, foreseen cost, market share)
- Safety, health, environment and social aspects
- Technical feasibility

Chapter 3 provides for each stage (gate) details on what information should be provided and in particular what experimental proof is needed for each stage gate to be able to judge the technical feasibility of the innovation. Specific evaluation methods for economics, health and safety, environmental, social, and sustainable development are provided in Chapter 10–13.

7.6.4 Reporting

Any design consists of words, symbols, drawings, and pictures to report the design result to others. So reporting orally and in writing is extremely important. Especially in the discovery stage this truth is sometimes forgotten. A brilliant idea may be mentioned in a discussion in the corridor. If it is not written down and reported to others it will be lost. Chapter 13 elaborates extensively on the importance of reporting designs and provides guidelines for doing so for each part of a design in each stage.

References and further reading

[1] Verkerk MJ, et.al. Philosophy of Technology, London: Routledge, 2007.
[2] Tassoul M. Creative facilitation. Delft: VSSD, 2009.
[3] Korevaar G. Sustainable Chemical Processes and Products: New Design Methodology and Design Tools, Delft: Eburon, 2004.
[4] Altshuller G. TRIZ keys to Technical Innovation, Worcester USA: Technical Innovation Center, 2005.
[5] Harmsen GJ. Chewing gum that does not foul the streets, oral presentation and abstract Proceedings ECCE 2007, Copenhagen, 2007, Conf CD ISBN 978-87-91435-57-9.
[6] Bermingham SK. A design procedure and predictive models for solution crystallisation processes. Development and application, PhD thesis, Delft: Delft University Press, 2003.
[7] Bruin S. Phase equilibria for food product and process design, Fluid phase equilibria, Vol. 158–160, 657–671, 1999.
[8] Wesdorp L. Good food requires engineering?! Simultaneous product and process development, Keynote lecture on ESCAPE-12 Symposium, The Hague, on CD-ROM supplement in: Grievink J, van Schijndel J (eds.). Computer-Aided Chemical Engineering, Vol. 10, Amsterdam: Elsevier, 2002.

[9] Cussler EL, Moggridge GD. Chemical Product Design, Cambridge: Cambridge University Press, 2001.

[10] Wesselingh JA, et.al. Design and Development of Biological, Chemical, Food and Pharmaceutical Products, Chichester, John Wiley & Sons, 2007.

[11] Seider WD, et al., Product and Process Design Principles, Hoboken NJ: John Wiley & Sons, 2009.

[12] Stankiewicz AJ, Moulijn JA (eds.). Re-engineering the Chemical Processing Plant, New York: Dekker, 2000.

[13] Green DW, Perry RH (eds.). Perry's Chemical Engineers' Handbook, Eighth Edition, New York: McGraw-Hill, 2008.

[14] Levenspiel O. Chemical Reaction engineering, 3rd edn, New York: J. Wiley& Sons, 1999.

[15] Douglas JM. Conceptual Design of Chemical Processes. New York: McGraw-Hill International editions, 1988.

[16] Schembecker G. Process Synthesis Course, Breda, Process Design Center, 2000.

[17] Harmsen GJ. Reactive Distillation: The frontrunner of Industrial Process Intensification: A full review of commercial applications, research, scale-up, design and operation, Chemical Engineering & Processing: Process Intensification, 46(9), 2007, 774 780.

[18] Harmsen J. Economics and Environmental Impact of Process Intensification: An assessment for the petrochemical, pharmaceutical and fine chemicals industries, Chapter 14, Process Intensification for Green Chemistry, KVK Boodhoo and AP Harvey (eds.), Chichester: John Wiley & Sons, (2013).

[19] Schwalbe T. Chemical Synthesis in Microreactors, Chimia, 56, 2002, 636–646.

[20] Diwekar U. Batch Processing Modeling and Design, New York: Taylor & Francis Inc, Abingdon, 2014.

[21] Sharratt PN. Handbook-of-batch-process-design, New York: Springer, 1997.

[22] Korovessi E, Linninger AA. Batch Processes, Boca Raton: CRC Press, 2005

[23] Kahneman D. Thinking, Fast and Slow, London: Penguin Books, 2012.

[24] Jonker G, Harmsen J. Engineering for sustainability A practical guide for sustainable design. Amsterdam, Elsevier, 2012.

[25] Kepner CH, Tregoe BB. The rational manager, Princeton NJ: Kepner-Tregoe Inc., 1981.

[26] Harmsen J. How to choose a process route: From the lab to production, NPT Procestechnologie, 2015, 16–17.

[27] Harmsen J. Best Renewable Routes to Adipic Acid, Acrylic Acid and Propionic Acid, Oral and abstract, AIChE Spring Meeting & 12th Global Congress on Process Safety, Houston, USA, 2016, 10–14.

[28] Moore WJ. Physical Chemistry, 4th edn, Englewood Cliffs N.J.: Longmans Green and Co., 1962.

[29] Perry RH, Green D, Perry's Chemical Engineers' Handbook, Eighth Edition, New York: McGraw-Hill, 2008.

8 Product modeling and optimization

In this chapter, not only the role of mathematical models in the design of products is described, but also the roles of other types of models: verbal, schematic, and physical. The information in this chapter is applied in the design levels and design steps of Delft Design Map (DDM) as indicated in Table 8.1.

Tab. 8.1: Delft Design Map (DDM) parts related to modeling and optimization of Chapter 8

Stage-Gate™:		Concept				Feasibility			Development			
	Framing	Supply chain imbedding				Process technology			Process engineering			Final
Design levels ⇒	PF	CW-PQ	PQ-PF	I-O	SP	TN	UN	PI	ED	OI	FO	(Final)
⇓ Design steps/ activities												
1 Scope												
2 Knowledge												
3 Synthesize												
4 Analyze												
5 Evaluate												
6 Select												
7 Report												

8.1 Verbal, schematic, mathematical, and physical models

In Section 4.6 the role of (mathematical) modeling in design is described. This section also emphasizes the use of nonmathematical types of models for both product and process design activities. Before going into these different model types, the creation of models by designers is the heart of the design activities, as all these models together make up/describe the design alternatives and the final design (be it from different perspectives).

This has been characterized by Van Overveld [1] (lecturer in general design methodology at TU Eindhoven) and in his very inspirational way has taught this in the TU Delft and TU/Eindhoven post-Master PDEng designer programs (see Section 14.2.4). Van Overveld characterized 'designing' and 'a design' as:

- "Designing is: making decisions to make stakeholders happy, according to an intentional plan, leading to a new 'artefact', an improvement to an existing artefact or a new or improved part of an artefact."
- "'A design' is a representation of the (new/improved) artefact, in the form of one or more models, that are mutually consistent. Each model focuses on one aspect

https://doi.org/10.1515/9783110467741-008

of the artefact (geometry, physics, chemistry, operation, ...) and communicates the artefact with the stakeholders, documents the decisions, helps the realization and aids in the use of the artefact."

Different model types are created and used also in the (bio)chemical product and process design field. These models describe, in their own way, parts of the design results and decisions. The most used model types are also classified in general systems theory by e.g., Skyttner [2]. Next to mathematical models, iconic/physical models, schematic models, and verbal models are also used. Engineers get well exposed to mathematical models and schematic models like graphs and charts, but receive less education and training in creating schematic and physical/iconic models. Reporting on design context, scope, knowledge, analysis, evaluation, and selection decisions is not recognized as constituting a model as well. All these models serve different purposes and are used during and after the design process. For communication purposes the physical and schematic models play a paramount role.
Viewing the design as a collection of (mutually) consistent models also helps in defining the scope and deliverables for the design team (in specifying what types of models and to what level of detail these models will be delivered).

Examples of these four model types will be given in Sections 8.2 and 8.3 and in Tables 8.5 and 8.6, and references are given of the examples of these models used in this book.

- **Physical (or iconic) models**. These models:
  - Look like the real system they represent (car, building, manufacturing plant, product prototypes, mock-ups)
  - Can be small scale or full-scale
- **Schematic models**. These models are pictorial representations of conceptual relationships of the design:
  - Molecular structures drawings, (bio)chemical reaction pathways
  - Product structure drawings (spatial distribution of phases with each individual size, shape, form)
  - Product structure drawing in different stages of manufacturing or use
  - Graphs showing relationships (suggested, measured) between product and process variables (reaction rate versus concentrations, heat transfer rate versus temperature differences, payback time versus capital investment)
  - Tables (containing symbols and data); pure component properties, stream specification, process stream summary
  - Diagrams (House of Quality, process block diagram, process flow diagram)

- **Verbal models.** These models contain words (organized in lines of text or tables) that represent systems or situations that (may) exist in reality. These words provide:
  - Design problem context
  - Relevant knowledge to solve a problem, analyze it, and evaluate performance
  - Assistance in construction of (parts) of the design (equipment CAD drawing)
  - Assistance in operating, maintaining, and repairing manuals
  - Justify design decisions and recommendations
  - Organized in lines, bullet lists, (text) tables, paragraphs, chapters

And finally:
- **Mathematical models:**
  - Are a reduced representation of reality
  - Most abstract and precise models in a structural language
  - Such models are systematic and precise indeed, but they will often be restricted with respect to their domain of valid application. I.e., the underlying modeling assumptions must be applicable in a particular domain and, ideally, an experimental validation of the model or of its critical submodels must have taken place.

These models of an artefact (verbal, iconic, schematic, mathematic) are representations of (parts of) the original artefact. A model has a correspondence to the actual artefact and is created/synthesized to solve design issues/problems of the real artefact. The design issues can be analyzed/evaluated by manipulating the model (because manipulating the actual artefact is not possible, as it is not there yet). The models should be so good a representation of (part of) the real artefact, that results of the manipulation of the model can be mapped back to the artefact. The complete set of models represents the artefact's design.

8.2 Process design schematic and mathematical models useful for product design

In (bio)chemical product and process design all four types of models are used, but only some of them have become standard tools. Chemical or process engineers – historically educated and trained in process design and mathematical process modeling and simulation – have been exposed to the use of a limited class of schematic models.

Chemical engineers value mathematical models very much as a tool in process design and are extensively using them in process and equipment design and have so for a long time. The mathematical modeling of manufacturing processes (composed of unit operations like chemical reactions and separations, including thermodynamic, kinetics, and transfer phenomena models) has a long history and is extensively used in process simulators and mathematical optimization software in all innovation phases. This will be extensively discussed in Chapter 9. In the initial innovation phases, mathematical models start in a simple form (e.g., mass balances based on reaction stoichiometry and conversion/yield assumptions) and develop in complexity for those processes that show potential to succeed and warrant more detailed and extensive modeling. Said mathematical process models are often linked to engineering diagrams or drawings like process block schemes (PBS), process flow diagrams/schemes (PFD/PFS), and piping and instrumentation diagrams (P&ID). These drawings – often produced to internal and/or external standards, may form the basis for mathematical modeling or a vehicle to capture the mathematical models' quantitative results.

Examples of the PBS and PFD of the TAME (Tert-Amyl Methyl Ether) production process are presented in Figures 8.1 and 8.2. 3D CAD process and equipment drawings are also used widely by process design and engineering companies.

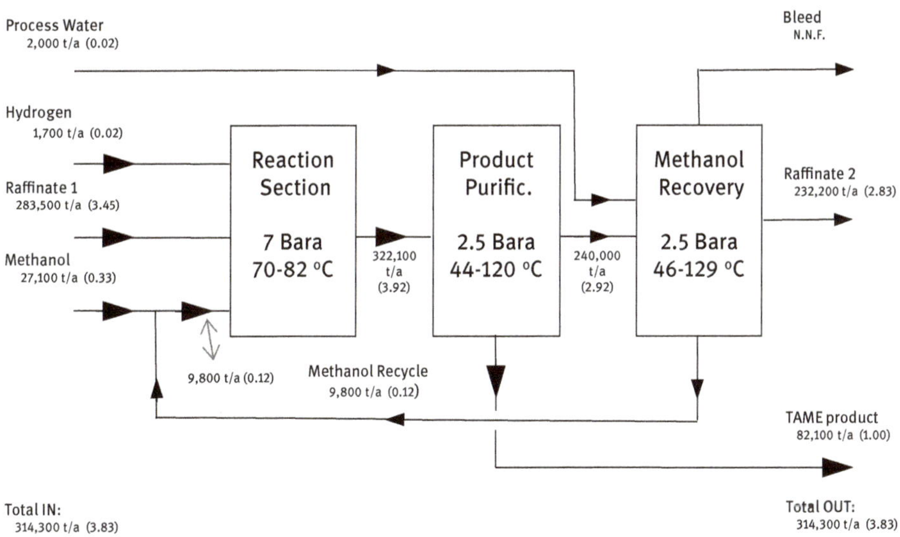

Please note:
- Values between brackets () are t/t values.
- Blocks and character fonts could have still smaller size.
- Values normally are round off, however without loosing consistency on balance.
- All streams ENTER at the LEFT hand side of the block scheme.
- ALL streams LEAVE at the RIGHT hand side of the block scheme.
- N.N.F. or NNF = Normally No Flow.
- Indicate temperature and pressure inside blocks and for all streams entering (LEFT) and leaving (RIGHT) the block scheme (not done in the above scheme).
- Indicate stream numbers for each stream (cross-referencing with Process Stream Summary).
- Never put arrows at the end of the connecting lines; for better readability of the scheme put the arrows at the middle of the line.

Fig. 8.1: Process Block Scheme (PBS) example – TAME (Tert-Amyl Methyl Ether) production process [3].

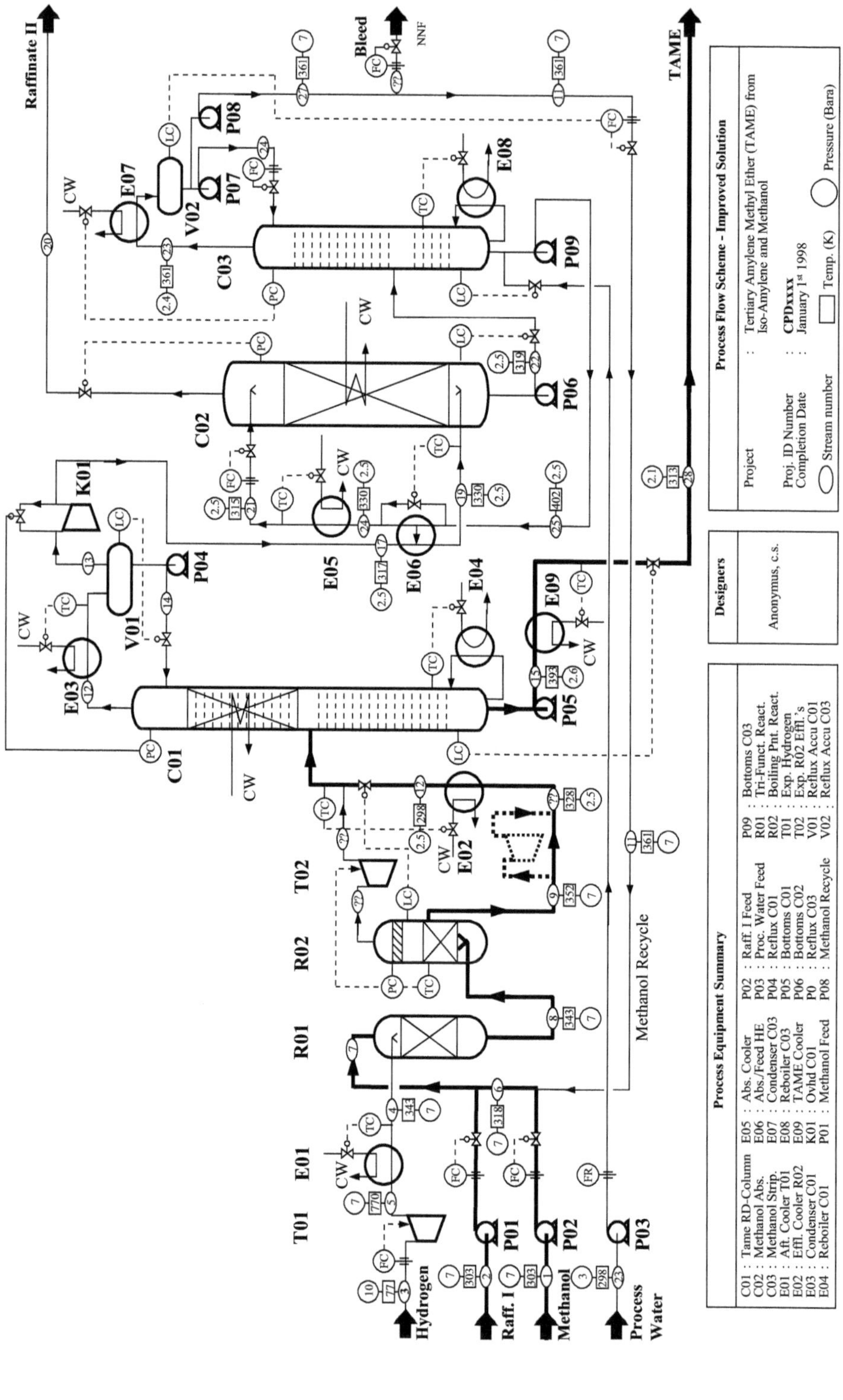
Raffinate II
Bleed
NNF
TAME
Hydrogen
Raff. I
Methanol
Process Water
Methanol Recycle
Process Equipment Summary
C01 : Tame RD-Column
C02 : Methanol Abs.
C03 : Methanol Strip.
E01 : Aft. Cooler T01
E02 : Effl. Cooler R02
E03 : Condenser C01
E04 : Reboiler C01
E05 : Abs. Cooler
E06 : Abs./Feed HE
E07 : Condenser C03
E08 : Reboiler C03
E09 : TAME Cooler
K01 : Ovhd C01
P01 : Methanol Feed
P02 : Raff. I Feed
P03 : Proc. Water Feed
P04 : Reflux C01
P05 : Bottoms C01
P06 : Bottoms C02
P0 : Reflux C03
P08 : Methanol Recycle
P09 : Bottoms C03
R01 : Tri-Funct. React.
R02 : Boiling Pnt. React.
T01 : Exp. Hydrogen
T02 : Exp. R02 Effl.'s
V01 : Reflux Accu C01
V02 : Reflux Accu C03
Designers
Anonymus, c.s.
Process Flow Scheme - Improved Solution
Project : Tertiary Amylene Methyl Ether (TAME) from Iso-Amylene and Methanol
Proj. ID Number : CPDxxxx
Completion Date : January 1st 1998
Stream number
Temp. (K)
Pressure (Bara)

◂ Fig. 8.2: Process Flow Diagram/Scheme (PFD/PFS) example – TAME (Tert-Amyl Methyl Ether) production process [4].

These engineering diagrams and drawings also play a key communication tool within the design team, but also to other stakeholders outside the team. They form key input documents for preliminary economic analysis (gross margin calculations based on product production and raw materials usage, see Chapter 10), for equipment lists and equipment data sheets (PFD) and for HAZOP studies (P&ID) (see Chapter 11) to name a few. The more detailed 3D CAD equipment and plant drawings are used for layout optimization and equipment and plant construction purposes. As a communication tool often between different companies (brand owner, engineering and procurement companies, equipment manufacturing and plant installation companies) these standardized drawings are widely spread.

The standardization of engineering drawings and their standardized use have not yet cascaded down to the design of chemical products. Chemical products that are process-like devices (fuel cells, lab-on-chip, sensor-on-a-chip) are also easily represented by the above tools, as the device function and performance delivery resemble a chemical process.

For other products (like biofuels, engine lubricants, coatings, laundry detergents, flexible packaging, etc.) product drawings are much less standardized (as they are often only used in a limited range of stakeholders) and are, therefore, not very widely applied.

For product design teams, drawings form a very important – but often vastly underestimated – tool for new product design and development. It is mainly in the design knowledge capture and the communication thereof, that the roles of drawings/sketches are paramount. Drawings are key as the first mental models representing the interactions of the product (and its components) with its surroundings during its entire life cycle, not only for visualizing these key product component interactions, but also as a basis for mathematical modeling and communication, an analysis and synthesis vehicle for generating the optimal product design. Therefore, the skill of making drawings/sketches in product design is viewed as a key competence every designer should learn to master.

8.3 Product design schematic models

8.3.1 House of Quality model for consumer function and property function

In reviewing the different product and process design phases (as presented in Section 4.3), the use of product related drawings/sketches and their links to mathematical models will be shown here.

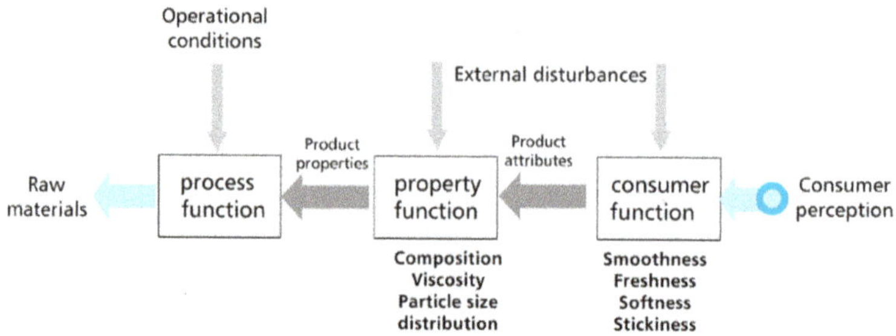

Fig. 8.3: Consumer and property functions in the consumer wants and product attribute design spaces [5].

The first design levels after the project framing level are: 'Consumer Wants–Product Concept' and 'Product Concept – Product Function.' In the 'Consumer Wants– Product Concept' level the translation of consumer needs and wants (voice of customer) into quantifiable product attributes takes place. These needs, wants and attributes can be vastly different for any (chemical) product. The Quality Function Deployment (QFD) method using the House of Quality (HoQ) design tool has been applied very successfully for this.

Almeida et al. [5] present a nice overview of the relationship between consumer wants/perception (great taste, scoopable directly from freezer, healthy) to the property function for an ice cream example (see Figure 8.3).

In this diagram, these consumer preferences for ice cream are first translated by a 'consumer (or quality) function' into quantifiable product attributes (e.g., smoothness, freshness, softness, stickiness). The consumer liking/perceptions of a product are mostly obtained through questionnaires and interviews. Careful preparation of these customers' interviews requires a good understanding of the product in use. Especially for product innovations it is good practice to visualize the various customer/product interactions. This starts, for instance, with the purchase of the ice cream in the supermarket, the transport and storage prior to consumption, the consumption activities, and final waste disposal. The product attributes are scored against an arbitrary scale that represents the relative performance of the attribute against consumer expectations.

In the second step these quantifiable attributes are then linked by a 'property function' to measurable product properties based on the composition and product microstructure. For the ice cream case study, product properties include: water crystal size and volume fraction, fat crystal size and volume fraction, air cell sizes, and volume fraction, etc. A schematic representation of an ice cream's microstructure is depicted in Figure 8.4.

The flow of design decisions runs from consumer product qualities to the process function and the selection of raw material feeds. This flow is the reverse of the cause and effect chain of physical events when making a product. While the logic of the flow

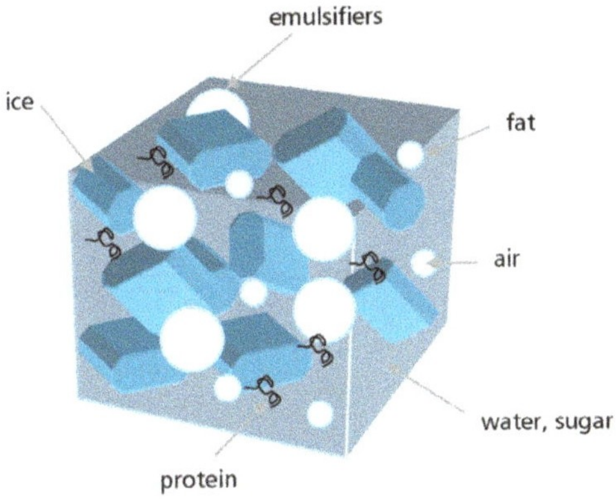

Fig. 8.4: Schematic representation of the microstructure of ice cream [5].

of design decisions is in itself impeccable, a practical information gap problem arises. This problem is due to the ever-increasing degree of freedom when going with the design decision flow, i.e., in every step of determining functions (consumer ⇒ property ⇒ process) there are relatively few specifications of what the outcomes of a function should be, while there is an excess of inputs to a function. Consequently, there is an abundance of degrees of freedom. The question arises how one can handle or fix this degrees of freedom excess? This information gap problem can be solved in a meaningful way when a lot of side information is added per function. That is, additional restrictive requirements on the function inputs must be imposed, along with an optimization of the performance of a function. Finding meaningful restrictions and setting suitable optimization goals are essential parts of the design process.

The quality function deployment (QFD) method can be used to translate consumer wants (consumer voice) into product attributes requirements. As presented by Almeida et al. [5], this translation occurs at the initial matrix of the QFD, termed House of Quality (HoQ)-1 (Figure 8.5). In Figure 8.6 the two subsequent steps (HoQ-2 and HoQ-3) translating consumer wants via product attributes and (physical) product properties to the process function requirements are shown. House of Quality templates for Microsoft Excel are available at the QFD ONLINE website [6]. Seider et al. [7] also provide good insight in how to apply the QFD/HoQ methodology.

8.3.2 Understanding the product application/use process: visualizing systems splits

In support of the House of Quality QFD method, it greatly helps to visualize and draw the product in the environment(s) where the product will be applied or used. The mouthfeel, creaminess, and iciness (sensory) product attributes should be met and not changed during the life cycle phases prior to consumption. More importantly, by quantifying the different conditions across the manufacturing and in-use processes

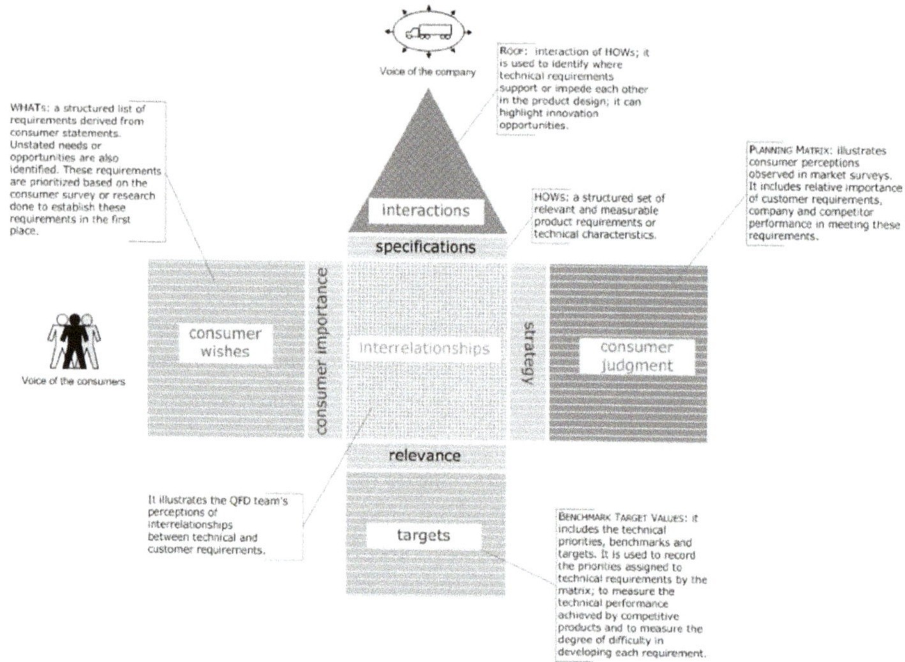

Fig. 8.5: Schematic representation of the use of House of Quality (HoQ) 1 in the Quality Function Deployment (QFD) methodology translating consumer wants (voice of consumers) into product attribute requirements [5].

(temperature and relative humidity during storage, transport, display, shopping, home storage, and consumption, including time of exposure, etc.) potential risks for product quality can be identified and modeled. If prolonged exposures to higher temperatures prior to the ice cream consumption are identified, product/environmental interaction drawings and mathematical models can be used to quantify, substantiate, and mitigate this risk. Transport phenomena models involving the ice cream material, the ice cream wrap, carton packaging box and the heat transport models can be used. Relevant physical transport and material properties (heat conductivity in moist air, ice cream, packaging material, density and specific heat, heat of melting, etc.) come into the equation.

By applying the approaches in this section and Section 8.4.1, the understanding and quantification of the key product application process mechanism can greatly enhance product (re-)design.

Wesselingh et al. [8] emphasize the essentials of product/systems interaction and promote drawing system splits to start the modeling approach. He advocates making different splits of the product/system interactions to get an overview of the product in use and during the different phases of its lifetime. He uses the example of a laundry washing machine and a laundry detergent product. Different systems can be chosen to get different viewpoints on this system and to make simple mathematical models. An

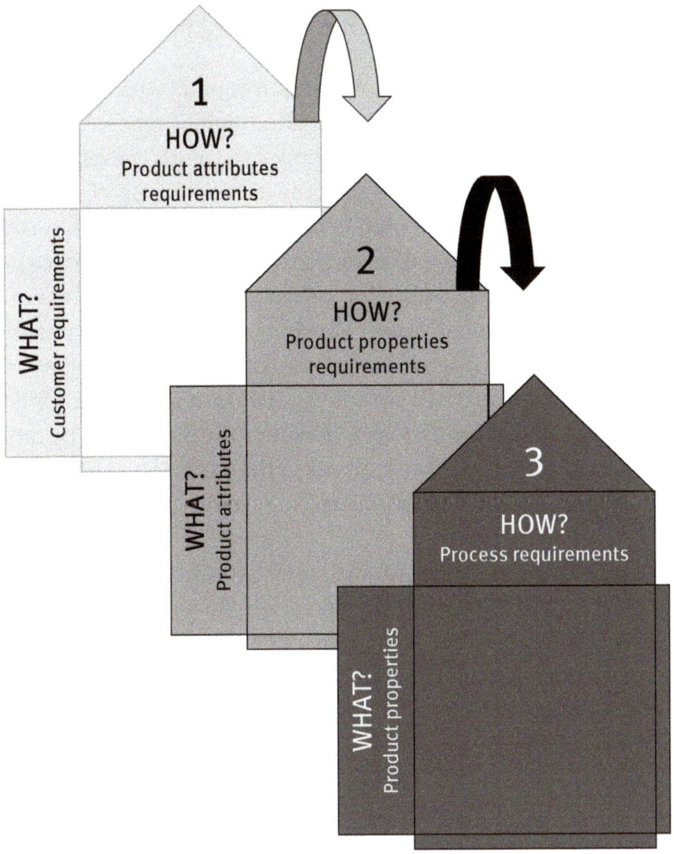

Fig. 8.6: Schematic representation of the use of Houses of Quality (HoQ) 1, 2 and 3 for translating consumer wants into product attributes, product properties, and process requirements.

input/output mass and energy balance analysis of washing machine (inputs: soiled dry textile, clean cold water, laundry detergent, electricity power; outputs: wet clean textile, dirty water [containing soil and detergent components], heat losses).
Another way to split this system is to subdivide the laundry process into process steps, such as: add laundry, add detergent into dispenser, add water, heat-up, wash, rinse, centrifuge, and take out laundry. The system can be quantified by taking simple heat and mass balance equations and using system variables like mass and temperature of textile, water, detergent, required washing temperature, power supply, heat-up rate, washing time, rinsing time, and centrifugation.

A closer look can be taken to examine the textile composition and structure. How is the textile structure: solid fibers, spun into yarns, or woven into textile? What are the volume fractions of voids in the yarns and between the yarns in the woven textile? One can quantify the fiber density and diameter, the yarn density and the total surface area of the fibers that needs to be wetted and covered with surfactant molecules for the efficient removal of soil particles or stains.

The interaction of the detergent product in water and in contact with the textile in the washing machine can also be visualized. The textile tumbles repeatedly into the water, and water (with dissolved surface active components) (partly) refreshes the liquid between the fibers and yarns. Through the bending and stretching of the textile the replacement of dirty liquid by fresh liquid is taking place, and through the surface-active activity of the detergent ingredients soil is removed from the surface of the fibers and is kept suspended in water. Antiredepositing ingredients avoid the resettling of the soil on the fibers. The liquid water refreshing rate can be computed/estimated based on a simple transport phenomena model for the convective flow of water in/out of the textile between the yarns (by bending), and the much slower transport of water/dirt/chemicals into the yarns (between the fibers) by diffusion. The rate determining step for the overall process is the diffusion in the yarns to the fibers (as can be derived from regime analysis and characteristic times estimations (covered in Section 8.4.1).

The soil, stains, and the detergent product and structure itself can also be sketched and modeled. They can be split into different stain types (soluble and insoluble particles, fat/oil, anionic surfactants when not scavenged). Identification of molecular identity, quantities, and dissolution behavior in water are all important. The sketches/drawings can capture and visualize the key phenomena to be considered in the use and life cycle phases of the product.

8.4 Mathematical models for consumer and property functions

In this section, different approaches to create mathematical models for consumer function and product property function are presented.

8.4.1 Characteristic times and regime analysis

A very convenient way of quantifying the different processes/mechanisms taking place in applying the product and finally delivering the performance is to compute the so-called characteristic times (or time constants) for these mechanisms. In the 1960s this practice of developing time constants for rates of changes in chemical engineering systems developed at the disciplinary interface of physical transport phenomena and process dynamics. In the 1980s this methodology of determining characteristic times within a process was applied to the scale-up of bioreactors, by using the 'scale-up by scale-down' approach. This method was used to calculate the various characteristic times for mechanisms (mixing, mass and heat transfer, oxygen depletion, product formation, and biomass growth) in stirred aerated fermentors

(Oosterhuis [9], Sweere [10]) and to establish the rate-limiting mechanism(s). The rate-limiting mechanism determines the regime such as the oxygen or substrate limiting regime in (aerobic) fermentations. The process was researched and optimized in a scaled-down fermentor in which the rate-limiting mechanism(s) on this laboratory scale were the same as in the large-scale fermentor to be designed.

This approach has been further developed and taught over the years by Kossen, Luyben, Heijnen, and Van der Lans of TU Delft's biotechnology department, and is still taught to MSc and PDEng students in Delft University of Technology's biotechnology and chemical engineering fields.

This regime analysis method using characteristic times is also very versatile for understanding and improving product application/use processes and the design of the microstructure and composition of the product.

Characteristic times or time constants can be defined as the ratio of a capacity over a corresponding rate, where capacity and rate are to be used in a rather general

Tab. 8.2: Characteristic times for transport phenomena and (chemical) kinetics.

Time constant [s]	Nominator	Denominator
Travel time: $\tau = L/v$	L [m]: travel length	v [m/s]: travel velocity
Residence time: $\tau = V/F$	V [m^3]: volume	F [m^3/s]: volume flow rate
Diffusion time: $t_d = L^2/D$	L [m]: diffusion length	D [m^2/s]: diffusion coefficient
Conversion time: General:		
– $t_c = (CV)/(r_c V)$, $t_c = C/r_c$	V [m^3]: volume C [kmol/m^3]: – reactant concentration	V [m^3]: volume r_c [kmol/m^3/s]: – conversion rate per volume
Zero order: $t_c = C/k$	C [kmol/m^3]	k [kmol/m^3/s]: – 0-order rate constant
1st order: $t_c = 1/k$	1 [–]	k [1/s]: – 1st order rate constant
Mass transfer time From gas bubble into liquid:		
– $t_{mt} = 1/(k_l a)$	1 [–]	$k_l a$ [s^{-1}] or [(m/s) · (m^2/m^3)]: – k_l: liquid side mass transfer coefficient [m/s] – a: specific surface area [m^2/m^3]
Depletion of gas bubble: – $t_{mt} = m/k_l a$	m [–]: ratio C_g^*/C_l^* C_g^*: reactant concentration in gas corresponding to the reactant saturation concentration in liquid (C_l^*)	
Heat transfer (equilibration) time: – $t_{ht} = V\rho c_p/(UA)$	$V\rho c_p$ [J/K] [(m^3 · kg/m^3) · (J/(kg/K))]	UA [W/K] [(W/(m^2 · K)) · (m^2)] ($1/U = 1/\alpha_l + (\rho c_p)/\lambda + 1/\alpha_e$)
Momentum transfer time (shear stress relaxation): – $t_{shear} = (L^2\rho)/\eta$	L [m]: shear layer thickness ρ [kg/m^3]: density	η: [kg/(ms)] viscosity

way. Some examples of time constants that can be derived from differential mass and energy balances are presented in Table 8.2.

Characteristic times for the laundry detergent example in the previous section can be calculated. For the steps involved in the washing process characteristic times can be estimated for: heating up water, heating up textile, disintegration and dissolution of the detergent powder into the water, diffusion time of chemicals/water into yarns, time for convective transport of water/chemicals into and out of the woven yarns, enzymatic conversion of fat, proteins, bleach reactions. Based on results and equations, the rate-limiting step(s) can be identified. By experimentation – in a properly designed scaled-down experimental setup (operating at the regime where the same step is rate-limiting) the 'rate' constants can be determined. The experimental results and regime insight can be used to modify the product design (composition and/or structure), and the in-use process of the product. It may prevent making noneffective changes (and cost) in products and processes that are not limiting.

It helps greatly to use schematic 'drawing to scale' where possible, so that the designers appreciate the different length scales of the product and systems it interacts with. Order of magnitude estimations for the different system mechanisms' time constants/characteristic times are sufficient. By determining the slowest or rate determining process(es), one can focus on these to improve the product and product application process, and assume that mechanisms with the faster order of magnitude take place 'instantaneously'. This improves the understanding of the system and also simplifies the mathematical modeling required. In [11] some further examples of product/system splits and modeling are presented:

- Detergent requirements (p. 38)
- Flow properties of toothpaste (p. 64–66)
- Running an oil tanker (p. 77–80)
- Injection needle (p. 95–97)
- Designing a capsule for controlled release (p. 119–120)
- Design of Rockwool (p. 170–172)

A very useful source for transport phenomena data and models historically dispersed over many literature sources was compiled and published by Janssen [12]. It provides a good starting point for preliminary estimation of the rate key (transport) processes in the application of products, depending on structure and size.

A comparable approach to quickly estimating material and component properties for component substitution in existing products is also presented by Cussler and Moggridge [13]. They present examples of component substitutions based on thermodynamics (like solubilities) and kinetics (chemical conversion rates and mass and heat transfer rates).

8.4.1.1 Estimation skills – Fermi problems

Estimation skills like those just mentioned are key for a quick appreciation of the important and less important phenomena to be considered in the design of the product and its application process/use.

Another very nice way for designers to train and apply estimation skills (or 'back-of-the-envelope' calculations) is to solve so-called 'Fermi problems', named after the physician Enrico Fermi. Fermi estimates deliver order of magnitude estimates. Information with this accuracy is in many cases often sufficient to make a decision – for example to neglect a phenomenon or to study it in more detail.

8.4.1.2 'Systemic' time constants – eigenvalues

Estimation of time constants of the separate rate phenomena can give much local insight into a particular phenomenon. Unfortunately, having a full set of time constants as such is hardly indicative of the behavior of the overall system (product/process) in which various phenomena interact. Suppose there are N resources in a process (e.g., species mass, energy) which can interact with each other and each interaction represents a physical phenomenon with a rate effect (reaction, phase transfer, temperature change, ...). Then $\frac{1}{2} \cdot N \cdot (N-1)$ time constants can be obtained related to the rate phenomena. The alternative is to order the rate parameters (from which the time constants were derived) in a (square) matrix. An example is the Jacobian matrix (an ordered set of linear derivatives of an equation with respect to all variables for all equations) of a set of conservation equations for the resources in a process (unit). Such a matrix has $N \times N$ entries and each entry can be turned into an equivalent time constant. However, mathematical analysis shows that the overall behavior of the process/product system is determined by the eigenvalues of such a matrix. One obtains only N eigenvalues. The joint effects of $\frac{1}{2} \cdot N \cdot (N-1)$ time constants of the system phenomena are condensed into a generalized N or "systemic" time constants. The benefit of computing and analyzing such eigenvalues extends the knowledge that can be inferred from the full set of rate related time constants. For instance, eigenvalues will reveal:

- Any linear dependency between resources, meaning that a resource varies as a linear combination of some other resources (occurrence of zero-value eigenvalue[s]);
- Stiffness: some resources respond very swiftly to changes while others are lagging (very high ratio of the largest eigenvalue over the smallest nonzero one);
- Oscillatory behavior (occurring for an even number of complex-valued eigenvalues with nonzero imaginary parts)
- Instability of the process/product system
  (some eigenvalues have positive real parts causing exponential growth)

These patterns of behavior are hard to fathom from just seeing a set of real-valued time constants.

8.4.2 Data driven nonlinear product modeling: artificial neural networks

Due to the absence of fundamental relationships between the consumer perception of the product's sensory attributes (e.g., smoothness, softness, stickiness, etc.) and product physical attributes (composition, ice crystal size distribution, air cell sizes, etc.), the 'product property function' can be formulated by means of data mining techniques.

In 2008 Bongers [14] presented a method for relating ice cream consumer perception to product physical attributes. This approach was based on large sets of experimental data (and Artificial Neural Networks [ANN]). The product property function involved a reduced set of physical product properties that could predict the consumer perception of the ice cream product. Based on this work Unilever has been able to introduce an innovative lower caloric ice cream with superior consumer perception by reducing fat content and controlling ice crystal and air bubble sizes by improved processing.

A similar approach was adopted by Almeida (2007) [33] in the rationalization of the ingredients in the manufacture of a mayonnaise-like emulsified product. By means of data mining techniques (partial least square and neural network techniques), only relevant product attributes were classified in sensorial and analytical attributes. These attributes were then used to create alternatives for the product's microstructure and ingredient composition.

The neural networks technique was also used by Dubbelboer et al. [15] in the formulation of the property function for high-internal-phase emulsions.

Another class of chemical product design tools – for studying and modeling product property functions – are DOE (Design of Experiments) software tools MODDE 11 (Umetrics) and Design-Expert®. Next to their capability to design experiments, these tools also perform results analysis (model building and regression) and are increasingly used to build product property models.

8.4.3 Scientific models for product state and behavior

8.4.3.1 Constituents

Before embarking on a description of the possible constituents of a product, it is helpful to introduce a common distinction between two different types of products. Following this distinction, the class of products to be considered for design will be further refined. Two product types can be discerned in relation to their extensive form:

- 'Matter', being continuously extensive (like any fluid)
- 'Device', a discrete entity with predetermined dimensions (like solid objects)

Fluid phase products are continuously extensive. Single molecular gases or liquids qualify, but blended single phase fluids and formulations do as well, in addition to multiphase systems like emulsions, foams, and gels (see Table 8.4). This is independent of how the product is made (batch-wise or continuous manufacturing) or how the product is packaged. With respect to solid phase products both forms can occur. If the size of the particulates of the solid phase is very small and there are numerous ones that can move more or less freely, the product becomes fluid-like again, such as a crystalline powder. It is not an individual particulate that matters in the use of the product. The particulates are used collectively as a kind of bulk material. Therefore, the product is classified as extensive.

However, if the solid phase product has to be handled on the basis of individual pieces rather than as a bulk, and the solid phase product has sufficiently large (macroscopic) dimensions, say, well above one millimeter, the product will classify as discrete. A slab of steel and an electronic chip are examples of a discrete product, even when the size is adjustable in the manufacturing process. Also, the multiple solid state layers on a wafer will be considered as discrete components in a device-like product.

The characterization of a structured product in this paragraph is written primarily with the continuously extensive product in mind. At the end of the paragraph some supplementary remarks will be made concerning device-like, discrete products. Here, the following assumptions are made, that the product:

- Is continuously extensive
- Is made up of several chemical and/or biological species
- Has one (pseudo)continuous thermodynamic phase
- Is dispersed when more than one (fluid) phase is present
- Has an interface between each set of two phases that acts as a distinguishable entity having other properties than those in the bulk of the adjacent phases
- Has diffusion of species and thermal energy in and between the phases
- May have chemical reactions between species

The buildup of constituents in a multiphase structured product is visualized in Figure 8.7, where the relation between phase composition, the multiphase structure of the mixture, and performance is shown.

8.4.3.2 Product structure

Concerning the geometric aspects of the bulk structure, it is observed that in many products the bulk is made up of a dispersion of one or more thermodynamic phases with a multilevel distribution; see Figure 8.8.

Examples of such dispersed systems are emulsions (liquid phase in another liquid phase), foams (large volume fraction of gas bubbles in a dispersing liquid), bubbly liquids (dilute suspension of gas bubbles in liquid), and gels (semirigid dispersion of solids in a fluid). Dispersion may repeat itself at increasingly smaller scales. Usually, one level of dispersion is considered in chemical equipment modeling, but for structured products like food, cosmetics and pharmaceuticals, the use of multiple

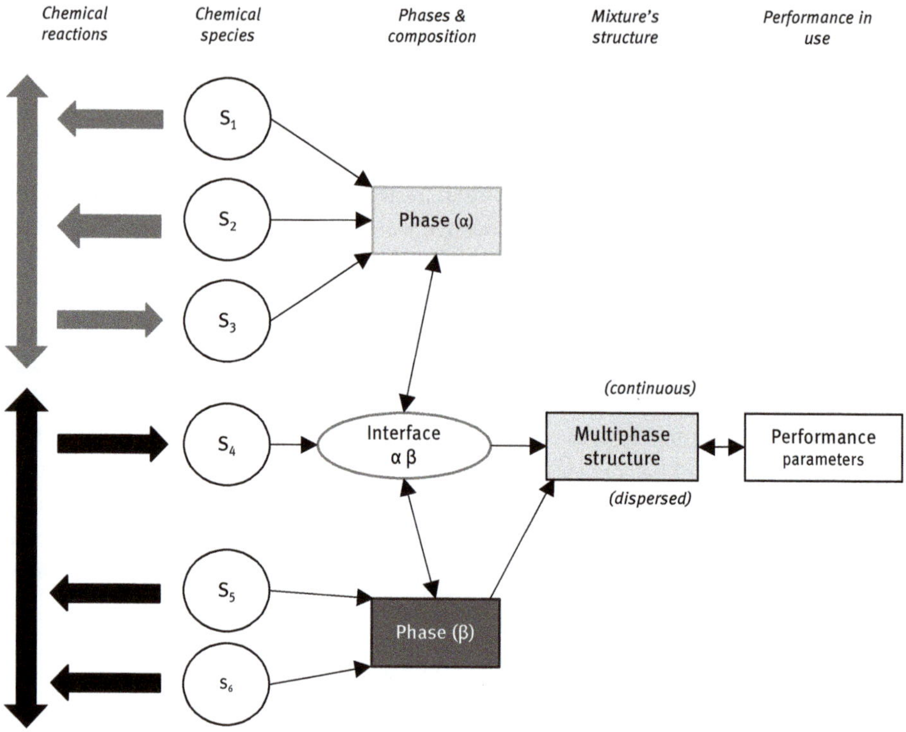

Fig. 8.7: Product composition, structure and performance for fluid phases.

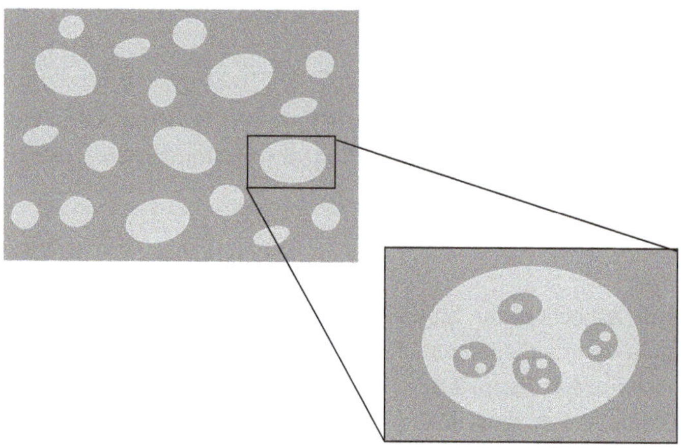

Fig. 8.8: Multilevel/repetitive phase dispersions (oil/water/oil/water).

Tab. 8.3: A matrix representation of the phase ratios.

Continuous phase	Distributed phase α	Distributed phase β	Distributed phase γ
Phase α	continuous	$\varepsilon_{\beta}^{(\alpha)} = 0.3$	$\varepsilon_{\gamma}^{(\alpha)} = 0.1$
Phase β	$\varepsilon_{\alpha}^{(\beta)} = 0.2$	continuous	$\varepsilon_{\gamma}^{(\beta)} = 0.3$
Phase γ	$\varepsilon_{\alpha}^{(\gamma)} = 0.1$	$\varepsilon_{\beta}^{(\gamma)} = 0.4$	continuous

levels may be relevant. For repetitive dispersions, the spatial complexity of the system will further increase. A discrete volume entity can act as a continuous phase in which even smaller discrete volume entities are spatially distributed. E.g., a droplet of a dispersed phase (β) is contained in a continuous phase (α), which is in turn contained in a droplet of the dispersed phase (β).

However, a droplet of the dispersed phase (β) may contain even smaller droplets of the phase (α) and of another phase (γ). This dispersion of phases might repeat itself at successive levels of increasing detail. The extreme case of a multilevel spatial distribution of dispersions (or even a continuous distribution) is not considered. The repetitive occurrence of dispersed phases has implications for the calculation of the overall phase ratios in the product. This is explained by means of a small example.

The first column in Table 8.3 indicates the continuous phase. The matrix row pertaining to a continuous phase shows which mass fractions of the other (two) phases are present (as dispersed phases) in that continuous phase. E.g., when a volume element of phase β is the continuous phase it will by example contain a 0.2 mass fraction of phase α and a 0.3 mass fraction of phase (γ). This information allows for computing the overall phase ratios in a simple way. For example, consider a two-level dispersion between three phases, where at the first level of dispersion, phase α is the continuous phase, while the other two phases (β) and (γ) are dispersed. At the second level of dispersion droplets of phase (β) contain dispersions of phases (α) and (γ), as well as droplets of phase (γ) containing dispersions of phases (α) and (β).

Computation of relative amounts yields:

$$\varepsilon_{\alpha}^{\text{overall}} = [1 - \varepsilon_{\beta}^{(\alpha)} - \varepsilon_{\gamma}^{(\alpha)}] + \varepsilon_{\beta}^{(\alpha)} \cdot \varepsilon_{\alpha}^{(\beta)} + \varepsilon_{\gamma}^{(\alpha)} \cdot \varepsilon_{\alpha}^{(\gamma)} = 0.67 \tag{8.1}$$

$$\varepsilon_{\beta}^{\text{overall}} = \varepsilon_{\beta}^{(\alpha)} \cdot [1 - \varepsilon_{\alpha}^{(\beta)} - \varepsilon_{\gamma}^{(\beta)}] + \varepsilon_{\gamma}^{(\alpha)} \cdot \varepsilon_{\beta}^{(\gamma)} = 0.19 \tag{8.2}$$

$$\varepsilon_{\gamma}^{\text{overall}} = \varepsilon_{\beta}^{(\alpha)} \cdot \varepsilon_{\gamma}^{(\beta)} + \varepsilon_{\gamma}^{(\alpha)} \cdot [1 - \varepsilon_{\alpha}^{(\gamma)} - \varepsilon_{\beta}^{(\gamma)}] = 0.14 \tag{8.3}$$

Note that the overall phase ratios do sum up to unity.

8.4.3.3 Interfaces between dispersed phases

Each volume body of a phase will have bulk-like features, (e.g., in the bulk of the dispersed entities, like droplets) as well as interfacial features for the contact surfaces between phases. When describing interfaces between two thermodynamic phases, one can adopt two viewpoints. In the "macroscale" view the surface is considered to be an idealized two-dimensional, smooth object of negligible thickness. In the 'microscale' view it is seen as a curved three-dimensional body with a finite thickness (see [16]).

A special interface is the exterior surface of a product. The exterior shape of a fluid-like product with low or moderate viscosity takes the form of its containment vessel. There are usually no exterior surface functions of such products expected. However, a fixated exterior surface will occur for solid phase products or fluid products with a very high viscosity. Then the form of the surface and surface functions to be performed are made part of the design.

8.4.4 Product structure matrix

One can think of a multitude of combinations of thermodynamics phases that are possible for creating multiphase products. Table 8.4 provides a schematic overview of the various options.

A representative example of the complexity of a multiphase dispersed system is ice cream. Ice cream is an emulsion, suspension and foam at the same time. The continuous phase is water with a high concentration of sugars. The liquid water contains a population of small ice crystals. The ice crystals form the first dispersed phase. The second dispersed phase is air, which is foam-like and dispersed in the form of small bubbles in the continuous liquid phase. The third dispersed phase is fat, distributed as small droplets in the water phase, so-called fat globules. The stability of this emulsion is controlled by means of emulsifiers. Actually, emulsifiers are being used to partially destabilize the fat emulsion and so cause partial coalescence. These partially

Tab. 8.4: Product structure matrix for multiphase chemical products *(adapted version, original from Prof. A.A. Broekhuis, RU Groningen, Netherlands).*

Dispersed phase ⇒ **⇓ Continuous phase**	**Gas**	**Liquid**	**Solid**
Gas	Mixtures	Aerosols / Mist	Dust
Liquid (organics, water, metals, polymers)	Solutions Bubbles Foams	Solutions Emulsions Gels	Solutions Colloids / sols Suspensions Pastes
Solid (organics, metals, ceramics, polymers	Adsorbents Foams	Adsorbents Blends Compounds	Blends, alloys Powders, granules composites **Device-like (discrete):** Solid films, coating, multiple stacked layers

coalesced fat droplets tend to collect at the waterside of the interface with the air bubbles and so help to stabilize the foam of air in water. The coalesced fat can form a spatial 3D network.

A vast amount of scientific knowledge and detailed scientific models have been developed for describing the behavior of (chemical) products and their components.

A very nice overview of the chemical product design related theories is presented in Bröckel et al. [17], members of the European Federation of Chemical Engineers (EFCE), section group "Product Design and Engineering." The following product design related theories and basics are presented:

- Interaction forces between particles (by Schubert)
  - From forces between atoms and molecules, to adhesion forces between particles in both gaseous and liquid environments
- Fundamentals of crystallization (by Ulrich, Jones)
- Emulsification techniques for the formulation of emulsions and suspensions (by Schuchmann)
- Characterization of dispersed systems (by Polke, Schäfer)
- Modeling of chemical systems to predict product properties (by Gani, Abildskov)

Wesselingh [18] presents an elegant approach for using simple, but very illustrative and useful models to quantify various product physical structures (in the nm to mm range). These models can be used for a first order of magnitude quantification of product structures in terms of volume fractions, surface area of dispersed phases, number of surfactant molecules needed to form monolayers on the interface, etc. Adsorption mechanisms and rheology are also explained and quantified.

Bröckel et al. present in [19] a number of illustrative examples of product design, like:

- Biodiesel (by Meier)
- Fats, oils and waxes (by Reiz and Kleinebudde)
- Starch and starch-based products (by Leeb and Schuchmann)
- Gelatin (for food, pharmaceutical and technical applications – by Babel)
- Sugars (by Häusler)
- Synthetic precipitated silicas (by Schmoll)
- Heterogeneous catalysts (by Kunz et al.)
- Peroral products (by Mäder)
- Carotenoid products (by Leuenberger)
- Coffee-based beverages (by Schuchmann)
- Laundry powder components (by Boerefijn)
- Agrochemical products (by Frank)

In 2013 Bröckel et al. [20] extended the product design and engineering series even further with a focus on the formulation of gels and pastes. Again, many examples from food, personal care, and ionic liquid applications are discussed, together with the fundamentals of the rheology of dispersed systems, and how to apply these fundamentals for the formulation of gels and pastes products.

8.5 Relations between product and process modeling

8.5.1 Causal flow of information on process – product modeling

The purpose of any manufacturing process is to convert the given feeds into products with target performances. For product and process design one can make models of the feeds, the process and the products. The product state is an output from a behavioral model of the product. This state is also important in predicting the performance metrics of a product by means of a performance model of a product. This can be done on the basis of the physical-chemical state of the product as well as the processing conditions that prevail under a product application. It is an issue in product and process modeling how to consistently link the behavioral models of products and processes. The behavioral models of the feeds, the process, and of the products must match, i.e., be mutually consistent. For *simple* feeds and products, the situation is simple indeed. A simple product has a single phase with a homogeneous composition. This fully matches with the common characterization of a stream in a process flowsheet model by its composition and flow rate. However, a complex product (and/or feed) has a single-phase product of a distributive nature (e.g., a crystal size distribution) or it has a multiphase composite structure, possibly also with one or more phase with distributive properties. This was covered already in the preceding paragraph (Section 8.4). This section will focus on connecting models of the product and its manufacturing process.

For product design and improvement, it is desirable to have clear, causal quantitative links between:

(a) the structure and composition of a product at the (supra)molecular scale as resulting from the product manufacturing process,
(b) the macroscopic physical properties of a product, and
(c) the qualities of a product and associated performance for users.

Therefore, one needs three models in a cascade to generate this flow of information from product ingredients in the feeds via the product manufacturing process to product properties and the customer qualities in product use. This is the causal flow of information (physical cause ⇒ effect), shown in Figure 8.9.

Each of these models will be highlighted. The positioning of the three models and the flow of information between the models is in line with Bernardo (2016) [21].

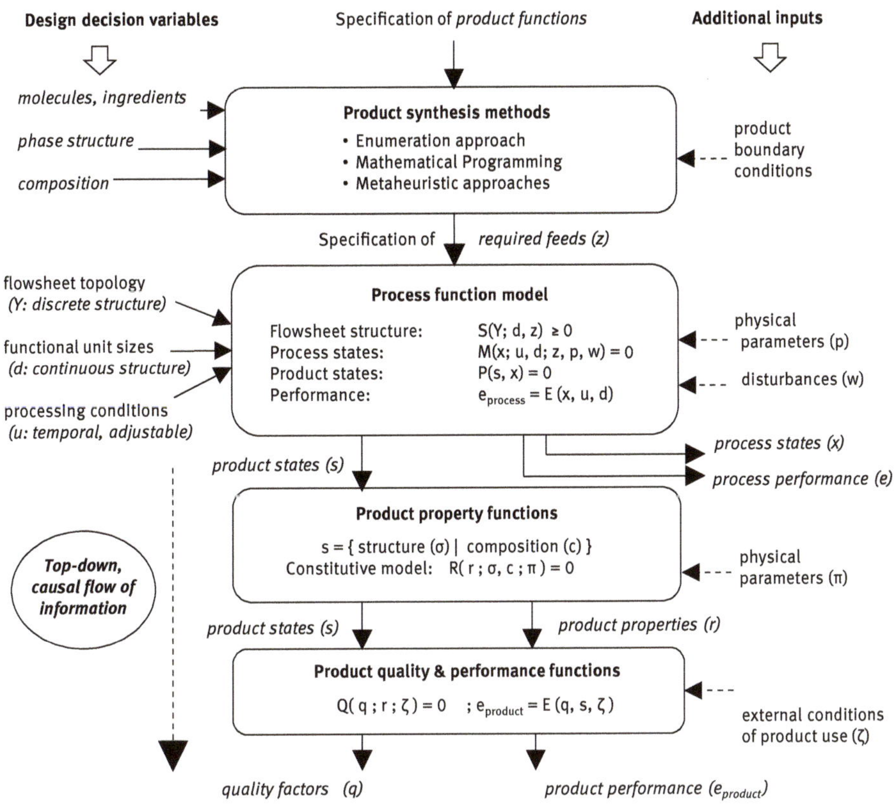

Fig. 8.9: Causal flow of information through function models from manufacturing process to product qualities and performance.

8.5.1.1 The product manufacturing ⟹ product structure and state model

The structure and the physical state of a product come into existence in a manufacturing process, starting from product ingredients or precursors in its raw feed streams. A model of a product manufacturing process has to be able to describe the structure and states of the process as well as the physical state of its product(s). Hence, a mathematical representation of matter in process streams and unit operations should allow for the formation and transport of a structured product. Making a proper representation of product structure and state(s) is to be considered an integral part of the process modeling. However, current process modeling practice is a bit reluctant in this respect, dealing with simple (single phase) products only. Most commercial process flowsheet simulation programs do not offer facilities to model and simulate the formation of structured products, except as an add-on. Dedicated models of product formation may be interfaced with a process simulator. Equation-oriented model building software is well suited to include such models (e.g., *gPROMS* [22], *JACOBIAN* [23]). Dedicated product-process modeling tools are becoming available for classes of structured products; e.g., *gPROMS-FORMULATE*, suitable for dealing with particle populations with distributive properties [24].

8.5.1.2 A product structure and state ⟹ product properties model

This model derives physical and chemical product properties of the (bulk of the) product (rheological properties, heat capacity, conductivity, pH,...) from its underlying physical (micro)structures and molecular compositions. Such structures involve two or more thermodynamic phases, their distributive properties and the interfacing between the phases. The chemical composition per thermodynamic phase must be known in order to derive the macroscopic properties from suitable constitutive equations. The latter links molecular information to macroscopic transport properties. During recent years, very nice progress has been made through the developments of several product modeling case studies by the research groups of Prof. R. Gani, Prof. K.M. Ng, and Prof. M. Eden. This has recently led to the first 'chemical product simulation and design tool' (*ProCAPD*), that is offered and supported by the new company 'PSE for SPEED', started by Gani in 2017 [25]. Product properties can be simulated and optimized based on the products' molecular composition and structure. The tool uses very extensive databases and has been demonstrated for design-analysis of products as: single molecular products (solvents, refrigerants), mixtures/blends (fuels, lubricants), and liquid formulated products (cosmetics, paints). TU Delft hosts each year a workshop by Prof. R. Gani for PhD students and PDEng trainees (see Section 14.3.4.1).

8.5.1.3 The product property ⟹ customer qualities (attributes) model

This model links customer related qualities and performance indicators with the physical properties of a product. Customer qualities are almost always empirically observable quantities, including the conditions under which a product is used or applied. Thus, the conditions in a product application process should be taken into account in a customer qualities model. Thus, such qualities are rooted in both product properties and product application conditions. Rigorous models based on physical first principles of product use are hard to derive or simply too expensive. Therefore, a customer qualities model is mostly an empirical, data driven, black-box model, either a statistical one or an Artificial Neural Network (ANN) – already discussed in Section 8.4.2. The model creates a data-supported bridge between the product properties, the conditions of (external) use of a product, and the product qualities and performance metrics, as appreciated by customers.

8.5.1.4 Inter-connectivity between product and process models

The causal flow of information through these three models assumes that information is already available on the required ingredients for the product(s) in the process from raw feeds to manufacturing. This has to be generated by a preceding product synthesis effort. It is stressed that the design of a product and its manufacturing process runs conceptually in an opposite direction. First one identifies the desired product qualities and the associated performance metrics (SHEETS related). Then, one works towards a suitable set of product properties and onto product composition and structure. This allows for the identification of suitable product ingredients. Then one can think of a conceptual design of a manufacturing process. The flow from product qualities to product ingredients and a process design generates a tremendous increase in information. This increase comes from a vast range of design decisions, particularly on the process manufacturing side:

- Process topology involves the selection and ordering of processing functions along with their streams connections in a process flowsheet. These are discrete decisions, initially man-made, sometimes leading to superstructures over which one can optimize a design.
- Type and sizes (volumes, exchange surfaces, length...) of processing units have to be derived from specified functional targets for the units (e.g., degree of conversion, of separation, heat transfer duties, ratio of pressure increase, etc.).
- Operational conditions in the process units (pressure, temperature, preferred flow regimes, phase ratios, etc.)

8.5.1.5 Mathematical model-based product-process optimization

Many product-process design cases are supported by repetitive process – product model simulations in which human designers vary design decisions while aiming for the target product qualities. Such iterative approaches often rely on serial design decision making. Such iterative, serial approaches lead to feasible designs but not necessarily optimum ones. Therefore, more comprehensive optimization approaches are developed. These approaches rely on a computing-driven automated search over the joint spaces of the design decision variables for the optimum performance. The objective is to optimize some SHEETS performance metrics of process and product. The search for a feasible optimum should remain within the validated domains of the process models, particularly taking care to stay within the regions of applicability of constitutive equations.

Model-based optimization of product-process designs experiences difficulties of four generic kinds:

- The model equations do not comply with the mathematical requirements for the existence of an extremum (e.g., convexity), undermining a theoretical guarantee for an optimum.
- Even if a model would satisfy all conditions for the existence of an optimum there may be multiple ones and the true extremum may be hard to find within the set of local extrema.

- Exhaustive computational efforts during the search for the optimum due to the high dimensionality of the search space as well as nonlinearity of the process and product models
- In addition to using real-valued continuous variables (e.g., P, T, x) it can be necessary to use discrete variables with binary {0, 1} or integer {0, 1, 2, 3, ...} values. Such integers may enumerate a class of chemical groups; e.g., {1 = $-CH_2-$; 2 = CH_3-; 3 = $-CH{=}HC-$; 4 = $HO-$}.

To cope with the hardships in a comprehensive optimization of product and process designs one can apply some form of problem decomposition. Two kind of decompositions are mentioned:
- First optimize the product structure, composition, and performance. One can arrange this outcome as a tight set of product related constraints to be obeyed in a process optimization. However, the outcome is likely to be an expensive process, making the perfect product at a negative profit, if the process is even feasible. One may work toward a compromise by relaxing the product optimization and sacrificing slightly product qualities that tend to make a manufacturing process more expensive. By means of repeated simulations one may find a feasible solution for both product and process designs. It is hard to say if the combined solution is the optimum one.
- A sequential approach to the product design can ease the computational burden. In computer-aided molecular design (CAMD), discrete variables occur to represent chemical groups that can be synthesized into myriads of different molecular structures for the product. Based on a molecular structure, the continuous, real-valued product properties can be determined. The search for such discrete set of variables for molecular structure creation is too demanding in combination with process synthesis optimization. An approximation would be to work with continuous product variables only, creating a hypothetical product with optimized product properties and qualities. Then, the first cycle is to jointly optimize the designs of process and (hypothetical) product. The next cycle of optimization is to search in the discrete space of chemical groups for a combination that forms a (real) product with properties and qualities close to those of the optimized hypothetical one. Having found the closest real product with near optimal properties and qualities, one can re-optimize the process design for this real product. This two-step approach is developed and applied to the design of a solvent based process in [26], and it is also an approach that is applied in case studies by Gani [25]. First, the designs of the process and of a hypothetical solvent (for internal use in the process) are jointly optimized. Such optimization is possible by using only continuous variables in the process and the product models. Having once identified the ideal hypothetical solvent, a search is made for a small set of real solvents closely matching the identified ideal solvent properties. From this small set of near-optimum real solvents, the best performing candidate is selected by re-

optimizing the process for each solvent. This design approach accounts for the interactions between process and solvent designs in an affordable optimization procedure.

8.5.2 Notes on first-principles models of formation of a structured product

Structured products are made up of multiple thermodynamic phases. The continuous phase of the product may contain one or more dispersed phases of gas bubbles, liquid droplets, or solid particles such as crystals. The particulates of a dispersed phase are often distributed in size (nonuniformly) and may have internally different physical and mechanical states and chemical composition. Furthermore, a product will have a geometric extension in up to three dimensions. The distribution of the phases and their properties can vary with the spatial position. The structure of a product involves the distributions of phase properties as well as the spatial distributions. Last but not least, the structure of a product may change over time due to internal physical and chemical events in the phases and between the phases, such as chemical reactions, heat and mass transfer, and flow of phases. A visual abstraction of a structured product and an outline of a generic mathematical model of a structured product are given in Figure 8.10.

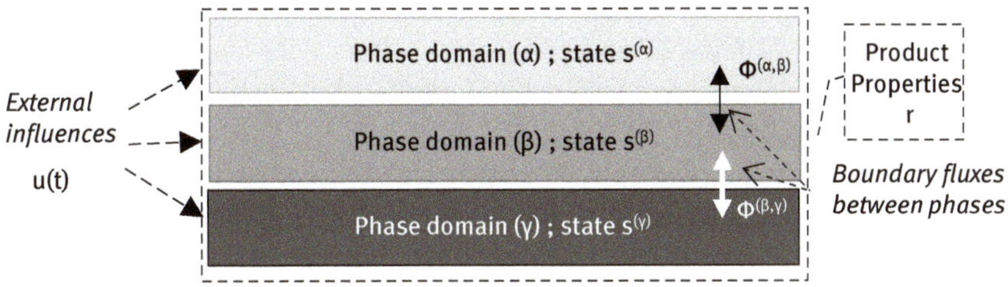

Behavioral model per phase for composition & structure:

$$\partial s^{(*)} / \partial t = M\,(s^{(*)}, \nabla_{z^*} s^{(*)}, \Phi^{(*,\#)}, u(t); p^{(*)})$$

Fluxes of mass & energy across interfaces: $[*, \# = \{\alpha, \beta, \gamma\}]$

$$B(\Phi^{(*,\#)}, \nabla_{z^*} s^{(*)}, \nabla_{z\#} s^{(\#)}) = 0$$

Product structure – properties relationships:

$$R(s^{(\alpha)}, s^{(\beta)}, s^{(\gamma)}, r) = 0$$

State of a phase: $s^{(*)} = \{$temp., press., compos., size distribution, mass hold-up$\}$

Fig. 8.10: Summary of a generic model of a multiphase, structured product.

In a structured product, each phase has its own internal physical state (s) that evolves in time due to internal events in the phase domain as well as to interactions with the neighboring phases by fluxes (φ) and with the external world (u). Heat, mass, and momentum fluxes between adjacent phases are governed by the states of those phases. The properties of a multiphase product (r) are defined by the integral effect of the states (s) of the phases domains $\{\alpha, \beta, \dots\}$. Each of the phases in a structured product has to be premanufactured in an upstream part of the process as a 'simple' product, involving a single phase characterized by its physical state (temperature, pressure, and composition) and physical properties (density, heat capacity, viscosity, thermal conductivity, and surface tension). The various simple products will go as ingredients into a product formation unit, such as an extruder or a colloid mill, to form the desired structured product.

Process modeling and simulation support the design and analysis of manufacturing and product application processes. For simple products, there is a coherent body of modeling knowledge for a broad class of manufacturing processes. This knowledge is captured in the common commercial flowsheet simulators for processes dealing with locally uniform fluid phases (gases, liquids). Such generic modeling is far less common for processes which handle structured products. This is due to the complexity and diversity of structured products and the (labor) cost of modeling. Products leaving a manufacturing process and entering a product application process will also set the level of complexity for the associated process modeling. This complexity can be effectively dealt with by means of first principles, equation-oriented modeling, combined with the numerical power of modern equation solvers. Such modeling platforms are under development and already offered for formulated products and their manufacturing processes [24, 25].

To end with a challenging example, let's have a look at the modeling of freezing ice cream. Ice cream has a microstructure consisting of ice crystals, air bubbles, liquid fat, and a sugary solution (see Figure 8.4). Starting with the liquid feed and air, ice cream is formed in a freezer. The freezer is an externally cooled extruder with a scraped surface heat exchanger (Figure 8.11). Such units and models are topics of study in the chemical engineering literature; e.g., [5, 27, 28].

The water in the sugary solution freezes at the cooled surface of the barrel, resulting in the growth of a layer of ice on this surface. This layer is periodically scraped off by rotating blades and dispersed as a population of ice crystals of different forms and sizes in the liquid – air mixture. The mixture of liquid/air/ice is axially transported by the rotor. Fundamental modeling of the ice cream formation is of a staggering complexity. It is staggering because *in principle* one has to deal with:

- Eight phases: ice crystals, ice layer, air bubbles, fat droplets, aqueous sugar solution, rotor, wall, coolant
- Three spatial coordinates (even a spiraling one) and time
- Distributions of particulate size (or mass) for the ice crystals, air bubbles, fat droplets

- Kinetics of growth, agglomeration, breakage of crystals, bubbles, and droplets
- Mass, energy, and momentum balances for the phases in the ice cream mixture
- Energy balances for rotor, wall, and coolant as well as a momentum balance for the rotor
- Physical transport phenomena (mass, energy, momentum) in and transfers between phases
- Thermodynamic equilibrium conditions at interfaces between phases
- Physical properties for thermodynamics and transport (with non-Newtonian flow)

In view of this overwhelming complexity, current ice cream freezing modeling efforts are directed at pragmatically developing very much simplified, yet realistic models of this process unit for design and control [27, 28].

Another parallel modeling effort is required for the application process of the structured product. This is even more challenging. For instance, the application process of ice cream is human consumption and digestion. Currently, mechanistic models of the mouth-feel, melting behavior, flavor development and sensing, digestion of ice cream in intestines, and of the impact of ice cream ingredients on body condition and long term health are hard to find. Yet, new developments in food engineering are on the way by studying food digestion in the human gastrointestinal track as a multiscale process, in analogy to multiscale chemical processes [29].

In summary to the relation process-product modeling and computing one can notice distinct developments towards models of (micro)structured products based on first principles as well as data driven. First principles product models have to account for much structural information (spatially and distributive properties) leading to complicated sets of (partial) different equations on multiple domains with changing boundaries. This is in contrast to much of current process modeling and computing where large sets of nonlinear algebraic equations prevail. Combining and integrat-

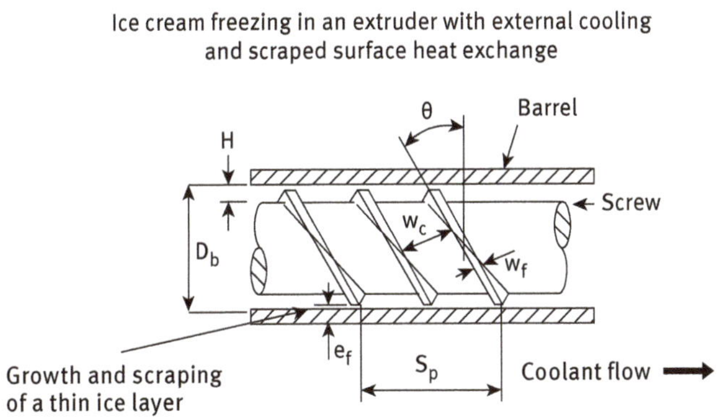

Fig. 8.11: Ice cream freezing in an extruder [5, 28].

ing product and process modeling is needed for further advancements in chemical engineering applications.

8.6 Product models: overview

For product design in the (bio)chemical engineering field, the different types of models that are created and used in designing and discussed in this chapter are listed in Tables 8.5 and 8.6. The focus in these tables is on the first levels of the Delft Design Map to show the large number of types of models created and used in a design process.

Tab. 8.5: Schematic overview of various model types created and used during the DDM's project framing and product concept-consumer wants (quality function) design levels (see also Section 4.3.2).

Design levels	**Models (physical, schematic, verbal, mathematical) (examples only, not complete list)**
PF: Project Framing	– Business context graphs – *Development/prediction in time: market volume, sales volume, price level product, price level raw materials, energy, population growth, etc.* – Supply chain block diagram – Value chain diagram – Product's life cycle diagram – Schematic showing design driver and design type – BOSCARD table – Project organization chart (members and accountabilities) – Project team Belbin roles diagram – Project and time planning chart – ...
CW-PC: Consumer Wants – Product Concept: (quality function)	– Product application/use process block diagram – Product life cycle block diagram – Brainwriting collage or table with identified stakeholders from all life cycle phases – Storyboard showing various consumer interactions with product – Storyboard showing other stakeholders interactions with product – House of Quality (1) diagram – *Consumer wants to product (quantitative) attributes* – Product attributes quantitative specifications table – ...

Tab. 8.6: Schematic overview of various model types: created and used during the DDM's product concept (product) property function design levels (see also Section 4.3.2).

Design Levels	**Models (physical, schematic, verbal, mathematical) (examples only, not complete list)**

PC-PF: Product Concept: (product) Property Function	– Schematic drawing representing alternative product structures
	  – *Schematic drawings of the phases (solid/liquid/gas), their size (distribution), volume fraction, spatial distribution in the product, composition of each phase*
	  – *See Figure 8.4 for ice cream product*
	– Pure component properties table (See Table A13.3)
	– Mathematical models of alternative products and states
	  – *Composition, structure, and physical properties (crystal size and distribution, droplet size and distribution, volume, specific surface area, particle density, powder bed density, viscosity); see Section 8.3.2.*
	  – *Steady state but may be extended to the change of product state over its life cycle (manufacturing, storage, transport, application/use, end-of-use), or dynamic models*
	– Block diagram showing product application/use processes
	  – *Application/use of: transportation fuels, lubricants, paint, laundry detergents, medication administering, ice cream consumption*
	– Schematics showing key mechanism of the application/use processes: systems splits (Section 8.3.2)
	  – *Mass transfer: mixing, dispersion, dissolution, heat transfer, (bio)chemical reaction*
	– Time constant equations of the product application/use process mechanisms
	– Artificial Neural Networks models
	– Mathematical models of product's (components') behavior during product's application/use
	  – *Also in storage, transport, collection, recycling*
	  – *(Bio)chemical reactions (stoichiometry, byproducts, waste products production)*
	  – *Physical changes: phase changes, dispersion, dissolution, heat and mass transfer (energy use in product application/use)*
	  – *Simulating effects of different product alternatives*
	– House of Quality (2) diagram
	  – *Product (quantitative) attributes translated into product properties (composition/structure, application/use conditions)*
	– Brainwriting diagram of alternative (bio)chemical components
	– Molecular formula / structure
	– Reaction equations
	  – *Stoichiometry, byproducts, waste products produced in application/use*
	– Supply chain block diagram update
	– Preliminary 'SHEETS' performance models (verbal, mathematical)
	  – *SHEETS: Safety, Health, Environment, Economy, Technology, Social*
	  – *Safety, health, environmental: effects of raw material and products transport and logistics, during waste production during product application/use*
	  – *Economy: raw material costs, energy use in product application/use*

8.7 Modeling for 'Safety, Health, Environment, Economy, Technology, Social (SHEETS)'

For schematic and mathematical modeling of products in relation to the 'SHEETS' evaluation criteria the reader is referred to:
- **S**afety Chapter 11
- **H**ealth Chapter 11
- **E**nvironment Chapter 12
- **E**conomy Chapter 10
- **T**echnology this chapter and Chapter 9 (for process modeling)
- **S**ocial Chapter 12

List of symbols and abbreviations

Symbols

A	[m^2]	surface area (e.g., for heat transfer or mass transfer)
a	[m^2/m^3]	specific surface area
c	[*]	generic composition variable in product property model
C	[kmol/m^3]	reactant concentration
C_g^*	[kmol/m^3]	reactant concentration in gas corresponding to C_l^*
C_l^*	[kmol/m^3]	saturation reactant concentration in the liquid corresponding to C_g^*
c_p	[J/(kg · K)]	specific heat capacity
d	[*]	generic size (length, area, volume) parameter of a process unit
D	[m^2/s]	diffusion coefficient
$e_{process}$	[*]	generic process performance indicator in process model
$e_{product}$	[*]	generic product performance indicator in product quality model
F	[m^3/s]	volume flow rate
k	[kmol/(m^3 s)]	0-order rate constant
k	[1/s]	1st order rate constant
k_l	[m/s]	liquid side mass transfer coefficient
L	[m]	travel length, diffusion length
m	[–]	equilibrium concentration ratio (gas/liquid)
p	[*]	generic parameter in a process/product model
q	[*]	generic quality variable in a product quality model
r	[*]	generic product property variable
r_c	[kmol/(m^3 s)]	conversion rate
s	[*]	generic state variable of a product model

t_c	[s]	conversion time
t_d	[s]	diffusion time
t_{ht}	[s]	heat transfer time
t_{mt}	[s]	mass transfer time
t_{shear}	[s]	shear stress relaxation time
u	[*]	generic input variable to a process/product model
U	$[W/(m^2\,K)]$	overall heat transfer coefficient
V	$[m^3]$	volume
v	[m/s]	travel velocity
w	[*]	generic disturbance variable of a process/product model
x	[*]	generic state variable in process model
Y	[–]	binary variable
z	[*]	variable of a feed stream to a process model
α_i	$[W/(m^2\,K)]$	internal heat transfer coefficient
α_e	$[W/(m^2\,K)]$	external heat transfer coefficient
$\varepsilon_\beta^{(\alpha)}$	[–]	phase mass ratio (of phase β dispersed in continuous phase α)
$\varepsilon_\gamma^{(\alpha)}$	[–]	phase mass ratio (phase γ dispersed in continuous phase α)
$\varepsilon_\alpha^{(\beta)}$	[–]	phase mass ratio (phase α dispersed in continuous phase β)
$\varepsilon_\gamma^{(\beta)}$	[–]	phase mass ratio (phase γ dispersed in continuous phase β)
$\varepsilon_\alpha^{(\gamma)}$	[–]	phase mass ratio (phase α dispersed in continuous phase γ)
$\varepsilon_\beta^{(\gamma)}$	[–]	phase mass ratio (phase β dispersed in continuous phase γ)
$\varepsilon_\alpha^{overall}$	[–]	overall phase mass ratio of phase α in total system
$\varepsilon_\beta^{overall}$	[–]	overall phase mass ratio of phase β in total system
$\varepsilon_\gamma^{overall}$	[–]	overall phase mass ratio of phase γ in total system
ζ	[*]	generic external condition of use in product quality model
η	[kg/(ms)]	viscosity
λ	[W/(mK)]	heat conductivity
π	[*]	generic physical parameter for a product model
ρ	$[kg/m^3]$	specific density
σ	[*]	generic structure variable in a product property model
Φ	$[*/(m^2\,s)]$	generic flux of a physical quantity
τ	[s]	travel time, residence time

Abbreviations

3D	3-Dimensional
ANN	Artificial Neural Network
BOSCARD	Background, Objectives, Scope, Constraints, Assumptions, Reporting, Deliverables
CAMD	Computer-Aided Molecular Design

CAD	Computer Aided Design
CW-PC	Consumer Wants – Product Concept
DDM	Delft Design Map
DOE	Design Of Experiments
EFCE	European Federation of Chemical Engineers
HAZOP	HAZard & OPerability study
HoQ	House of Quality
P&ID	Piping & Instrumentation Diagram
PBS	Process Block Scheme
PC-PF	Product Concept – Property Function
PF	Project Framing
PDEng	Professional Doctorate in Engineering
PFD	Process Flow Diagram
PFS	Process Flow Scheme
QFD	Quality Function Deployment
SHEETS	Safety, Health, Environment, Economy, Technology, Social
TAME	Tert-Amyl Methyl Ether

References and further reading

[1] Van Overveld CWAM. PowerPoint Slides guest lecture in PDEng course Advanced Principles of Product and Process Design (ST6064) at TU Delft, 2010.
[2] Skyttner L. General systems theory – problems, perspectives, practice 2nd edn, World Scientific Publishing Co. Pte. Ltd., Singapore, 2005, 97–99.
[3] De Haan AB, Swinkels PLJ, de Koning PJ. Instruction manual conceptual design project CH3843 – from idea to design, Appendix-18, Department of Chemical Engineering, Delft University of Technology, Delft, Netherlands, 2016.
[4] De Haan AB, Swinkels PLJ, de Koning PJ. Instruction manual conceptual design project CH3843 – from idea to design, Appendix-17, Department of Chemical Engineering, Delft University of Technology, Delft, Netherlands, 2016.
[5] Almeida-Rivera C, Bongers P, Zondervan E. A structured approach for product-driven process synthesis in foods manufacture. In: Martin M, Eden MR, Chemmangattuvalappil NG, eds. Computer Aided Chemical Engineering Vol.39 Tools for chemical product design – from consumer products to biomedicine, Amsterdam, Netherlands, Elsevier BV, 2017, 417–441.
[6] QFD Online Quality Function Deployment tools and information for real time application, free QFD Templates 2007–2010. Accessed October 10, 2017, at www.qfdonline.com/templates/.
[7] Seider WD, Seader JD, Lewin DR, Widagdo S. Process and Process Design Principles: Synthesis, Analysis and Evaluation, 3rd edn, Hoboken, NJ, USA, John Wiley & Sons Inc., 2010, 45–47.
[8] Wesselingh JA, Kill S, Vigild ME. Design & development of biological, chemical, food and pharmaceutical products, Hoboken NJ, USA, John Wiley & Sons Inc., 2007, 27–38.
[9] Oosterhuis NMG. PhD Thesis, Delft University of Technology, 1984. Accessed August 19, 2017, at https://repository.tudelft.nl/islandora/object/uuid:03a887b7-8c20-4052-8d6b-7fe76918d7ec/datastream/OBJ.

[10] Sweere APJ, Luyben KChAM, Kossen NWF. Regime analysis and scale-down: tools to investigate the performance of bioreactors. Enzyme Microb. Technol., 9, July, 1987, 386–398.
[11] Wesselingh JA, Kill S, Vigild ME. Design & development of biological, chemical, food and pharmaceutical products, Hoboken NJ, USA, John Wiley & Sons Inc., 2007, 38, 64–66, 77–80, 95–97, 119–120, 170–172.
[12] Janssen LPBM, Warmoeskerken MMCG. Transport phenomena data companion, VSSD, Delftse Uitgevers Maatschappij, Delft, 1987.
[13] Cussler EL, Moggridge GD. Chemical Product Design, New York, USA, Cambridge University Press, 2001, 75–85.
[14] Bongers PMM. Model of the product properties for process synthesis. In: Braunschweig B, Joulia X (eds.), Computer Aided Chemical Engineering 25, 18th European Symposium on Computer Aided Process Engineering – ESCAPE 18. Amsterdam, Netherlands Elsevier BV, 2008, 55–60.
[15] Dubbelboer A, Zondervan E, Meuldijk J, Hoogland H, Bongers PMM. A neural network application in the design of emulsion-based products. In: Lockhart Bogle ID, Fairweather M (eds.), Proceedings of the 22nd European Symposium on Computer Aided Process Engineering, 17–20 June 2012, London. Amsterdam, Netherlands, Elsevier BV, 2012, 692–696.
[16] Edwards DA, Brenner H, Wasan DT. Interfacial transport phenomena and rheology, Boston MA, USA, Butterworth-Heinemann, 1991.
[17] Bröckel U, Meier W, Wagner G, eds. Product design and engineering – best practices, volume 1: basics and technologies, Weinheim, Germany, Wiley VCH Verlag, 2007.
[18] Wesselingh JA, Kill S, Vigild ME. Design & development of biological, chemical, food and pharmaceutical products, Hoboken NJ, USA, John Wiley & Sons Inc., 2007, 259–293.
[19] Bröckel U, Meier W, Wagner G (eds.). Product design and engineering – best practices, volume 2: raw materials, additives and applications, Weinheim, Germany, Wiley VCH Verlag, 2007.
[20] Bröckel U, Meier W, Wagner G (eds.). Product design and engineering: formulation of gels and pastes, Weinheim, Germany, Wiley VCH Verlag, 2013.
[21] Bernardo FP. Integrated process and product design optimization, Ch. 12. In: Martin M, Eden MR, Chemmangattuvalappil NG, eds. Computer Aided Chemical Engineering Vol.39 Tools for chemical product design – from consumer products to biomedicine. Amsterdam, Netherlands, Elsevier BV, 2016, 347–371.
[22] Process Systems Enterprise (UK). gPROMS: advanced process modelling platform. Accessed October 5, 2017, at https://www.psenterprise.com/products/gproms.
[23] Process Systems Engineering Laboratory, Massachusetts Institute of Technology (USA), Process modelling & computing software JACOBIAN. Accessed October 5, 2017, at https://yoric.mit.edu/Jacobian.
[24] Process Systems Enterprise (UK). gPROMS|FORMULATE: gPROMS Formulated Products, platform for integrated digital design of robust formulated products and their manufacturing processes. Accessed October 5, 2017, at https://www.psenterprise.com/products/gproms/formulatedproducts.
[25] PSEforSPEED Company Ltd., Computer-aided application of process systems engineering (PSE) for sustainable product-process engineering, evaluation & design (SPEED). Accessed October 5, 2017, at http://www.pseforspeed.com.
[26] Bardow A, Steur K, Gross J. Continuous-Molecular Targeting for Integrated Solvent and Process Design, Industrial & Engineering Chemistry Research, 49, 2010, 2834–2840.
[27] Arellano M, Benkhelifa H, Alvarez G, Flick D. Coupling population balance and residence time distribution for the ice crystallization modeling in a scraped surface heat exchanger, Chemical Engineering Science, 102, 2013, 502–513.
[28] Dorneanu B. Model Reduction in Chemical Engineering. Case Studies Applied to Process Analysis, Design and Operation. PhD Thesis, Delft University of Technology, 2011. Accessed Octo-

ber 5, 2017, at https://repository.tudelft.nl/islandora/object/uuid:afecbacd-0ac1-4a01-b7ab-db1ba0c42f50?collection=research.
[29] Bornhorst GM, Gouseti O, Wickham MSJ, Bakalis S. Engineering Digestion: Multiscale Processes of Food Digestion, Journal of Food Science, 81, 2016, R534–543.
[30] Ng KM, Gani R, Dam-Johansen K (eds.). Chemical product design: towards a perspective through case studies, Computer Aided Chemical Engineering – Volume 23, Amsterdam, Netherlands, Elsevier BV, 2007.
[31] Khalloufi S, Kharaghani A, Almeida-Rivera C, Nijsse J, Dalen G, Tsotsas E. Monitoring of initial porosity and new pores formation during drying: A scientific debate and a technical challenge, Trends in Food Science & Technology, 45, 2015, 179–186.
[32] Khalloufi S, Almeida-Rivera C, Janssen J, Van der Vaart M, Bongers P. Mathematical model for simulating the springback effect of gel matrixes during drying processes and its experimental validation, Drying Technology 29, 2011, 1972–1980.
[33] Almeida-Rivera C, Jain P, Bruin S, Bongers P. Integrated product and process design approach for rationalization of food products. In: Pleşu V, Şerban Agachi P (eds.), Computer Aided Chemical Engineering – Volume 24, Amsterdam, Netherlands, Elsevier BV, 2007, 449–454.
[34] Linke P, Papadopoulos, AI, Seferlis P. Systematic Methods for Working Fluid Selection and the Design, Integration and Control of Organic Rankine Cycles – A Review. ENERGIES, 8, 2015, 4755–4801.
[35] Papadopoulos AI, Stijepovic M, Linke P, Seferlis P, Voutetakis S. Toward Optimum Working Fluid Mixtures for Organic Rankine Cycles using Molecular Design and Sensitivity Analysis. Industrial & engineering chemistry research, 52(12), 2013, 116–133.
[36] Schilling J, Tillmanns D, Lampe M, Hopp M, Gross J, Bardow A. From molecules to dollars: integrating molecular design into thermo-economic process design using consistent thermodynamic modeling, Molecular systems design & engineering, 2, 2017, 301–320.

Part C: **Design optimization**

9 Process modeling and optimization

Synopsis

This chapter deals with modeling in support of three successive innovation stages of product. Linear process modeling is associated with applications in the concept stage. Nonlinear process modeling is linked with the feasibility and development stages. Process simulations will find applications in the development stage, while the feasibility stage benefits from optimization applications, which scout the entire design space and optimize over it. The modeling approach chosen here has a practical inclination. Many academic textbooks on modeling cover the common modeling steps, such as development of model equations, mathematical model analysis and numerical means for computing solutions. Therefore, these steps are largely skipped here. Instead the contextual and application sides of modeling in process engineering will get more attention. For instance, how does one get to a relevant model for a process engineering problem? Furthermore, how does one analyze and evaluate the outcomes from computational model applications and how are clues derived for process innovations? Aspects of sensitivity analysis, uncertainties and identification of technological constraints from model based solutions will also be discussed.

9.1 Justification and objectives of process modeling

Modeling objectives for process synthesis, analysis and optimization

The purpose of process modeling and computing is to offer effective, quantitative means to synthesize, analyze and evaluate process designs. The main goal of this modeling and computing chapter is to bring out the conceptual aspects of modeling, simulation and optimization for process design applications. In view of limited text space this emphasis will go at the expense of the underlying mathematical and numerical aspects of model building and equation solving. Here we will simply assume that:

- Models will be properly formulated from mathematical and computational perspectives.
- Their domain of applicability (containing feasible physical solutions) is well understood.
- An applied numerical solution procedure is consistent with the model and sufficiently accurate.

The focus will be on the context and practice of developing model applications for the concept, feasibility and development stages in support of product and process designs; see also Chapter 4. Attention will also be given on how to extract relevant infor-

https://doi.org/10.1515/9783110467741-009

mation from model applications for purpose of improvement and innovation. There is an established industrial practice of supporting process design case studies with commercial steady state process flow sheet simulators. This vast area of computer aided process simulations has been thoroughly covered in [1], and therefore will be skipped here. Dynamic modeling and model reductions for process control and operations are also omitted.

Further background reading on process modeling for simulation and optimization in process development, design and operations is offered in a number of high quality text books; regarding modeling [2–8]; flow sheeting simulations for design [1, 9–12] and optimization [13–15] and [16].

Conceptual representation of a process: *network features*

A process is an open system that internally consists of a network interconnected and interacting unit operations. The system is in interplay with its physical, economic and social surroundings. The system performs specified production functions; i.e., turning physical feeds into products with a chemical functionality. The production is guided by process and product control information. The system may operate in different modes over time. Each mode is characterized by a distinct internal state of the process.

A process is connected to its surroundings by:

- Feed, product and waste streams (the latter to waste treatment facilities).
- Control set points for its operational conditions and modes as supervised by plant operators.
- The environmental conditions.
- The cash flow it generates for the process owner.

The behavior and performance of a process is codetermined by the interactions between a process and its surroundings. This includes the imposed modes of operation. Each mode can be seen as the realization of an external scenario, under which the process must make product(s) while performing well enough. Development of scenarios is part of the design; these scenarios are used for testing candidate processes under design. Consequently, the set of process models must be able to handle all scenarios. This explains why next to steady state models also dynamic models are required.

Sharing a generic view on process modeling and computing

Design models carry a lot of process information on behavior and performance. The common steps in the modeling work procedure are outlined in Table 9.1. These steps are presented in a sequential fashion. This is *not* how it works in practice. There is a lot of going forward and backward over the steps during the modeling activity. One may also visualize these activities as placed on the perimeter of a circle with easy crisscrossing between activities.

Tab. 9.1: Common steps in a modeling work procedure.

Step	*Activity*
1	Select the *object of modeling*, and justify a modeling effort from *anticipated benefits*.
2	Agree on *goal(s) of model* and intended application domain(s) for future model use.
3	Find a suitable *conceptual representation* of the object to be modeled.
4	Develop a *mathematical model* of the object for its structure/behavior/performance.
5	Code the mathematical model in *software* with *verification* of correctness of coding.
6	Select a *numerical solver* with enough numerical precision for intended applications.
7	*Validate* the coded model by comparing computed answers with experimental data.
8	*Select case studies* based on *interaction scenarios* between object and its surroundings.
9	*Apply model* to *case studies* and *analyze* computed results on achieved accuracies.
10	*Evaluate* computed results *on significance* for goals and new insights.
11	*Report* model, case studies, numerical results and the evaluation of results.

The next sections will deal with the nature of model applications in the three design stages: concept, feasibility and development. The successive stages of process development and design call for models of different functionalities and capabilities. Three classes of models are considered:

- Linear models for process input-output blocks: for use in the *concept* stage.
- Nonlinear models for *process simulation*: for use in the *development* stage.
- Nonlinear models for *process optimization:* for use in the *feasibility* stage.

For each stage the main modeling aspects generic to that stage will be discussed, involving:

- Consistent model formulation with proper degree of freedom analysis.
- Significance of solutions and sensitivity analysis.
- Analysis of active inequality constraints in case of a process optimization.
- Identification of technological limitations in optimized solutions.

Having dealt with the generic aspects, a number of (small scale) cases are presented as illustrations of certain modeling aspects. These examples are small scale on purpose in order to illustrate a way of thinking and acting in modeling. There is a good reason to keep them small. The many facilities of process modeling software with associated computing power often tend to overwhelm novice modelers. Therefore, it is preferable to develop basic model synthesis and analysis skills first by means of small scale problems. These problems can even be solved by hand; they serve as vehicles to transfer ways of thinking in model synthesis, analysis and evaluation. Each case is concisely presented in this chapter by means of motivation and objective, modeling approach and main outcomes. Details of the cases and results are collected in the Appendix to this Chapter A9.

Before embarking on modeling for the three design stages it is expedient to make a remark about a restrictive meaning of some terms in this chapter. The terms behavior and performance will be used in a more focused way than commonly done in chemical process engineering. The term behavior of a process or product will relate to patterns of change in its (bio) chemical and physical quantities. The term performance is intentionally reserved for factors that express some kind of value and appreciation by human society, including customers of a product, operators and owners of a process and/or legal authority. Such performance factors are related to safety, health, economy and ecology.

9.2 Contributions to a concept stage with linear modeling

9.2.1 Modeling for a concept stage

Let us move from general modeling aspect to the specifics of modeling for process development and design. In this chapter a distinction is made between three generic main items of any process model:

- A sub model describing the physical/chemical/biological behavior of a process. This sub model deals with changes in physical and chemical resources (species mass, energy, momentum, electrical charge, etc.) as they go through a process.
- Another sub model quantifying some indicators for process performance. These indicators reflect the degree of appreciation by process owners of what the process (design) achieves in terms of economy and ecology under safe operation.
- Scenarios defining pertinent influences of the surroundings of a process on the process itself. One must specify external scenarios under which a process is supposed to perform its function(s) in order to compute and analyze process behavior and performance.

A process interacts with its environment and this feature must be represented in a process model. The common, causal flow of information to and from a process model is shown in Figure 9.1.

It represents the flow of physical quantities from process inputs to its outputs. The inputs cover the physical feed streams, the selected operating conditions (u), physical parameters (p) for the process units, as well as other known external influences (w) acting on the process. The fourth class of inputs involves the geometric design variables (d) of the process units (e.g., lengths, areas, volumes). Their numerical values will have been established in the preceding design phase. The outputs (x) are the collective set of physical states of units and streams in the process. In this Figure, it is assumed that the types of unit operation and the ways in which such units can be connected to an internal network structure have been decided upon by preceding conceptual design decisions. Figure 9.1 mentions the three main sub models: behavioral,

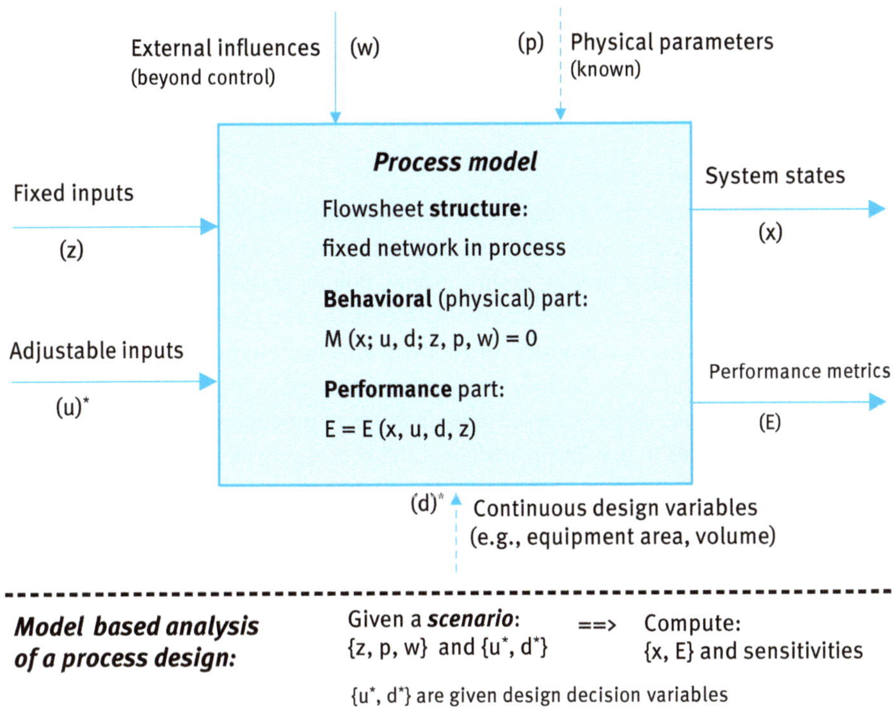

Fig. 9.1: Information flow through a process model for analysis of a process design.

performance and the external scenario(s). Here, the structure part is assumed fixed. This assumption needs to be relaxed when modeling for process synthesis studies that specifically deal with alternative process structures. Consequently, the model setup in Figure 9.1 is suitable for the *analysis of a design* only. The inputs must all be given by the designer/modeler; there is no automatic adjustment of the inputs as to target for process states or to optimize process performance metrics.

Design leads to a reversal of information flow to a model

In actual design situations the geometric design variables will be unknowns rather than given quantities. These unknowns add to the degrees of freedom in a design situation. So, to take away the additional degrees of freedom, the model needs to be augmented by as many design specifications as there are unknown geometric design variables. These new design specifications are often given as specified targets for some key physical outputs of the process, e.g., nominal production rate and product qualities as well as constraints on waste streams (sizes and waste content). There can also be specifications of duties of process units, such as degree of conversion and selectivity in a reactor and a desired degree of separation or purities in a separation column. The general rule is that targets on physical outputs are set, while having the required geometry of the equipment as an outcome to meet such targets. The flow of design information from targets to geometric features can be reverse to the causal (physical) flow of information.

Modeling for the concept stage

A concept stage is focused on creating alternative, competing product concepts. In the background, though, one must also keep in mind the basic processing principles for eventually making such a product from the available feeds. It is prudent to think already about a process block diagram which outlines the key processing steps to make a product. The viability of a product concept will also be determined by its manufacturing complexity and costs. At this concept stage, process models and computations should be kept simple, addressing the main features of processing functions only. For this reason, it suffices to use *linear* models of the processing steps in a process block diagram for a first exploration. A model is linear when each equation in the model is linear in all variables. There are five common forms of deterministic models, summarized in Appendix A9. The application cases in this section deal with linear algebraic and differential equations with real variables. The exception is the modeling of the structure of a block diagram/flow sheet involving a linear algebraic model with binary variables. Six cases will be given to illustrate some aspects of modeling.

9.2.2 Cases for concept stage modeling

Case 9.1: Exploring the *stoichiometric space* for an innovative *chemical conversion*

Motivation and objective

Chemical processes have a conversion section in which the main chemical species in the feed streams are converted into target species for the product(s). A conversion is characterized at the chemical level by (a) the reaction stoichiometry; (b) associated thermodynamic reaction equilibrium; (c) reaction kinetics; and (d) level of catalytic activity and catalyst stability over time. The considerations of the associated physical transport phenomena, phase contacting patterns and hydrodynamic flow regimes are part of the subsequent physical reactor selection and design. The reaction kinetics can be seen as an option for optimization of product yield over the reaction conditions and over the ratios for reactant feed concentrations. Stoichiometry is often assumed to be given and fixed. Yet stoichiometry can also be a degree of freedom in the design of a process concept. A simple example will be given of a search over a stoichiometric space to find the best reaction serving a particular conversion goal. In this case the objective is the chemical conversion of *carbon-dioxide* as a greenhouse gas into *carbon-monoxide* as a chemical reactant in *syngas*. The latter is, among others, a feed for methanol synthesis and *Fischer–Tropsch synthesis* for liquid hydrocarbon fuels. The reuse of carbon in chemical products helps in closing the carbon cycle.

Approach

Linear algebra analysis of the *stoichiometric matrix* is applied to extend the range of *dry reforming* reactions.

Outcome
The outcome of this reaction modeling example shows how a systematic exploration of the space of reaction stoichiometry can lead to an optimized yield of a target product. The derived stoichiometric model enables the exploration of potentially interesting and useful reactions in a given species space. The *super dry reforming reaction* is the highest performing reaction for CO_2 to CO conversion. An experimental study of this reaction and its inventive underlying catalytic scheme was reported in [17]. The stoichiometric analysis also revealed the high dry and medium dry reforming reactions as interesting side reaction. In principle, these reactions may occur simultaneously with super dry reforming. In a reflection on uncertainties and limitations playing a role in this modeling, one should consider that the outcome space is determined by the chosen set of chemical species. One could argue that solid carbon (C_{solid}) should be added to the chemical species set so as to allow for more reactions, such as the Boudouard reaction ($2\,CO \Leftrightarrow CO_2 + C_{solid}$).

Case 9.2: A *conversion – separation – recycle* structure using lumped units

Motivation and objective of modeling
When moving from concept to feasibility stage, one must have some ideas about the use of species mass resources in a future manufacturing process for the product under consideration. Quite a few product options may have to be explored, so simple process modeling and fast computing are preferred. The simplest mathematical models in chemical engineering are linear ones, which can be formulated and solved easily. As an example, a *linear algebraic model* of a common *reactor-separator-recycle* processing structure is derived and solved. This case may serve as a leading example on how to tackle such cases in general.

Approach
A *process block diagram* is generated with a unique labelling of the process units, the streams connecting the units as well as the chemical species in the streams. The units in the process are treated in a lumped, input-output way. The functional task(s) of a unit are simply characterized by means of model parameters, such as degree of conversion, selectivity and separation factors. Linear component mass balances will be generated over all process units and integrated to balances over the entire process. A degree of freedom analysis reveals how many external conditions need to be specified as part of a scenario under which a process operates.

Outcome
The main outcome is a process block diagram (Figure 9.2) represented by a square linear algebraic model $\mathbf{A} \cdot \underline{\mathbf{x}} = \underline{\mathbf{b}}$. The vector, $\underline{\mathbf{x}}$, represents the species mass flows in the process. Hints are given on performing sensitivity analyses regarding how these mass flows depend on the chosen values for the process unit parameters. An energy balance is also of importance but it is bilinear (mass flow multiplied with temperature). However, by knowing the species mass flows and by adding some reasonable physical approximations, an energy balance is reduced to a linear form and easily solved.

Case 9.3: Specification of *connectivity* in a block diagram by use of a *structure model*

Motivation and objective of modeling
The units in the process *block diagram* of Case 9.2 are connected by streams (see Figure 9.2). The models of the process units include the species flows in the connecting streams. If units A and B are directly connected by stream (l), the models for unit A and B make use of the same flow variables. The behavioral models for the units account for the connections as well. There is no separate model

to describe the connectivity of units by means of streams. The behavioral and structural models have been lumped together. However, suppose one would want to rearrange a process block diagram with different connections between process units using alternative streams. In such a case, the model equations of the units involved in a reshuffle needs to be rewritten because ingoing and outgoing streams are relabeled. Such rewriting is error prone and may involve quite some bookkeeping effort. Furthermore, it is conceptually unpleasant that a change in the connectivity of units must result in changes of the model equations describing unit behavior. It would be better to see instead a change in the connections by streams. In other words, a separation between the behavioral models of the units and a process flow sheet structure model is preferred.

Approach
Each process unit is given slots for the incoming and outgoing flows, where each slot gets its own set of flow variables. This explicit definition of slot related input and output flow variables allows for a formulation of the process unit model that is independent of the actual connections occurring in a process block diagram. The flow sheet structure is used to link stream slots of the process units. Each process stream must connect an output slot of a unit with an input slot of another unit.

Outcome
A decoupling arises between the behavioral models and structural connectivity of units by streams in the flow sheet. This preferred way of modeling can be realized at a slight expense of having more linear connectivity equations in the model. The full set of the connectivity equations forms the *structure model* of the block diagram. To cover optional alternative connections between process units, a formal extension is made to include binary variables to switch such connections on and off. The price of the additional computing efforts due to the solving of additional linear connectivity equations is very small; it pales in comparison to the benefit of transparent flow sheet models.

Case 9.4: Optimization of process performance by linear programming (LP)

Motivation and objective of modeling
The species balances over the process units in a process block diagram give rise to a linear model with some residual degrees of freedom. The latter are taken away by specifying an external scenario under which the process must function. Such a scenario can set some intake rates of the feeds or targets for the product output rates. When the design problem is fully specified the model can be solved. One often adds a performance metric for the design; e.g., an economic potential function and/or chemical element efficiency can be computed. The latter indicates how much of a key element (say carbon) in the feed streams ends up in the products; it is often expressed as a fraction.

A full specification of a design problem gives a 'point' solution only. Practically it is more relevant to explore the design decision space for the best performance of a design. To stay in the linear domain, the performance metric must also be linear in the (unknown) process variables.

Approach
For linear process design optimization one has to do five things.
(a) To open up some *degrees of freedom* in the design model for optimization.
(b) To select a *linear performance metric.*
(c) To introduce *linear inequality constraints* on process variables to demarcate the physical domain in which a process design remains physically feasible.
(d) To select a reference production scenario and compute its performance.
(e) To compute the optimum solution and assess its relevance over the reference solution.

Outcome
The example of the conversion-separation-recycle process is used to illustrate an optimization study. An economic profit function is added to the design model as well as a target output for the main product with an upper limit on the maximum amount of side product that can be sold. The optimized outcome is compared with the reference (nominal) situation. The gain in profit can be explained and visualized in a feed intake diagram. Furthermore, sensitivities of the profit with respect to key processing parameters (selectivity and a separation factor) and a sales parameter are determined. This *sensitivity analysis* shows that a future profit enhancement would benefit most from an improvement in the selectivity factor among these three parameters. While this numerical example is fictitious, it is the analysis method that is practically relevant to the optimization outcome.

Case 9.5: Inverse use of linear models for process targeting and estimation

Motivation and objective of modeling
The use of a square linear process model, $\mathbf{A}\underline{\mathbf{x}} = \underline{\mathbf{b}}$, has been covered in the preceding modeling cases. The process features are parametric captured in matrix **A** and vector $\underline{\mathbf{b}}$, while the unknown process variables in vector $\underline{\mathbf{x}}$ can be computed. That is, if matrix **A** is nonsingular. However, there are more complicated situations of practical relevance in process design and operations where linear process models play a role.

Let $\underline{\mathbf{u}}$ represent a vector of inputs to a process and $\underline{\mathbf{y}}$ represent the process outputs. These are related by a linear process model $\mathbf{A}\underline{\mathbf{u}} = \underline{\mathbf{y}}$. The number of inputs and outputs will be unequal in general. Such a nonsquare linear model can be used in three different ways:

(1) Forward: given **A** and $\underline{\mathbf{u}}$, compute $\underline{\mathbf{y}}$. This case is straightforward and will be skipped.
(2) *Targeting*: given A and targets for $\underline{\mathbf{y}}$, compute $\underline{\mathbf{u}}$.
(3) *Estimation*: given a sequence of experimental observations of inputs $\{\underline{\mathbf{u}}_1, \underline{\mathbf{u}}_2, \ldots, \underline{\mathbf{u}}_K\}$ and of outputs $\{\underline{\mathbf{y}}_1, \underline{\mathbf{y}}_2, \ldots, \underline{\mathbf{y}}_K\}$ find the connecting process parameters which are the elements of matrix **A**.

Targeting and estimation problems will become complicated if the numbers of inputs and outputs are different. The objective is to offer approximate solutions to the targeting and estimation problems.

The practical question is when can these methods be applied in chemical product and process engineering? In terms of product engineering, one can think of the outputs $\underline{\mathbf{y}}$ representing the empirical variables for product performance testing, and $\underline{\mathbf{u}}$ representing the internal chemical and phase composition variables (e.g., for fuels, lubricants, food and healthcare products). In terms of process engineering, one can think of the relations between the inputs to a process unit and its outputs. One may also think of an identification of a process control matrix as linking steady state relationships between inputs and outputs around an operational working point. The practical restriction is in the assumption of linear relationships between inputs and outputs.

Approach
One wants to find solutions to match in the best possible way, with the linearized model. However, the 'best possible way' has to be defined more specifically. This can be done by means of an additional function that aims for minimization of deviations from model linearity. For this purpose one often uses quadratic functions over the model residuals: $\underline{\boldsymbol{\varepsilon}} = \mathbf{A} \cdot \underline{\mathbf{u}} - \underline{\mathbf{y}}$. The benefit of using quadratic expressions as fitting functions in both targeting and estimation is that one can derive closed from linear analytical expressions as answers.

Outcome
Linear analytic expressions are derived for the following cases:
(a) Targeting when there are more adjustable inputs $\underline{\mathbf{u}}_N$ than output targets $\underline{\mathbf{y}}_M$, ($M < N$).
(b) Estimation, assuming measurement errors in outputs $\underline{\mathbf{y}}$ only.
(c) Estimation accounting for random measurement errors in both inputs and outputs

The full expressions are given in the Appendix to Chapter 9. These expressions are analytic in form, yet computationally demanding due to repeated matrix and vector operations. These are better done numerically; e.g., by use of *Matlab*™.

Regarding the resulting uncertainties in the computed model variables, it is also possible to derive estimates for their covariance matrices. Although, the expressions for such covariance matrices become quite involved.

Statistical targeting and estimation are complicated subjects and it is not hard to get into conceptual traps. It is recommended to consult a book on engineering statistics, e.g., [18], or expert statisticians. They can advise on suitable error distributions, fitting functions and on the interpretation of the computed results.

Case 9.6: Notes on the use of other types of linear models

Motivation and objective of modeling
To highlight the existence of some other types of linear process models beyond the algebraic ones.

Approach
Offering an overview of the analytical expressions for each type of model

Outcome
Linear models are given for:
(a) Dynamic evolution of process state variables.
(b) *Dynamic process* models with spatially distributed states in one dimension.
(c) Discrete time process dynamics: development of a *FMCG warehouse inventory*.
(d) Linear equations with integer coefficients and binary variables.

This overview completes the set of applications of deterministic linear models in the concept stage. These models are supposed to always contain a fair degree of knowledge of what is supposed to happen inside a process (gray box model). If it is relatively cheap and easy to collect experimental data there is an alternative of using fully data driven *linear black box* models. Such a model can be generated and used in an experimental optimization procedure. Data driven modeling is covered as an optimization case study in Section 9.4.

9.3 Nonlinear process model simulations: contributions to a development stage

9.3.1 Nonlinear modeling and simulation for a development stage

There are many introductory academic text books available on the principles of process modeling, associated mathematical analysis of the solutions and the numerical simulations in the context of process design and flow sheeting. For background reading one is referred to text books [1–16]. Much of this material is looking inside the model and how to structure and solve the model equations. This section aims at looking more outwardly at process modeling from a somewhat higher vantage point of model use. First some generic structural aspects of process models and their use will be reviewed. This review is supplemented with a few cases highlighting aspects of use of nonlinear process models.

9.3.1.1 Process representation

Any process model is based on a conceptual representation of a process (to be). A process is a network consisting of multiple nodes interacting with each other by exchanges of physical and information resources. Each node represents a unit operation. Within a unit operation, several physical and chemical tasks can be carried out in parallel; e.g., reactions and separations in a reactive distillation column. The connections between the units are specified as physical streams and as signals (for measurement and control) in a process flow sheet. Also, utility streams servicing process units will be considered as part of the process system.

Looking on a more detailed scale, the inside each unit operation is made up of a subnetwork consisting of interconnected compartments. Within these compartments, the states of the resources change because of physical and (bio) chemical phenomena (e.g., reactions, phase changes, fluid and particles flows, heat and mass transfers). There can be active agents facilitating the changes of states of physical entities, such as catalysts and mass separation agents such as membranes. The active agents for signals in process control and automation are sensors (physical quantities ⇒ information signal) and actuators, e.g., valves (signal ⇒ physical quantity). The process control software using dedicated control laws act as 'unit operations' for processing signals.

Obviously, a process is an open system having connections with the outer world:

- Physical streams entering the process as feeds and leaving as products and wastes.
- Physical disturbances acting on the process units, such as heat losses to the environment.
- Information flows from the process sensors to the plant operators.
- Process control set points are imposed on the process, collectively defining an operating mode.

The interactions between process and its outside world are partially defined by the operational scenario imposed on a process as well as by the external disturbances, and partly by the resulting state of the process.

9.3.1.2 Product modeling

In this subsection the focus will be on the modeling and simulation of a process without particular attention to the structural complexities of the product(s). The process stream delivering the product has no special status; i.e., in the sense of having some different modeling attributes. The product stream, such as any process stream, is made up of one or two coexisting fluidic thermodynamic phases. No provision is made for distributive properties of the product, such as a size distribution of the particulates in a phase or a chain length distribution of polymers. Also, it is assumed that the properties (e.g., chemical composition, geometric shape, porosity, etc.) of the active agents in a process (e.g., catalyst, solvents and membranes) are known and can be used in the process model. One may want to account for distributive properties of thermodynamic phases in the process units and in the formation of products. This capability requires a high level of integrated product and process modeling sophistication (see Chapter 8). Such a level is still beyond the capabilities of the current generations of process flow sheeting software.

9.3.1.3 Process equipment modeling

A process plant is made up of process equipment items in which processing operations take place. In other words, a process equipment piece hosts one or more of the nodes of the processing network. Regarding the function of process equipment, there is much more to it than just the process events inside. The complementary aspects are the integrities of its mechanical condition and its operational software for safety and control and its availability for the production function the equipment it was designed for. The latter aspects (its state of availability for normal production conditions, state of mechanical integrity and safety system aspects) are not part of the conventional process modeling in chemical process engineering. These aspects are modeled separately under the common assumption that actual physical state of the matter inside a process equipment item has rarely an immediate influence on its mechanical state and availability. Having outlined these aspects of process representation, the focus will narrow to the modeling of the physical and chemical events and states inside a process, along with the flow of resources.

9.3.1.4 Scope of a process model

A process model is an amalgamation of models of the various units in the process, as well as of the streams connecting these units including utility streams. In contrast, the unit operations models are an ensemble of compartment models, with sub models of the behavior of the active agents and of the transfer of the physical and chemical resources between compartments. It is practical to introduce a formal compact representation of the process model. Let process model $[M]$ represent the process system

[PS] at a suitable scale of detail for design purposes:

$$PS \Rightarrow [M] \qquad M(\underline{\mathbf{x}}; \underline{\mathbf{u}}, \underline{\mathbf{d}}; \underline{\mathbf{p}}, \underline{\mathbf{w}}) = 0$$

M: Set of all equations describing physical and chemical behavior inside the process.
$\underline{\mathbf{x}}$: State variables (physical, information) of the process units and internal and exit streams.
$\underline{\mathbf{u}}$: Input variables associated with *incoming* streams that can be manipulated.
$\underline{\mathbf{d}}$: Geometric design variables (fixed structures and sizes).
$\underline{\mathbf{p}}$: Physical parameters, having some uncertainty characterized by a covariance matrix.
$\underline{\mathbf{w}}$: Disturbances, acting more or less randomly on the process from outside or from inside.

The process model $[M]$ is deterministic while most of its model equations are nonlinear in the process state variables. The model equations are continuous in the variables and (at least once) differentiable in its variables.

The above model formulation $[M]$ may seem to suggest that this model structure is restricted in its validity to continuous processes with lumped process units, resulting in a set of nonlinear algebraic equations. What about distributed process units having an internal geometric coordinate, such as a tubular reactor? Or a batch process unit where time is imminent and such a unit is charged and discharged at specific times? Rather than having a continuous, steady flow in and out, it will be a time varying flow or even a big chunk of mass transferred at once from one process operation to the next. Fortunately, the same model format and reasoning applies. The only change is in a broader interpretation of a state variable x. In a distributed process unit, a state variable x may have a spatial derivative in direction z: $\nabla_z \underline{\mathbf{x}} = \partial \underline{\mathbf{x}}(z,t)/\partial z$. Similarly there can be a time derivative of a state variable: $\underline{\mathbf{x}}' = \partial \underline{\mathbf{x}}(z,t)/\partial t$. In these cases, one can extend a single state $\underline{\mathbf{x}}$ to its triple $\{\underline{\mathbf{x}}, \nabla_z \underline{\mathbf{x}}, \underline{\mathbf{x}}'\}$ and insert this triple in the overall state vector. Then of course, the process model must include the corresponding boundary and initial conditions for these derivatives.

The process model $[M]$ describes the internal physical chemical behavior of the process. However, the modeling problem is still open ended, leaving some degrees of freedom, because the process model itself does not yet account for the external conditions in its surroundings for which it has to be solved to find the process states $\underline{\mathbf{x}}$. This open ended nature is evident from the number of unspecified variables in the model $\{\underline{\mathbf{u}}, \underline{\mathbf{d}}; \underline{\mathbf{p}}, \underline{\mathbf{w}}\}$. A set of specifications for these variables is called a scenario. A scenario reflects the design and external conditions under which one wants to determine the physical and (bio) chemical process behavior. It is practical to make a distinction between the relatively simple process analysis cases and the more complicated process synthesis cases.

9.3.1.5 Process analysis scenarios

In such cases, a design of the process is available and the design variables d are known. For instance, one may want to analyze how state variables x (of internal process units and streams) will vary under variations of inputs (u) and design variables (d). For that purpose, the process model [M] must be complemented with an *analysis scenario*. A typical base case analysis scenario contains the following numerical specifications:

$$
\begin{aligned}
[S_{\text{Design}}]: &\quad \underline{\mathbf{d}} - \underline{\mathbf{d}}_{\text{ref}} = 0 \\
[S_{\text{Input}}]: &\quad \underline{\mathbf{u}} - \underline{\mathbf{u}}_{\text{ref}} = 0 \\
[S_{\text{param}}]: &\quad \underline{\mathbf{p}} - \underline{\mathbf{p}}_{\text{ref}} = 0 \\
[S_{\text{disturb}}]: &\quad \underline{\mathbf{w}} - \underline{\mathbf{w}}_{\text{average}} = 0
\end{aligned}
$$

Here, each specific quantity represents a particular numerical value, applicable in the base case. So, given specifications for all variables $\{\underline{\mathbf{u}}; \underline{\mathbf{d}}; \underline{\mathbf{p}}, \underline{\mathbf{w}}\}$ except the states $\underline{\mathbf{x}}$, which can be solved from M. The combination $\{\underline{\mathbf{x}}; M \text{ and } S\}$ is often a very large set of algebraic variables and equations in equal numbers that need to be solved numerically. In case of a dynamic process system, one must specify the trajectories of inputs and disturbances as functions of time.

9.3.1.6 Process performance evaluation metrics

The process model enables computation of its internal states and this model has an inward focus on what is happening in the process. There is also a complementary outward look on the performance of a process from a societal perspective. The process under design can be evaluated by means of different, complementary metrics. Here, three kinds of metrics are distinguished, representing different values to the process operator, owner and society:

- technical metrics: $E_{\text{tech}} = P_{\text{te}}(\underline{\mathbf{x}}, \underline{\mathbf{u}}, \underline{\mathbf{d}})$
- economic metrics: $E_{\text{econ}} = P_{\text{en}}(\underline{\mathbf{x}}, \underline{\mathbf{u}}, \underline{\mathbf{d}}; c_{\text{I}}, c_{\text{O}}, p_{\text{P}})$
- ecological metrics: $E_{\text{ecol}} = P_{\text{ec}}(\underline{\mathbf{x}}, \underline{\mathbf{u}}, \underline{\mathbf{d}}; e_{\text{I}}, e_{\text{O}})$

The technical metrics can cover product yields over feed intake(s); energy efficiency; element efficiency and risks (DF&EI – Dow Fire & Explosion Index). The economic ones account for size related investment costs (c_{I}), throughput related operating costs (c_{O}) and the proceeds from product sales (p_{p}). The time value of money can be accounted for by deriving net present value profiles. Ecological metrics will address sustainability issues. For instance, a process design can be assessed by second law analysis to determine its (irreversible) exergy losses. Once having solved a process model under a specified scenario with a process simulator, it is relatively a minor effort to compute the performance metrics as well and perform sensitivity analyses.

9.3.1.7 Sensitivity analyses

One can compute how much small changes in inputs $\underline{\mathbf{u}}$ and in design variables $\underline{\mathbf{d}}$ will influence the performance metrics (E) through the state variables $\underline{\mathbf{x}}$. This can be done by determining the scaled sensitivities of the performance metrics with respect to the states: $S_{Ex} = \{\partial ln(\underline{E})/\partial ln(\underline{\mathbf{x}})\}$ and of the process states with respect to the process inputs ($\underline{\mathbf{u}}$): $S_{xu} = \{\partial ln(\underline{\mathbf{x}})/\partial ln(\underline{\mathbf{u}})\}$ and with respect to the design variables ($\underline{\mathbf{d}}$): $S_{xd} = \{\partial ln(\underline{\mathbf{x}})/\partial ln(\underline{\mathbf{d}})\}$. Process simulators offer these facilities for such local sensitivity analyses. One may rank these sensitivities in order of decreasing magnitude (i.e., using absolute values of computed sensitivities) and see which inputs and design variables affect the process behavior most strongly. Similarly, one may compute and rank order the sensitivities with respect to the physical model parameters. The latter is particularly relevant if some of the model parameters have a substantial uncertainty margin in their determination from experimental data. An overview of sensitivity analyses for chemical systems is offered in [19].

Simulations for process analysis are conceptually simple, as a process model computes the states directly from the inputs and the design variables. It is a forward flow of information in the process model [M]. The situation becomes more complicated when a few of the key process states are fixed as design targets for process performance. Then, one must search for values of the design variables ($\underline{\mathbf{d}}$) which match the specified process performance. This is a so called 'inverse' problem as the flow of information from targets on a few states goes 'backwards' to the required design variables that generate these targets. For process models consisting only of algebraic equations, this reverse flow of information can usually be handled well without excessive computational efforts. For models containing (partial) differential equations, an inverse problem is much tougher to solve than the forward problem. That is because the integration of the equations follows the causal chain of events along the geometric and time coordinate(s) and there is no way of reversing it. Consequently, the models have to be repeatedly solved from inputs to outputs, searching for an input that hits a target output. This iterative procedure can be computationally very demanding.

9.3.1.8 Process design and synthesis cases with targets

In a design situation, one needs to find values for the unknown process design variables d. Hence, there are more unknown variables: both x and d. The degree of freedom in model [M] is now equal to the number of design variables, assuming there is an external scenario for setting the remaining variables $\{u; p; w\}$. This degree of freedom must be taken away by setting some targets $\{t\}$ for the performance of a design. The number of targets must be equal to the number of process design variables. Otherwise, the design is under, or over, specified. These targets can be imposed both on important performance indicators as well as on the geometric design variables. For instance, the yield of a product relative to its raw material, its production rate and its product quality can be design targets. Rather than giving direct specifications of geometric design variables certain heuristics for their proportions are applied; e.g., fixing the aspect ratio (diameter/length) of a vessel. The specifications can be set up in three categories: technical process performance, operational conditions of process units and constraints on geometric design variables. However, the total number of specs cannot exceed the number of design variables in process model; it must match exactly.

$$[S_{\text{perf}}]: \quad T_{\text{perf}}(\underline{\mathbf{x}}) - \underline{\mathbf{t}}_{\text{perf}} = 0$$

$$[S_{\text{oper}}]: \quad T_{\text{oper}}(\underline{\mathbf{x}}) - \underline{\mathbf{t}}_{\text{oper}} = 0$$

$$[S_{\text{design}}]: \quad T_{\text{design}}(\underline{\mathbf{d}}) - \underline{\mathbf{t}}_{\text{design}} = 0$$

T_{***}: Performance target expressed as a function of process states or design variables

t_{***}: Numerical value for a performance target

The technical process performance variables express the effectiveness of the process to turn inputs (feeds and utilities) into outputs (product qualities, yields and amount of waste). These performance variables must consistently relate to the Key Performance Indicators (KPIs) used in the higher management levels in a manufacturing organization.

Special attention is warranted to the occurrence of different modes of operations.

(a) Normal continuous production over a time period to make the main product grade(s).
(b) Normal continuous production over a time period to make another product grade(s).
(c) Dynamic transitions between two successive product grades.
(d) Transitions between different production output rates.
(e) Repetitive cleaning (e.g., for food production) and regeneration operations of process units.
(f) Startup from standby (with low holdup and internal circulation) to nominal production.
(g) Shutdown from high production to standby level.

Apart from the steady state production modes, all other modes of operations will requires dynamic modeling and simulation.

9.3.1.9 Additional specification of an external scenario for a design

Having set the targets, one must impose additional specifications for the other variables $\{\underline{\mathbf{u}}; \underline{\mathbf{p}}; \underline{\mathbf{w}}\}$ in the process model $[M]$. A process design scenario consists of specifications for the external inputs that can be manipulated, the model parameters and the external influences (beyond manipulation):

$$[S_{\text{input}}]: \qquad \underline{\mathbf{u}} - \underline{\mathbf{u}}_{\text{ref}} = 0$$
$$[S_{\text{param}}]: \qquad \underline{\mathbf{p}} - \underline{\mathbf{p}}_{\text{ref}} = 0$$
$$[S_{\text{disturb}}]: \qquad \underline{\mathbf{w}} - \underline{\mathbf{w}}_{\text{average}} = 0$$

$\underline{\mathbf{u}}_{\text{ref}}$: reference values for the process model input variables.
$\underline{\mathbf{p}}_{\text{ref}}$: reference values for the process model parameters.
$\underline{\mathbf{w}}_{\text{average}}$: reference value for the disturbances.

The input specifications will address the flow rates, compositions, temperatures and pressures of the raw material streams, as well as the conditions of the available utility streams. The amounts of utility consumed or produced can be made part of the process state vector as utility amounts depend on the inner workings of the process.

It may happen that a designer faces a situation where more targets need to be set than there are design degrees of freedom. The excess number of targets going beyond the number of design specifications can be handled by removing as many specifications on the inputs. The inputs that are set free will be manipulated in the model computations to meet the excess number of targets.

In a case of a dynamic mode of operation (e.g., a switching between production modes) the input specifications will be dynamic ones. These dynamic scenarios may challenge the design performance and call for readjustment of geometric design decision variables, such as lowering volumes and holdups as to increase the rates of transition (within safe bounds, of course).

The parameters in the process model will, in general, be the statistical averages resulting from a parameter estimation procedure using experimental data. In favorable circumstances one may even have a covariance matrix (V_{param}) describing the extents and patterns of uncertainties in the parameters. Specifying the external random disturbances can be conceptually hard. It often involves random (and noisy) phenomena with poorly specified or unknown probabilistic distributions. The disturbances can be, for instance, climate related quantities, such as the environmental temperature and humidity, affecting the processing states. It may also be a random variation or a gradual decline in the activity of a process agent, such as a catalyst. In design situations one often takes the expectation value (averaged value of a disturbance variable taken over a long time interval). It is already a boon to know the spread in the variation per disturbance variable. Getting a full covariance matrix (V_{disturb}) is practically ruled out.

9.3.1.10 Consistency in model formulation

When formulating a process model $[M]$ there is a challenge to be consistent in the selection of variables and equations within a physically feasible domain. In other words, one must avoid redundancies and contradictions in the model. For instance, the design specification imposed on a process design model should not violate thermodynamic feasibility and stability. To elaborate, it may happen that a process unit model in a flow sheet has two thermodynamic phases and the evolution of the states of the two phases in the model causes one of the two phases to lose its stability. In physical reality such phases would merge into a single phase. In the model, such spontaneous merging will not happen. Instead the physical properties of the two thermodynamic models will start to coincide, possibly leading to linear dependencies between equations with a resulting loss of model solvability.

The computational solving of a model $[M]$ requires it to be well posed:
- All equations are differentiable with respect to all variables $\{\underline{\mathbf{x}}; \underline{\mathbf{u}}, \underline{\mathbf{d}}; \underline{\mathbf{p}}, \underline{\mathbf{w}}\}$.
- The finite derivatives of equations are required to avoid numerical singularity of the Jacobian matrix J_M $(= \partial M/\partial x)$.
- Model $[M]$ has as many linear independent equations as state variables x. Linear dependent equations are combinations of other independent equations and therefore redundant.

It is assumed that all model equations can be solved simultaneously by means of a suitable numerical solver. In process simulation practice for novice modelers it happens that complaints arise about slow convergence to a weird solution or even total failure of a numerical solver to converge to a solution. The first resort is to attribute the responsibility for ill performance to the numerical solver. Upon closer inspection of the model it often appears there is a mathematical or physical inconsistency in the model. Such an inconsistency will challenge any solver to converge to an unphysical or nonexistent solution. The first case in this section (case 9.7) is a simple illustration of a locally ill conditioned model, violating the rule of bounded derivatives and causing trouble in model simulations. The next case (9.8) deals with the consequences of a design specification for a reactor violating thermodynamic reaction equilibrium.

9.3.1.11 Process model computations

Having set the design targets $\{\underline{\mathbf{t}}\}$ as well as the complementary design scenario specifications for $\{\underline{\mathbf{u}}, \underline{\mathbf{p}}, \underline{\mathbf{w}}\}$, one can jointly solve the process model $[M]$ and the target specifications [S] equations for $\underline{\mathbf{x}}$ and $\underline{\mathbf{d}}$. The solution that one obtains for a chosen reference scenario is called the base case solution. One can compute solutions to variations on this base case by varying the specified reference values for the inputs ($\underline{\mathbf{u}}$), physical parameters ($\underline{\mathbf{p}}$) and the external disturbances ($\underline{\mathbf{w}}$). Using the solutions in the perturbed cases in comparison with the reference solution, one can compute the sensitivities. These sensitivities can be ranked in order of (absolute) magnitude to find out which of the variables going into the process model determine the computed states the strongest. Figure 9.2 offers a visual overview of a simulation problem for process design, using design targets for process states.

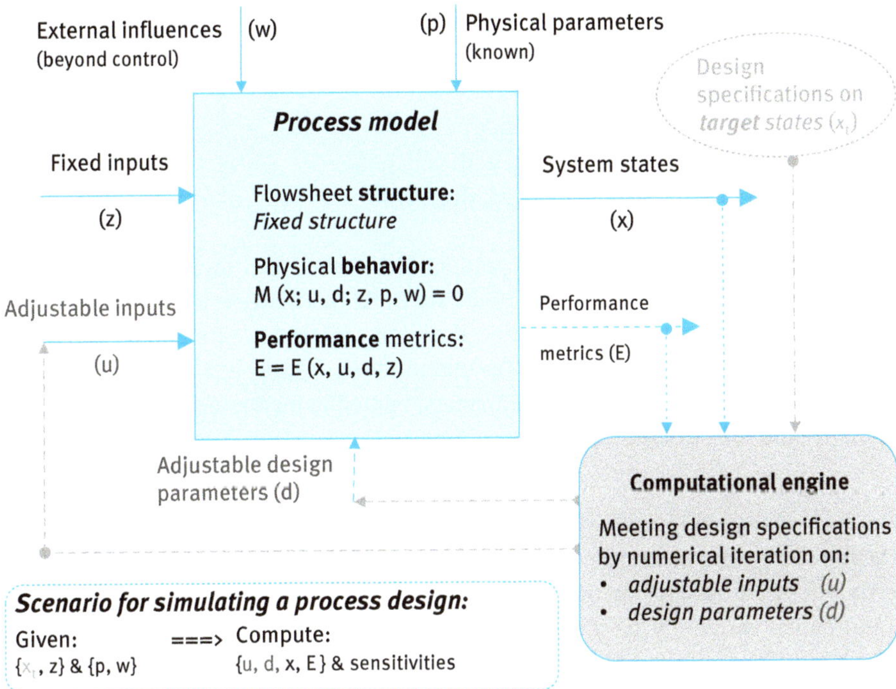

Fig. 9.2: Overview of process simulation for design.

9.3.1.12 Analysis of obtained model solutions

What kind of intricacies of model solutions can one expect when having solved a process model?

Though rather rare, one may occasionally find computationally multiple solutions. Theoretically, this can never be ruled out for nonlinear models. Remember that every polynomial of degree N has N mathematical solutions some of which are in the complex domain. Fortunately, in many chemical engineering systems there is often only one real solution among the mathematical set of (complex) solutions. Even if there would be a few real solutions, only one of these has the better physical stability characteristics; i.e., being more robust against external disturbances. Usually, it is this solution that is preferred as *the* design solution.

Having found more than one solution may also turn out to be a boon. The unexpected new solution may have a better performance than the anticipated solution. Moreover, it will make the designers aware that the multiplicity of solutions has implications for the dynamics and stability of the future process. E.g., there is a possibility of disturbance driven transitions between steady states.

9.3.1.13 Process performance evaluation metrics

Three kinds of metrics are considered, representing different values to the process operator, owner and society:
- technical metrics: $E_{\text{tech}} = P_{\text{te}}(\underline{\mathbf{x}}, \underline{\mathbf{u}}, \underline{\mathbf{d}})$
- economic metrics: $E_{\text{econ}} = P_{\text{en}}(\underline{\mathbf{x}}, \underline{\mathbf{u}}, \underline{\mathbf{d}}; c_{\text{I}}, c_{\text{O}}, p_{\text{P}})$
- ecological metrics: $E_{\text{ecol}} = P_{\text{ec}}(\underline{\mathbf{x}}, \underline{\mathbf{u}}, \underline{\mathbf{d}}; e_{\text{I}}, e_{\text{O}})$

Examples of technical metrics are: product yields over feed intake(s), energy efficiency, element efficiency and risks (e.g., by DF&EI). The technical metrics will be closely linked with the earlier introduced design performance functions (T_{perf}) and targets ($\underline{\mathbf{t}}_{\text{perf}}$) for the process design. The quantities $\{c_{\text{I}}, c_{\text{O}}, p_{\text{P}}, e_{\text{I}}, e_{\text{O}}\}$ in the economic and ecological metrics are model parameters, related to investments (I), cost of operation (O) and prices of products (P).

9.3.1.14 Using uncertainty information in a sensitivity analysis

Having obtained a base case design, one has still to consider the impact of the uncertainties in a design situation on the design outcome and the evaluation metrics. There can be variability in the specifications for operational and input variables, such as when there are different production modes. Usually these uncertainties are of a deterministic nature. A designer may know there can be multiple production modes with different feed intakes. These variations in the operational and input variables are often handled by defining a number of likely production modes, as part of specified external scenarios. The resulting designs for these modes can be compared and possibly reconciled to arrive at a robust master design that can handle all production modes. Apart from having distinct production modes, there can be uncertainty about the proper targets for some of the operational variables as well as about the input variables and streams. One can analyze the impacts of these uncertainties by means of model based sensitivity analysis. If necessary, a design can be adapted to attenuate the sensitivities. Fortunately, there is also a way of coping with remaining sensitivities to disturbances after completion of a design and after having put the process plant into operation. Then, one can still compensate for many arising additional uncertainties by adjusting the operational and input variables through process control to meet set performance targets.

In case of a rigorous model development with experimental validation and model parameter estimation, the uncertainties in the system parameters ($\underline{\mathbf{p}}$) can be coestimated. These uncertainties are expressed by a covariance matrix of the parameter estimates. Computationally, this covariance matrix comes for free along with the mean values of the parameter estimates as obtained in a statistical maximum likelihood parameter estimation procedure from experimental data. For further model uncertainty analysis, one may sample parameter values from the space spanned by the covariance matrix. Given such a set of sampled parameter values, one can compute and analyze the resulting variances in process design variables (E), states (x) and performance metrics (E).

Uncertainties in external disturbances are often harder to cope with. There is often hardly any information on the distributions of their magnitudes, if anything at all. These uncertainties are often handled after completing the base case design by means of:

- Application of experience based overdesign factors to process equipment. There is a risk of this approach by applying overly conservative factors. Historical factors may not account yet for the increased accuracy in equipment design possible by model based design simulations.
- Some form of a mathematical uncertainty analysis, based on probabilistic distributions of the uncertainties and reducing the chance that a design cannot cope with disturbances to a low acceptable level. Such global system analysis facilities are offered by advanced simulators, e.g., gPROMS process modeling platform [20].

The above outline completes the story on process modeling for simulation purposes. The application of process models for optimization of a process design is part of the next subsection.

9.3.2 Cases for development stage modeling

Three modeling cases are presented. The first two cases will highlight two common pitfalls in modeling while the third case (Case 9.9) offers a plea for model supported operability analysis in the design phase.

Case 9.7: Note on ill conditioned constitutive equations in design models

Motivation and objective of modeling

To highlight the detrimental effect of an ill conceived *constitutive equation* on the solvability of a process (unit) model.

Approach

To use a continuous flow well mixed tank reactor (CSTR) model with a commonly used *reaction rate* expression that has a local physical anomaly.

Outcome

The model has a region in which the solution will numerically fail due to an unbounded first derivative of the model equation. The implication of this example for modeling practice is that it pays to carefully check all constitutive equations in a design model on their asymptotic behavior before embarking on any process engineering computations.

Case 9.8: Consistency of design specifications with thermodynamics

Motivation and objective of modeling
In modeling for process design, it is rather easy to overlook the limits imposed by thermodynamics on setting feasible design performance targets. If such design targets are set beyond *thermodynamic equilibrium*, the computations with the process model will fail. The failure can be due to a lack of convergence in the computations or finding unphysical design outcomes after 'successful convergence'. The purpose of this case is showing such a situation through a reactor design with an analytical solution.

Approach
The design case is the computation of the required residence time to achieve a target conversion in a well mixed with a single, reversible first order reaction $A \Rightarrow B$ at isothermal conditions.

Outcome
Having specified a degree of conversion above the reaction equilibrium, the reactor design equations generate a numerical solution with a negative residence time as a displeasing feature.

The resulting design heuristic is obvious enough. Physical design targets should be located within a thermodynamically feasible domain. The preferred distance between the design target and the thermodynamic equilibrium boundary is negotiable in design.

Case 9.9 Model based analysis of operability aspects in process design

Motivation and objective of modeling
The operational success of a process plant depends, to a large extent, on the quality of its design. The operational characteristics of process should be taken into consideration in process design and modeling. The conventional model based process analysis is suitable to study the effect of the dominant phenomena on the behavior and performance of a process at steady states under normal operational modes. *Operability analysis* is needed to assess the effects of major deterministic and random changes outside and inside a process. Operability analysis of a process has become an extensive subfield of process engineering, linking process design, control and operations. It relies much on dynamic process modeling away from the usual steady states. Here, the aim is to create awareness of the main aspects of operability, while giving some references to pertaining literature for further study.

Approach and outcome
Following Marlin [21], eight topics of operability are briefly covered from the perspective of teaching such topics in process design. Dynamic process models are applicable in all of the aforementioned operability cases, except for reliability and safety. Reliability is an exception as it requires probabilistic models for equipment failure rates. Safety is another exception because it often involves process events that go far beyond the normal operating window. The conventional process models are not tuned to such exceptional conditions beyond the normal operating window. Safety requires dedicated models to study the physical behavior of a process unit under abnormal conditions, such as runaway or flow reversal effects.

9.4 Nonlinear process model optimization: contributions to a feasibility stage

9.4.1 Nonlinear modeling & optimization for a feasibility stage

An explanation is in order of why modeling in the feasibility stage is covered only after modeling for the development stage. Commonly, in process design, it is the feasibility stage that precedes the development stage. The reason for switching the order is rooted in the additional complexity of model use for optimization and its associated computational procedure. In a feasibility stage one may perform an optimization over a range of process design options using relatively simple process models to search for an optimized performance over a wide space of design options. A few of the more promising process options that are identified in the optimization procedure will be transferred to the subsequent process development stage. There, such options will be analyzed in more detail, requiring a more extensive process model along with strong process simulation capabilities. A process model in the feasibility stage may contain less detail and be simpler in general than the model for development. The paradox is that the computational framing of the models in the feasibility stage is seen as being more complicated due to its optimization features, relative to simulation only in the development stage.

The main benefit of optimization over simulation in design is that optimization offers a search over the entire solution space for the best outcome. A simulation results in a single point in the solution space. When, additionally, a simulation based global system analysis is applied [22], some more light is shed on the behavior in the vincinity of that single solution point. In system optimization, one can explore a predefined space for a point of optimum performance. It is because of this capability to explore a wider space that one wants to apply optimization in the feasibility phase to identify the more promising options within the entire solution space (for process/products). Furthermore, advanced optimization methods can also be used for the synthesis of process and/or product structures from a predefined set of building blocks (e.g., unit operations for a process phase building blocks for a product). In process simulation, the process structure is fixed *a priori*. Therefore, it is the combination of (i) creation of structure, (ii) the computing and simulation of physical behavior and (iii) optimization of system performance that makes optimization applications attractive in the feasibility phase. However, there are two restrictions to be accounted for. The search over a solution space is always a heavy computational task. Therefore the models of the building blocks for product/process synthesis must remain relatively simple and should focus on core features only. A similar restriction on models may emerge from a relative lack of process and product knowledge in the feasibility stage. The details of the mechanistic functioning of the building blocks may still be lacking and thus approximate model descriptions are used.

The price to pay for the enhanced capabilities by optimization is a much more extensive description of the design problem relative to a simulation case. It is necessary to describe the boundaries of the search space in an optimization as well as the domain of validity of the process model. In addition to the usual model equations, one has the following attributes in an optimization problem:

- Structure related: discrete (binary) variables (Y) to characterize the presence of building blocks, the types of building blocks and the connectivity of blocks.
- Structure related: equations (equalities and inequalities) to model flow sheet structure.
- Domain related: valid domain of model application in the optimization.
- Feasible design targets: ranges for variation of process, design and performance variables.
- Objective function(s): one performance metric or two metrics for a Pareto tradeoff.

9.4.1.1 Overview of common elements in an optimization frame

The elements of a design optimization problem are given:

Process variables:		$\{x, u, d, Y; p, w\}$			
Structure model:	$[S]$:	$S(Y, x)$			≥ 0
Process model:	$[M]$:	$M(x; u, d, Y; p, w)$			$= 0$
Model application domain:	$[A_M]$:	$A_M(x, u, d)$			≥ 0
Design cotargets:	$[S_{design}]$:	$T_{design}(d)$	$-$	t_{design}	≥ 0
Process design scenario:	$[S_{input}]$:	u_{fixed}	$-$	u_{ref}	$= 0$
	$[S_{param}]$:	p	$-$	p_{ref}	$= 0$
	$[S_{disturb}]$:	w	$-$	$w_{average}$	$= 0$

Process performance metrics for optimization:

- Technical metrics: $\max.\ E_{tech} = P_{te}(x, u, d)$
- EconomicEconomic metrics: $\max.\ E_{econ} = P_{en}(x, u, d; c_I, c_O, p_P)$
- Ecological metrics: $\max.\ E_{ecol} = P_{ec}(x, u, d; e_I, e_O)$

Degrees of freedom: $\dim\{d\} + \dim\{Y\} + \dim\{u_{adjust}\}$

One may wonder about the role of the design cotargets. Why not optimize all design variables at once? It may happen that a design optimization is part of a retrofit of an existing process, where the design of certain parts of the process should be retained. The use of design cotargets allows keeping a portion of the design variables fixed. In a similar fashion, if one does not want to reinvest in new equipment, one can keep the existing equipment, fix the design variables and still optimize the process by varying the operating conditions such as pressure, temperature and internal recirculation flows. There is often some tradeoff possible between making changes in the geometric design variables or in the temporal ones, such as operating conditions (pressure, temperature and/or flow rates).

9.4.1.2 Mixed integer nonlinear programming format

When the physical meaning of the variables in a process model is ignored, one can write the design optimization in its usual compact mathematical form as a mixed integer, nonlinear optimization problem (MINLP):

$$\min_{\{x,Y\}} F(\underline{\mathbf{x}}, Y)$$

Subject to:

$$\underline{\mathbf{g}}(\underline{\mathbf{x}}, Y) = 0$$

$$\underline{\mathbf{h}}(\underline{\mathbf{x}}, Y) \geq 0$$

$$Y \in \{0, 1\} \quad \text{and} \quad \underline{\mathbf{x}} \in \mathbb{R}^N$$

The mathematical conditions for solving such a MINLP problem and the various numerical algorithms are covered in textbooks on optimization in chemical engineering [11–16].

9.4.1.3 Flow of information in the computational process

In a process model the causal flow of physical information runs from physical and information inputs (feed streams, preset operating conditions and design parameters) to physical process outputs. When targets are set on the outputs, being some states and performance metrics, one must start adjusting some inputs and design parameters as to meet these targets. In view of the nonlinearity of a process model as well as (numerical) stability considerations it is not possible to invert the model as done easily with linear algebra equations: $\mathbf{A} \cdot \underline{\mathbf{x}} = \underline{\mathbf{b}} \Rightarrow \underline{\mathbf{x}} = \mathbf{A}^{-1} \cdot \underline{\mathbf{b}}$. Stability considerations play another important role. For instance, if one solves a model of a tubular reactor by integrating the differential equations in the model from entrance to exit, both the analytical and numerical solutions are stable. Going the reverse way from exit back to entrance it is observed that the same solutions exhibit instable behavior, effectively prohibiting such a route. Consequently, process models must be solved in a forward manner. When aiming for targets on process states or searching for the optimum performance, the process model must be very many times solved in iterative moves towards the targets or the optimum performance. The causal flow of information in a process design optimization procedure is shown in Figure 9.3.

For a proper evaluation and interpretation of the solution of an optimization problem it is strongly recommended to obtain an basic understanding of the role of the Lagrangian function, of the necessary conditions for the existence of a local extremum (be it a minimum or a maximum) and of the sufficiency conditions for a local minimum. The necessary theory is presented in optimization text books focused on chemical processes, [11, 13–16].

What are practical issues arising in the application of MINLP to process optimization?

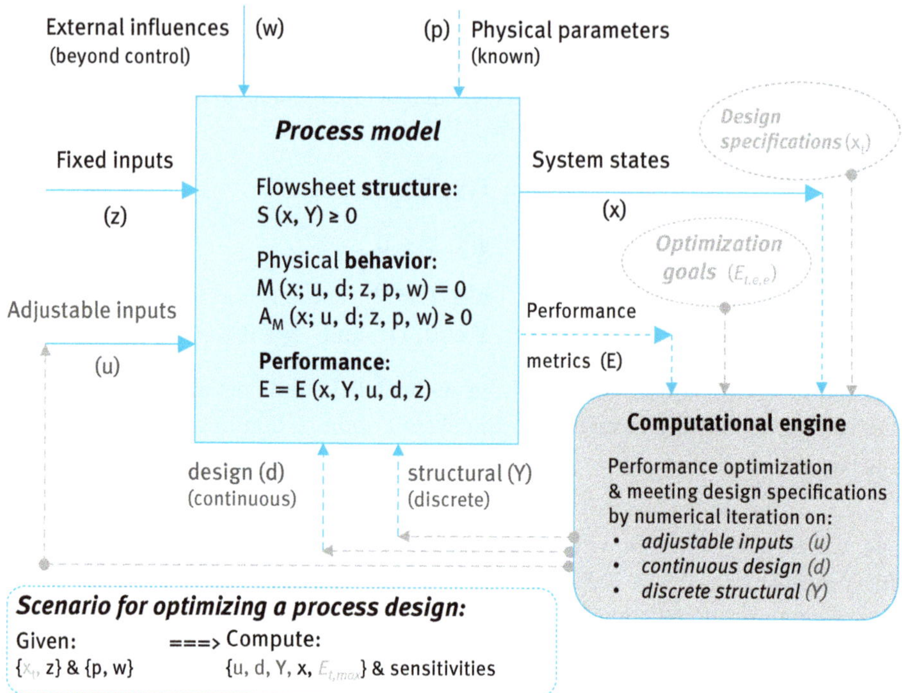

Fig. 9.3: Design optimization by solving an inverse problem.

(a) *Establishing the domain of model validity*
The derivation of the process model equations is similar to that of process simulation. The hardship is in getting a good set of inequalities to bound the solution and so prevent fictitious results outside the domain of model applicability. In practice it is often harder than expected to collect relevant information on the bounds of the domain of model applicability. The latter is the domain in which the model and, particularly, its constitutive equations can be applied with sufficient confidence. Commonly, one is well able to retrieve in the original documentation of a constitutive equation from which theoretical considerations and experimental data it was derived. However it is far less common to find, along with this equation, an explicit specification of the domain in which it can be applied with a particular degree of confidence. The delineation of such a domain is rarely given by, for example, a set of inequality constraints. Without such a specification, the numerical optimizer may move into regions where the performance seems to 'improve', while one or more constitutive equations in the process model have lost their significance. Such a loss yields a fictitious optimization result. Often the model developer has to go back to the original literature and look into the original sets of experimental data and infer over which ranges the experimental conditions were varied with the resulting ranges of the response variables.

(b) *Finding a representation of model uncertainty*
Similarly, it can be hard to find proper documentation of the statistical properties of the estimated parameters in the constitutive equations beyond their mean values, such as covariance matrices. Having such an uncertainty description would be helpful in assessing the significance of the computed optimization results in relation to underlying uncertainties in the process model.

(c) *Numerical conditioning of the model*
Scaling of process variables and equations is advised due to a wide range in magnitudes. Enthalpy flows and terms in an enthalpy balance may be of the order of 10^6 while molar fractions of trace components in mass balances can be of the order 10^{-6}. Applying some scaling will help and stabilize the numerical computations by the optimizer. It will also facilitate any internal local scaling by the optimizer.

(d) *Initialization of model variables to enhance convergence of iterations on model equations*
The assignments of starting values to the process model variables needs due attention in order to facilitate a smooth convergence to a solution of the model equations. Inconsiderate assignments of initial values may play havoc with the magnitudes of the nonlinear terms in the model equations. Such nonlinearity is often found in rate expressions (e.g., power-law reaction kinetics) and in thermodynamic constitutive equations (e.g., for activity coefficients). To get started with model initialization, it is found helpful, in practice, to artificially tone down the impact of the nonlinear rate terms in the balance equations. It is convenient to introduce into the model a common artificial scalar 'inhibition factor' as a multiplier to all strongly nonlinear terms in the model equations. The purpose of such an inhibition factor is to slow down the rates of change. Assigning initially a small value to this inhibition factor, a nonlinear model tends to behave as close to linear. Then it is easier to achieve initial convergence. Having obtained such a converged in between solution, one can use it as a starting point for a next round with a gradual ramping up of this inhibition factor. Stepping up the inhibition factor to unity will restore the original nonlinear features of the model.

(e) *Use of suitable modeling platforms for optimization of processes*
Modern process model building and computing platforms, based on equation oriented computations, offer compatible links to advanced MINLP optimization facilities. Examples are gPROMS [20] and AIMMS [23]. The conventional sequential, block oriented process flow sheet simulators are more restricted and offer some optimization facilities for process operating conditions (e.g., control set points). They do not offer optimization of flow sheet (super) structures.

These remarks complete the review of modeling for process (design) optimization. The next section will deal with four cases to illustrate some aspects of optimization applications.

9.4.2 Cases for feasibility stage modeling

Four cases are covered that highlight aspects of:
- constraint analysis to identify options for technological innovations
- Pareto tradeoffs in process design between two performance criteria
- synthesis of gas to liquid (GTL) processes
- experimental process optimization with data driven black box models

Case 9.10: Constraints analysis to identify options for technological innovations

Motivation and objective of modeling
Realistic process models should reflect current limitations the technologies underlying process unit operations in a process (conversion, separations and compression). In process design optimization studies such limitations should be expressed as *inequality constraints* in the process model. These inequalities delineate the operational domain in which a particular technology (e.g., catalyst, membrane) can trustfully be applied (e.g., validity range of reaction kinetics, maximum allowable temperature, maximum allowable pressure differences). The purpose of *constraint analysis* after an optimization of a design is to identify those active inequality constraints that predominantly limit process performance. Knowing how inequality constraints relate to underlying technologies one finds directions for improving process performance by means of innovations in the limiting processing technologies.

Approach
Model based process optimization delivers optimized process conditions and design variables. In addition it will also yield active set of inequality constraints constraining the process performance. The relative importance of these inequality constraints can be identified by means of their associated *Kuhn–Tucker multipliers*, coproduced as a result of the optimization. One can put the Kuhn–Tucker multipliers in order of decreasing magnitude. In that way one obtains the order by which active inequality constraints are restraining the objective function. This approach of finding the relative impact of active inequalities is explained by means of small reactor design example

Outcome
The result of the reactor design optimization shows that one inequality constraint is strongly dominating the optimum and could be a candidate for relaxation, while another inequality constraint has only a marginal influence on the performance in the optimum. An analysis of the active inequality constraints and their Kuhn–Tucker multipliers in an optimization outcome can help to find incentives and clues for further technology development and to enhance process performance.

In the same vein, an extensive, deep modeling study was reported in [24] for the conceptual design of a steam cracking reactor. This design was reoptimized with respect to olefin yields by lifting the current material induced temperature constraint by 100 K to a new anticipated level of 1300 K. The result is a very significant potential improvement in olefin yields (55 ⇒ 66% mass yield of ethylene).

Case 9.11: A simple example of a Pareto tradeoff between two performance metrics

Motivation and objective of modeling
Virtually every process design has to consider multiple performance criteria. The ranges of all design decision variables together form a high dimensional process design space. Each point in this

design space represents the complete array of design decision variables. In these points one can compute the process performance metrics. This is a low dimensional space of two or three dimensions. The overall objective of a design is to improve the performance criteria as much as possible. In certain regions of the design space one can move a design from one point to another point and all performance criteria improve indeed. However, often there appears a transition to another region where opposition between criteria arises. Making changes in a design can improve one performance metric while making anther metric worse. There is no way of improving both. The set of all designs where one cannot improve a performance metric without making at least one other metric worse is called a (noninferior) *Pareto set of design solutions*. If one takes all performance criteria equally serious, it is impossible to say which design within this Pareto set is the better one. All are equally qualified as 'best' designs.

Working with multiple performance functions in a design gives conceptually rise to *Multi Objective Nonlinear Programming* (MONLP). In process engineering practice one wants to keep using a normal single objective Mathematical Programming format and associated nonlinear programming (NLP) solver. Then there are three options to deal with multiple performance criteria. The assumption is that all criteria have to be minimized.

(1) To set a threshold level for the performance variable of each criterion, except one. The latter one is called the dominant criterion. The purpose of having thresholds is to relax from the minimization and stay well above the attainable minimum level of performance for each criterion. Each threshold will appear as an additional inequality constraint in the optimization. Next, the design is optimized with respect to the dominant criterion, while making sure that all other criteria meet their threshold levels. When a criterion is at its threshold level (i.e., becoming an active inequality), its Kuhn–Tucker multiplier indicates the sensitivity of the dominant criterion with respect to the criterion at threshold. The outcome may give reason to adapt the thresholds and repeat optimization till the results are "satisfactory" to the problem owners of a design. In other words, "satisfactory" depends on a reflective judgement by process engineering management, depending on societal and company cultures as well as on local site conditions where the process will be erected.

(2) To define a new master criterion as a linear weighted sum over all separate criteria. Often economic performance is used as the dominant criterion and all other criteria must be translated into monetary values. The weight factors in the linear sum expression represent the relative economic weights of the various criteria. The approach under 1) can help to establish reasonable estimates for such weight factors (≈ using Kuhn–Tucker multiplier). One can then minimize the new master criterion. If the results are not satisfactory to the problem owner of a design one may start iterative improvements in adjusting the relative weights in the master criterion.

(3) To create a Pareto set of design solutions, where each design in the set will have different performance values. Such a Pareto set can be generated by means of approach 1) where the selected threshold levels (initially high and relaxed) are systematically stepped down is severity. Till the level for that particular criterion is so low that optimization with respect to the dominant criterion becomes infeasible.

Approach

A Pareto optimization is demonstrated by means of a small scale reactor design problem with two performance criteria: profit and mass yield.

Outcome

The results indicate there is a region of design in which both performance criteria simultaneously improve. Adjacent is another region with a Pareto front: improvement of one criterion makes the other criterion worse. A similar result is obtained in an optimization of the design of a reactive – distillation column for *MTBE synthesis* [25]. An economic performance indicator is placed in a tradeoff with an ecological indicator, the exergy efficiency of this column.

Case 9.12: Optimization of designs of gas to liquid processes

Motivation and objective of modeling

The main purpose of GTL processes is to turn *natural gas (NG)* into liquid fuels. GTL processes tend to serve huge gas fields and are truly large scale. The capital investment is very high (up to several billions US$). There are many design decisions with interacting outcomes on process performance. Also, tradeoffs emerge between capital investment, carbon and energy efficiencies. Therefore, GTL processes are prime candidates for model based optimization *process synthesis*.

Approach

The feasibility of such an approach is shown in an industrial case study on model based, optimization driven synthesis of a large scale GTL process [26], using profitability and *carbon efficiency* as performance measures. The process flow sheet is represented by means of a block diagram with reactors and unit operations. The flow sheet is generalized to cover many possible synthesis options. The reactor models and unit operations are reduced to their core features to keep the model relatively small. An extensive set of inequality constraints is implemented to make sure the conditions in the reactors and the unit operations will stay in validated domain during optimization.

Outcome

An industrial case study [26] shows a design region in which both performance metrics can be improved simultaneously. However, after reaching the top in profitability there is a *Pareto frontier* with a tradeoff: an improvement in carbon efficiency leads to a decrease in profitability.

Few other industrial case studies of MINLP optimizations for process synthesis are found in the scientific literature. This is compensated for by a growing number of academic publications. Floudas and coworkers have applied optimization comprehensively to synthesize *XTL processes*. In XTL, the symbol X stands generically for either Biomass or Coal or Natural Gas or even all feeds in parallel. Also, in these studies the product slate has been extended from liquid hydrocarbon fuels to chemicals, including methanol, DME and olefins, [27]. It is even possible to extend optimizations to the synthesis of a national *supply chain* (for the USA) for the best locations for new GTL processes given local sources of NG and main customer markets for liquid fuels [28].

Case 9.13: Process performance optimization by model and experiments

Motivation and objective of modeling

Some new products, as well as the processes to make these products, are conceived experimentally in the concept stage. Some further optimization is warranted before moving to the next stages. Such an optimization is part of the feasibility stage. If a product or a process is complicated and only poorly understood, an effort to mechanistic modeling of its behavior and performance could be daunting and costly. In such situations an experimental optimization of the product/process performance with the help of data driven models can be more effective. Data driven process model development is sometimes cheaper and quicker than developing a (semi)mechanistic model. However, it is required that one can indeed measure all input and output variables of a product/process, relevant to the characterization of its behavior and performance.

Approach
A *data driven* modeling and optimization procedure is presented that applies equally well to the optimization of process and product performance. This story will focus on the conceptual outline of experimental process optimization, using a response surface approach.

Outcome
The outcome is a procedure leading to a parametric process model and its experimental validation. The model is repetitively improved through the following sequence:
(a) Making a computational prediction of the conditions of optimum performance.
(b) Experimental validation of this point.
(c) Adaptation: if there is still a significant deviation between the model predictions and the experimental outcomes, the most recent experimental data are used to refine model parameters.

In the end, the model has parameter estimates accurate enough to closely predict the optimum performance point. The strength of this successive approach of gradual refinement of parameter estimates is that model outputs and experimental data should become consistent in the region of optimum performance.

Another similar response surface approach is reported in [18, (Chapter 20.5)], on the modeling and optimization of a *coating process*. A design of experiments as well as the successive *response surface modeling* (regression) are extensively discussed and analyzed for statistical significance of the results. This application is an adaptation from a training exercise in E.I. du Pont de Nemours and Company (USA). It is possible to extend the response surface approach to *dynamic process systems* [29], such as (semi-)batch reactors. One applies a dynamic design of experiments. Upon collection of the corresponding experimental data, a black box or gray box model can be fitted to the data and applied for process operations (finding optimal trajectories) and control.

9.5 Concluding remarks

This chapter has highlighted approaches to model development and applications related to the concept, feasibility and development stages of product and process development and design. The emphasis here has been on making effective and inventive use of models for process innovation and optimization, illustrated by means of small scale examples. This amounts to making useful interpretations of modeling and computing results. The balance in the use of models has been in favor of steady state (algebraic) models, at the expense of modeling of process dynamics for control and (dynamic) optimization. However, the core thinking patterns in analyzing models and their results remain the same.

List of symbols, subscripts and abbreviations:

Symbols

$\mathbf{A}$	real (non)square matrix (real values as entries)
$\mathbf{b}$	real vector
c	cost related parameters in economic model expressions
d	symbol for design parameters (continuous) in a process model
E	set of quantitative performance metrics for a process design
N	dimension of a vector
p	generic symbol for a set of physical parameters in a process model
p_{P}	proceeds from product sales, in economic model equation
P	performance function
t	target value for a design quantity or performance function in process design
T	function expressing a metric (function) related to a target
u	generic symbol for inputs to a process model
w	disturbance variables in a process model
x	generic symbol for physical states in a process model
y	set of measured outputs of a process (model)
Y	generic symbol for structural design parameters (discrete) in a process model
z	generic symbol for a set of adjustable inputs to a process model
ε	deviation between computed and experimental value of a process output variable
$\mathbb{R}$	mathematical space of real variables
∇	spatial derivative

Subscripts

Average	averaged value of a quantity
design	design related
ecol	ecology related
ecom	economy related
oper	operations related
perf	performance related
ref	reference value of a quantity when making a comparison
tech	(process) technology related

Abbreviations

CSTR	Continuous flow, well stirred tank reactor
DF&EI	Dow Fire & Explosion Index
DME	Di-Methyl-Ether
GTL	Gas to liquid
KPI	Key performance indicator
LP	Linear programming
MINLP	Mixed integer nonlinear programming
MONLP	Multi objective nonlinear programming
MTBE	Methyl *tertiary* butyl ether
NG	Natural gas
NLP	Nonlinear programming
XTL	X = {biomass, coal, natural gas} to liquid fuels

References and further reading

[1] Dimian AC, Bildea CS, Kiss AA. Integrated design and simulation of chemical processes, Second edition, Elsevier, Computer-Aided Chemical Engineering, 35, 2014.
[2] Aris R. Mathematical modeling techniques, Research Notes in Mathematics, 24, Pitman Advanced Publishing Program, 1979.
[3] Bequette BW. Process Dynamics, Modeling, Analysis & Simulation, Prentice Hall, 1998.
[4] Aris R. Mathematical modeling, PSE Series Vol. 1, Academic Press, 1999.
[5] Hangos KM, Cameron IT. Process modeling & analysis, PSE Series Vol. 4, Academic Press, 2002.
[6] Cameron IT, Gani R. Product & process modeling: A case study approach, Elsevier, 2011.
[7] Rice RG, Do DD. Applied Mathematics & Modeling for Chemical Engineers, Second Edition, Wiley, 2012.
[8] Zondervan E. A Numerical Primer for the Chemical Engineer, Taylor & Francis (CRC Press), 2014.
[9] Rudd DF and Watson CC. Strategy of process engineering, John Wiley &Sons, Inc., 1968.
[10] Westerberg AW, Hutchison HP, Motard RL, Winter P. Process flowsheeting, Cambridge University Press, 1979.
[11] Biegler LT, Grossmann IE, Westerberg AW. Systematic methods of chemical process design, McGraw Hill, 1997.
[12] Seider WD, Seader JD, Lewin DR, Widagdo S. Product and Process Design Principles. Synthesis, Analysis, and Evaluation, 3rd edn, John Wiley & Sons, Inc., 2010.
[13] Floudas CA. Nonlinear and mixed-integer optimization. Fundamentals and Applications, Oxford University Press, 1995.
[14] Floudas CA. Deterministic Global Optimization: Theory, Methods and Applications, Kluwer Academic Publishers, 2000.
[15] Edgar TF, Himmelblau DM, Lasdon LS. Optimization of chemical processes, 2nd edn, McGraw-Hill, 2001.
[16] Biegler LT. Nonlinear Programming: Concepts, Algorithms and Applications to Chemical Processes, SIAM, 2010.
[17] Buelens LC, Galvita VV, Poelman H, Detavernier C, Marin GB. Super-dry reforming of methane intensifies CO2 utilization via Le Chatelier's principle, Science 354(6311), 2016, 449–452.

[18] Ogunnaike BA. Random Phenomena. Fundamentals and Engineering Applications of Probability and Statistics, CRC Press, Section 20.5, 879, 2010.

[19] Varma A, Morbidelli M, Wu H. Parametric sensitivity in chemical systems, Cambridge University Press, 1999.

[20] gPROMS process modeling platform of PSEnterprise (London, UK), https://www.psenterprise.com/products/gproms/platform.

[21] Marlin TE. Teaching operability in undergraduate chemical engineering design education. Computers & Chemical Engineering, 34, 2010, 1421–1431.

[22] Global system analysis in gPROMS process modeling platform of PS Enterprise, London, UK, https://www.psenterprise.com/products/gproms/technologies/global-system-analysis.

[23] AIMMS (Advanced Interactive Multidimensional Modeling System) of paragon decision technology, Haarlem, NL, https://aimms.com/english/developers/resources/manuals/optimization-modeling/.

[24] Van Goethem MWM, Barendregt S, Grievink J, Moulijn JA, Verheijen PJT. Model-based, thermo-physical optimization for high olefin yield in steam cracking reactors, Chemical Engineering Research and Design, 88, 2010, 1305–1319.

[25] Almeida-Rivera CP, Grievink J. Process Design Approach for Reactive Distillation Based on Economics, Exergy and Responsiveness Optimization, Industrial & Engineering Chemistry Research, 47, 2008, 51–65.

[26] Ellepola JE, Thijssen N, Grievink J, Baak G, Avhale A, van Schijndel J. Development of a synthesis tool for Gas-To-Liquid complexes, Computers and Chemical Engineering, 42, 2012, 2–14.

[27] Baliban RC, Elia JA, Weekman Vern, Floudas CA. Process synthesis of hybrid coal, biomass, and natural gas to liquids via Fischer–Tropsch synthesis, ZSM-5 catalytic conversion, methanol synthesis, methanol-to-gasoline, and methanol-to-olefins/distillate technologies, Computers and Chemical Engineering 47, 2012, 29–56.

[28] Elia JA, Baliban RC, Floudas CA. Nationwide, Regional, and Statewide Energy Supply Chain Optimization for Natural Gas to Liquid Transportation Fuel (GTL) Systems, Industrial & Engineering Chemistry Research, 53, 2014, 5366–5397.

[29] Klebanov N; Georgakis C. Dynamic response surface models: A Data driven approach for the analysis of time-varying process outputs, Industrial & Engineering Chemistry Research, 55, 2016, 4022–4034.

10 Evaluating economic performance

10.1 Introduction

Before companies agree to invest large amounts of capital on proposed projects, their management must be convinced that the selected project(s) will provide a sound return on investment (ROI) compared to alternatives. Therefore, it is crucial that economic evaluations are made before initiating a project, at various stages in its development, and before attempting the design of a process and plant. These evaluations determine whether a project should be initiated, abandoned, continued or taken to the scale up and implementation stages. Furthermore, economic evaluations pinpoint those parts of the project (product formulation, manufacturing and process) that require additional investigation. Even if the available technical information is insufficient for complete product/process design, economical evaluations must still be made to determine whether the project is economically (more profitable than competing projects) and financially (capital required for implementation can be raised) feasible. Although cost estimation is a specialized subject and a profession in its own, design engineers must be able to make rough cost estimates to support the decisions between project alternatives and optimize the design.

The purpose of most chemical engineering design projects is to provide the input information from which estimates of capital and manufacturing costs can be made and subsequently the profitability of a project can be assessed. Therefore, this chapter first introduces the principal methods used for comparing the economic performance and profitability of different projects. Subsequently the components comprising the manufacturing costs and the methods used for capital costs estimating are introduced and discussed.

10.2 Economic project evaluation

10.2.1 Project cash flow

During any project, cash initially flows out of the company to cover the engineering costs, equipment procurement, plant construction, and plant startup. After construction is completed the plant is put into operation and revenues from product sales start to flow into the company. The *net cash flow* (CF) is the difference between earnings and expenditures. A cash flow diagram, such as that shown in Figure 10.1, illustrates a typical cumulative net CF forecast over the life of a project. The CFs are based on estimates of investment, operating costs, sales volume and sales price for the project. The diagram can be divided into the following characteristic regions:

https://doi.org/10.1515/9783110467741-010

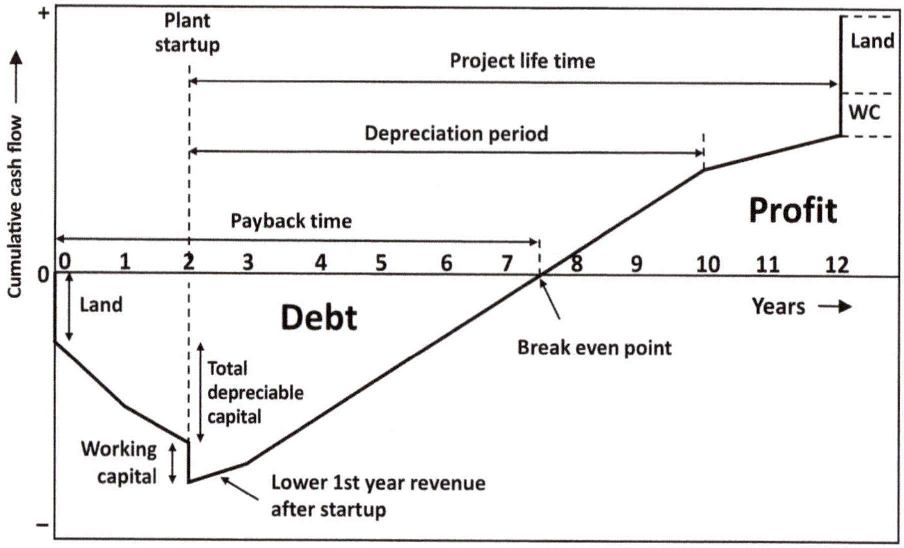

Fig. 10.1: Project cash flow (CF) diagram.

- Investment required to design, construct and provide funds for startup of the plant.
- Upturning cash flow curve due to a positive net cash flow as income is generated from sales. The cumulative amount remains negative until the investment is paid off, which is known as the *break even point*.
- Positive cumulative cash flow (CCF) where the project is earning a *return on the investment*.

The CF summarizes all the incoming and outgoing CFs through the operating year. Investments are considered a negative CF, while profits (after tax) plus depreciation are positive CFs. During the years of plant construction the CF for a particular year n is:

$$CF_n = -\varphi \mathrm{TDC} - \mathrm{WC} - C_{\mathrm{Land}} \tag{10.1}$$

The fraction of total depreciable capital (TDC) expended that year is φ, WC is the working capital and C_{Land} is the cost of land expended during that year. WC comprises the additional funds, on top of the costs to build the plant, to startup the plant and keep it in operation. It flows in and out of an existing operation and is usually assumed to be completely recoverable at the end of a project without loss. As illustrated in Figure 10.2, WC is divided into two main categories:

(1) *Current liabilities* that consist of bank loans and accounts payable (money owed to vendors for various purchases).

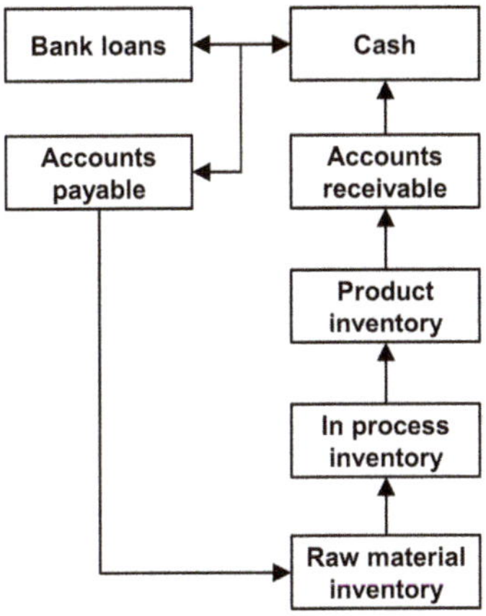

Fig. 10.2: Flow of working capital (adapted from [2]).

(2) *Current assets* that comprise
- available cash (salaries, raw material purchases, maintenance supplies and taxes).
- accounts receivable (products shipped but not yet paid and extended credit to customers).
- product inventory (material in storage tanks and bins).
- in process inventory (material contained in pipe lines and vessels).
- raw material inventory (material in storage tanks and bins).
- spare parts inventory.

WC can vary from as low as 5% of the fixed capital for a simple single product, process, with little or no finished product storage, to as high as 30% for a process producing a diverse range of product grades for a sophisticated market, such as synthetic fibers. A typical figure for general chemicals and petrochemical plants is 15% of the total permanent investment (TPI) as defined in Figure 10.7.

The CF for a particular year of plant operation is calculated from the *operating income* (OI) resulting from subtracting the operating expenses (variable and fixed costs) from the sales or revenue for that specific year:

$$\mathrm{OI}_n = S - \mathrm{VC} - \mathrm{FC} \tag{10.2}$$

Here, S is the annual sales, VC the annual variable costs and FC the annual fixed costs. The total annual product sales revenue is the sum of the unit price of each product multiplied by its sales quantity. When the depreciation (capital costs) are subtracted from the OI, the annual *gross profit* (GP) is obtained:

$$\mathrm{GP}_n = S - \mathrm{VC} - \mathrm{FC} - D \tag{10.3}$$

Where D is the annual depreciation calculated by dividing the total depreciable capital, C_{TDC} in Figure 10.7, by the depreciation period of N years, often chosen equal to

10 years. The annual *net profit* (NP) represents the net result in a specific year and is calculated by subtracting income taxes from the GP, where t is the taxation rate:

$$\mathrm{NP}_n = \mathrm{GP}_n \times (1 - t) \tag{10.4}$$

As depreciation is considered as an internal expense that is retained in the company, the CF from plant and product manufacturing operations for year n equals the after tax earnings plus the depreciation:

$$\mathrm{CF}_n = (1 - t)(S - \mathrm{VC} - \mathrm{FC} - D) + D \tag{10.5}$$

The cumulative cash flow (CCF) sums calculated CF for each year up to the point of the calculation:

$$\mathrm{CCF}_N = \sum_{n=0}^{N} \mathrm{CF}_n \tag{10.6}$$

Where CCF_n is the cumulative CF for a period of N operational years, CF_n the CF for year n, and n a positive integer for the operational year. The *discounted cash flow* (DCF) calculates the value of current money in the future considering that the value of money earned in the future is less than money in the present. This calculation is dependent on the discount rate, which further decreases the value of money when it increases. The discount rate used is generally the appropriate weighted average cost of capital (WACC) that reflects the risk of the CFs by incorporating the time value of money and a risk premium. Through the time value of money (risk-free rate) investors are compensated for having to wait until their investment is paid back. The risk premium reflects the extra return investors demand because they want to be compensated for the risk that the CF might not materialize after all. As a result the discount rate typically ranges from 10% for low risk projects up to 20% for high risk projects.

$$\mathrm{DCF}_n = \frac{\mathrm{CF}_n}{(1 + r)^n} \tag{10.7}$$

Where DCF_n is the discounted cash flow for year n and r the discount rate. The *cumulative discounted cash flow* (CDCF) calculation uses the calculated discounted cash flow for each year and sums them up to the point of the calculation. This number is indicative for the net result of a business case. When the CDCF equals to zero, the internal rate of return (IRR) is equal to the discount rate.

$$\mathrm{CDCF}_N = \sum_{n=0}^{N} \mathrm{DCF}_n \tag{10.8}$$

CDCF_n is the cumulative discounted cash flow for a period of N years.

10.2.2 Economic potential project evaluation method

Before even starting a design, it is crucial to verify that the target product(s) at least provides a value that is higher than the minimal consumption of raw materials. Such evaluation requires nothing more than first indications of possible sales prices, raw material prices and targeted production volume, which are typically obtained from market data. For important chemicals, market data and prices are recorded and published by various organizations such as ICIS (http://www.icispricing.com), IHS (www.ihs.com/industry/chemical) which also publishes the Chemical Economics Handbook containing market overviews for >300 compounds, Nexant (www.nexant.com/industries/chemicals), Orbichem (www.orbichem.com) and Intratec (www.intratec.us) or can be obtained via online brokers and suppliers (www.business.com/directory/chemicals, www.alibaba.com). For specialty chemicals and new products it is often more difficult to obtain reliable price data. In such cases consultants can be involved to provide economic and marketing information, or estimates of attainable price levels need to be made based on envisioned product performance and related/competing products in the targeted market(s).

Based on the obtained price data indications and calculated raw material consumptions, a first estimate of the *economic potential* (EP) of a project can be made by simply substracting the annual cost of the raw materials from the expected annual sales revenues:

$$\mathrm{EP} = S - \mathrm{RMC} \tag{10.9}$$

S is the annual sales and RMC is the the annual raw material costs. The sales can constitute only the main product but also the known byproducts. The main benefit of this first high level evaluation is that nothing more than a black-box input-output diagram and overall mass balance is required to assess the difference in the EP of the various design alternatives. For most projects, this is a valuable approach as raw material costs are the major part (50–90%) of the total manufacturing costs. When required, this EP estimation can be refined to include utility costs, but then more detailed information on the energy balances is also required.

10.2.3 Simple project evaluation methods

This section introduces some simple economic indicators that can be calculated quickly once the project investment and CFs are estimated. These indicators are widely used for preliminary screening of project attractiveness. By dividing the total capital investment (TCI) (fixed capital plus WC; see Figure 10.7) by the average annual CF, the *pay back time* (PBT) is obtained:

$$\mathrm{PBT} = \frac{C_{\mathrm{TCI}}}{\mathrm{CF}} \tag{10.10}$$

Here, C_{TCI} is the total capital investment and CF the average annual CF. This is not the same as the breakeven point indicated by the cash flow diagram, as it assumes that all the investment is made in year zero and revenues begin immediately. For most chemical plant projects, this is not realistic as investments are typically spread out over one to three years and revenues may not reach 100% of design basis until the second year of operation. The simple PBT is strictly based on a CF but, for simplicity, taxes and depreciation are often neglected and the average annual income is used instead of CF. Another simple indicator of economic performance is return on investment (ROI), which is defined as:

$$\mathrm{ROI} = \frac{\mathrm{NP}_n}{C_{\mathrm{TCI}}} \times 100\% \tag{10.11}$$

This is generally stated as a percentage per year. The required ROI will typically depend on the degree of risk associated with a project. For low risk projects, a ROI of 10% might be sufficient while for high risk ventures (new products, new markets and new technology, for example) the required ROI might be as high as 50%.

10.2.4 Present value project evaluation methods

The above simple economic indicators do not capture the time dependency of CFs during the project. CF timing is crucial to investors because not all capital must be financed immediately and money earned can be reinvested and start to earn a return as soon as it is available. For that reason money earned in the early project years is more valuable than that earned later. This *time value of money* can be accounted for by bringing the net CF in each year of the project to its 'present value' at the start of the project through discounting it at some internally agreed discount rate, which is set at the WACC in most companies.

The *net present value* (NPV) is one way to value a project in terms of CFs and the value of money. It states how much the business is worth at a certain point of time; typically a 10 year period is taken. It includes future revenues, expenses and converts those CFs into a positive or negative value.

$$\mathrm{NPV}_N = \sum_{n=1}^{N} \frac{\mathrm{CF}_n}{(1+r)^n} \tag{10.12}$$

The internal rate of return (IRR) is directly linked to both the discount rate and the NPV. It says something about the rate of growth, which a project is expected to generate. By calculating the NPV at various interest rates, it is possible to find an interest rate at which the cumulative NPV at the end of the project is zero:

$$\mathrm{NPV}_N = \sum_{n=1}^{N} \frac{\mathrm{CF}_n}{(1+\mathrm{IRR})^n} \tag{10.13}$$

This particular rate is called the IRR and is a measure of the maximum interest rate that the project could pay while still breaking even by the end of the project life. The more positive the IRR is, the more favorable the project is in terms of ROI. Often, this value is used for the comparison of different projects, as it is relatively easy to calculate and compare. Companies usually demand IRR levels higher than the internally agreed discount rate before they invest in a project.

10.3 Manufacturing costs

To determine the financial attractiveness of a project, the manufacturing (production) costs of a product are needed. Figure 10.3 separates the total production costs into three main categories: direct costs, indirect costs, and general costs. Direct (variable) costs are proportional to the production rate, whereas the indirect and general costs (fixed and plant overhead costs) tend to be independent of the production rate. General costs cover management of the firm, marketing the product and research and development.

Feedstocks, catalysts, solvents	**Raw materials**	**Direct costs**
Steam, electricity, fuel, refrigeration, water, waste treatment, landfill	**Utilities**	
Supplies: operating, maintenance Operating labor, supervision Maintenance labor, supervision Laboratory charges Royalties	**Operational costs**	
General plant overhead Safety and protection Medical, salvage services Payroll overhead Packaging, shipping, storage facilities Restaurant, recreation facilities Control laboratories Maintenance facilities	**Plant overhead costs**	**Indirect cost**
Depreciation Property taxes, financing, insurance, rent	**Fixed costs**	
Executive, clerical, engineering, legal, communications	**Administration**	**General cost**
Sales (or transfer), advertising, product distribution, technical sales service	**Marketing costs**	
Research and development		

Fig. 10.3: Breakdown of total production costs. (Adapted from [2]).

10.3.1 Direct costs

In calculating the *direct costs*, the annual cost of each raw material is found by multiplying the annual consumption by the price. Raw material costs often dominate the overall manufacturing costs. Consumables such as acids, bases, absorbents, membranes, solvents and catalysts are depleted or degraded over time and require replacement. Utilities comprise of steam, electricity, fuel, cooling water, process water, compressed air, refrigeration, waste treatment and landfill. Utility equipment is usually located outside the process area (OSBL) and may supply several processes. Analogous to raw materials the annual cost of each utility is determined by multiplying its annual consumption by its price.

Chemical plants require operating labor to produce a chemical, and maintenance labor to maintain the process. There is also indirect labor, needed to operate and maintain facilities and services. When developing a new process the quantity of operating labor can often be estimated in preliminary cost analysis from company experience with similar processes or published information on comparable processes. If a flowsheet or process flow diagram (PFD) is available, the operating labor can be estimated from the number of operations based on previous experience:

$$\text{Annual Labor Cost} = \sum \left(\text{Equipment type} \times \frac{\text{Operator/Shift}}{\text{Unit}} \right) \times (\text{number shifts}) \times \left(\frac{\text{cost/year}}{\text{operator}} \right) \quad (10.14)$$

Table 10.1 indicates some typical labor requirements for various types of process equipment. The number of shifts varies between countries but is generally 5 for 24/7 operating continuous plants and 4 for 24/5 operating batch plants. The required amount of operating supervision is closely related to the total amount of operating labor and typically averages about 15% of the operating labor costs.

Tab. 10.1: Typical labor requirements for process equipment [9].

Equipment type	Operators/unit/shift
Compressors and blowers	0.15
Centrifuge	0.35
Crystallizer	0.20
Dryer	0.50
Evaporator	0.25
Filter (continuous/batch)	0.15/1.0
Heat exchanger	0.10
Vessel, tower	0.2–0.5
Reactor (continuous/batch)	0.5/1.0

Maintenance costs consist of materials, labor, and supervision and may sometimes be higher than the cost of operating labor. Although maintenance costs increase as a plant ages, for economical estimates, an average value over the plant life is assumed. The maintenance supplies (materials and services) are typically estimated as a freacton of the TDC investment and range from 3.5% for a fluids processing plant to 5% for a solids processing plant and 4.5% for solids-fluids processing. Maintenance labor

Tab. 10.2: Breakdown of direct operational and indirect costs [4].

Operational costs	
Operating labor	See equation (10.14)
Operating supervision	15% of operating labor
Operating supplies	15% of maintenance supplies
Maintenance supplies	
– Fluid handling	3.5% of total depreciable capital (C_{TDC})
– Fluid-solids handling	4.5% of total depreciable capital (C_{TDC})
– Solids handling	5.0% of TDC (C_{TDC})
Maintenance labor, supervision	120% of maintenance supplies
Laboratory charges	15% of operating labor
Royalties	4% of total production costs without depreciation
Indirect costs	
Overall plant overhead	22% of operating + maintenance labor and supervision
Fixed costs	
– Depreciation	10% of total depreciable capital (C_{TDC}, chosen lifetime dependent)
– Property taxes	2% of TPI (C_{TPI})
– Insurance	1% of total depreciable capital (C_{TDC})
– Interest	Excluded from CF when profitability calculation is the main objective
– Rent	0% of total permanent investment (C_{TPI})
General costs	
– Administration	2% of sales
– Marketing & sales	3% of sales (transfer 1% of sales)
– Research & development	5% of sales (2–10% depending on business)

wages and benefits are typically of the same magninitude as the mainenance supplies costs. The costs for engineers, supervision and overhead are estimated at 30% of the direct maintenance labor. Custodial supplies, safety items, tools, column packing, and uniforms, which are not raw materials or maintenance supplies, are considered as *operating supplies*. A typical cost value for these types of supplies is 15% of the maintenance supplies.

In order to meet the specifications of the products, regular analysis of process streams is mandatory to determine and control their quality. As initial estimates, these *laboratory charges* for quality control typically range from 10 to 20% of the operating labor. The amount of *royalties* (licensing fees) depends heavily on the uniqueness of the process and can range from 0 to 6% of the total production cost, or alternatively, 0 to 5% of product sales. All these direct costs components are summarized in Table 10.2.

10.3.2 Indirect and general costs

As depicted in Figure 10.3, *indirect costs* consisting of fixed and plant overhead costs, change only little or not at all with the amount of production (plant capacity) because they are indirectly related to the production rate.

Fixed costs include depreciation, property taxes, insurance, interest and rent. These expenses are a direct function of the capital investment and financing structure. Rent is usually not taken into account in a preliminary estimate. The initial investment is paid back by charging *depreciation* as a manufacturing expense. Although there are several depreciation methods, using a constant yearly depreciation rate for a fixed period in preliminary economic evaluations is commonly accepted. A manufacturing plant or individual equipment has an economic, a physical, and a tax life. The economic life is the period until a plant becomes obsolete, the physical life when maintenance of a plant becomes too costly, and the tax life is fixed by the government. The plant life is usually set between ten and twenty years. Depreciable capital includes all incurred costs for building a plant up to the point where the plant is ready to produce, except land and site development. The magnitude of local *property taxes* depends on the particular location of the plant and may be initially be estimated at 2% of the TDC investment. Property insurance rates, typically 1% of the TDC per year, depend on the actual operation carried out in the manufacturing plant and the extent of protective measures taken. Interest is only a cost when it is necessary to borrow the capital needed to invest in the plant. However, in most cases the capital originates from the company and/or investors making project profitability calculation the main evaluation objective. Therefore, *interest* is normally excluded from the cash flow. *Plant overhead* are the costs involved for services and facilities to make the complete plant function as an efficient unit. These costs are closely related to the costs for all labor directly connected with the production operation and amount to about 22% of the total direct labor costs (operating, maintenance and supervision).

In addition to manufacturing costs, other *general costs* are associated with management of a plant. Main categories to be included are administrative, marketing, sales, research and development. *Administrative costs* are those expenses connected to executive and administrative activities. For a preliminary estimate they may be approximated by 2% of the sales. *Marketing and sales* costs vary widely for different types of manufacturing plants. For larger volume products 3% of the product sales is considered a reasonable estimate. When the product is only transferred 'over the fence', 1% of sales is a suitable approximation.

Finally, the process and product need to be continuously improved. Therefore, the cost of *research and development* should be added to the production cost. In the chemical industry these costs range from 1 to 10 percent of the total sales depending on the business. Table 10.2 summarizes the indirect and general cost components.

	Bill of Materials			Annual production	xxxxx ton/year	
	Raw materials	Consumption (unit/kg product)	Price (€/unit raw material)	Cost (€/kg product)	Annual consumption (unit/year)	Anual costs (€/year)
	Reactant A					
	Reactant B					
	Catalyst					
	Solvent					
	Carbon					
	...					
BOM (Bill of Materials)	**Total raw materials**					
	Utilities					
	Steam (ton)		25			
	Electricity (kWh)		0,07			
	Natural gas (GJ)		1,9			
	Process water (m^3)		0,35			
	Cooling water (m^3)		0,05			
	Waste water (m^3)		1			
	Landfill (ton)		20			
	Other waste (ton)		300			
	...					
	Total utilities					
	Total variable costs					
+	**Operational cost**					
	Operating cost					
	Labor					
	Supervision					
	Supplies					
Direct cost	Maintenance cost					
	Supplies					
	Labor, supervision					
	Laboratory charges					
	Royalties					
	Total operational cost					
+	**Indirect cost**					
	Plant overhead					
Indirect cost	Depreciation					
	Other					
	Total indirect cost					
	Manufacturing cost					
+	**General cost**					
	Administration					
General cost	Marketing & sales					
	R&D					
	Total general cost					
	Total production cost					

Fig. 10.4: Cost sheet outline.

10.3.3 Cost sheet

For economic project evaluation, accuracy is crucial in estimating total production cost. The most important factor is to make sure all costs associated with making and selling the product are included. These are included in the cost sheet depicted in Figure 10.4. In this cost sheet the raw materials and utilities comprise of the so-called *bill of materials* (BOM), indicating exactly what is needed as input to produce one kilogram or metric ton of product and therefore a main means of communication between the technical and financial people within a company.

ISBL	Major equipment, spare parts, surplus equipment freight charges, taxes, insurance, duties, startup allowance	**Delivered equipment**	**Direct costs**
ISBL	Placing equipment, paint, insulation, foundations, structural steel	**Equipment installation**	**Direct costs**
ISBL	Instrumentation purchase, calibration, installation	**Instrumentation**	**Direct costs**
ISBL	Process piping, pipe hangers, fittings, valves, insulation	**Piping**	**Direct costs**
ISBL	Electrical equipment, materials, installation	**Electrical**	**Direct costs**
OSBL	Process buildings, maintenance shops, warehouses, buildings for services, garages, steel structures, laboratories, medical, cafeteria	**Buildings**	**Direct costs**
OSBL	Site development, site clearing, grading, roads, walkways, railroads, fences, parking, landscaping	**Yard improvements**	**Direct costs**
OSBL	Utilities, waste treatment, receiving, shipping, packaging, storage, lighting, communications	**Service facilities**	**Direct costs**
OSBL	Property cost, surveys, fees	**Land**	**Direct costs**
	Administration, process design, general engineering, cost engineering, drafting, purchasing, expediting, inspection, supervision, reproduction, communications, travel	**Engineering**	**Indirect cost**
	Temporary facilities, offices, roads, parking, railroads, electrical, piping, fencing, equipment, supervision, accounting, purchasing, timekeeping, expediting	**Construction**	**Indirect cost**
		Legal expenses Contingency Contractor's fee	**Indirect cost**

Fig. 10.5: Components of the total depreciable capitalTDC for a chemical manufacturing plant. (Adapted from [2]).

10.4 Estimation of capital costs

10.4.1 Capital cost components

The total depreciable captital (TDC) involves the total cost for the design, construction and installation of a new plant or the revamp of an existing plant. It also includes the associated modifications required for the preparation of the plant site. A new manufacturing plant may be an addition to an existing site or a *grassroots* plant without nearby auxiliary facilities. It is customary to separate the directly associated processing equipment from the auxiliary facilities by an imaginary border named *battery limits*. Utilizing this division, the total capital investment (TCI) is made up of direct costs:

(1) *Inside battery limits (ISBL)* investment for the plant itself.

(2) *Outside battery limits (OSBL)* investment auxiliary facility/site modifications and improvements.

The TCI is also made up of indirect costs:

(3) Engineering and construction.

(4) Contingency charges.

As summarized in Figure 10.5, the *ISBL investment* includes the purchasing and installation cost of all the equipment items that make up the new plant. These *direct plant costs (DPC)* consist of:
- *Purchased equipment costs (PEC)* for all major process equipment including field fabrication and testing if necessary, such as:
  - vessels, reactors, columns
  - furnaces, heat exchangers, coolers
  - pumps, compressors, motors, fans, turbines
  - filters, centrifuges, driers
- *Installation costs* to integrate all equipment items into an operating plant, such as:
  - foundations, structures, roads, piling, process buildings, sewers
  - piping, valves, insulation, paint
  - wiring, instrumentation
  - solvents, catalysts
  - Installation labor and supervision, including construction costs such as construction equipment rental, temporary construction (rigging, trailers, etc.), temporary water and power, construction workshops

It is extremely important to define the ISBL scope of a project carefully in the early stages of a project, as other project costs are often estimated from the ISBL cost. With a poorly defined ISBL scope, the overall project economics can be badly miscalculated. After discussing the other capital cost components the following sections of this chapter provide several methods, from very global to highly detailed, for estimating ISBL costs.

OSBL investment includes the costs of all additions, improvements and modifications that must be made to the site infrastructure to accommodate adding a new plant or revamp an existing plant. Figure 10.6 illustrates typical offsite auxiliary facilities that are associated to the construction of a grassroots plant. Depending on the extent of needed/available offsite facilities, they can be a significant contribution to the TDC. Typical offsite investments include:
- Power generation plants, turbine engines, standby generators, electric main substations, transformers, switchgear and powerlines.
- Boilers, steam lines, condensate lines, boiler feed water treatment plant and supply pumps.
- Cooling towers, circulation pumps, cooling water lines and cooling water treatment.
- Water supply pipes, demineralization plant, waste water treatment, site drainage and sewers.
- Air separation plants (to provide site nitrogen for inert gas) and nitrogen lines.
- Dryers and blowers for instrument air and instrument air lines.
- Pipe bridges, feed and product pipelines, tanker farms, loading facilities, silos, conveyors, docks, warehouses, railroads and lift trucks.

- Laboratories, offices, canteens, changing rooms, central control rooms and maintenance workshops.
- Emergency services such as firefighting equipment, fire hydrants and medical facilities.
- Site security, fencing, gatehouses and landscaping.

OSBL investment is typically estimated as a proportion of the ISBL investment in the early stages of design. Offsite costs are usually in the range of 10% to 100% of ISBL costs, depending on the project scope and its impact on site infrastructure. For typical chemicals projects, offsite costs are usually between 20% and 50% of ISBL cost, and 40% is usually used as an initial estimate if no details of the site are known. For an established site with well developed infrastructure ('brownfield' sites), offsite costs will generally be lower. On the other hand, if the site infrastructure is in need of repair or upgrading to meet new regulations, or if the plant is built on a completely new site (a 'greenfield' site), then offsite costs will be higher. Table 10.3 provides some guidelines to make approximate estimates of OSBL investment as a function of process complexity and site conditions.

The *engineering and construction costs* comprise the costs of detailed design of the plant and other engineering services required to carry out the project:

- Detailed design engineering of process equipment, piping systems, control systems and off sites, plant layout, drafting, cost engineering, scale models, and civil engineering.

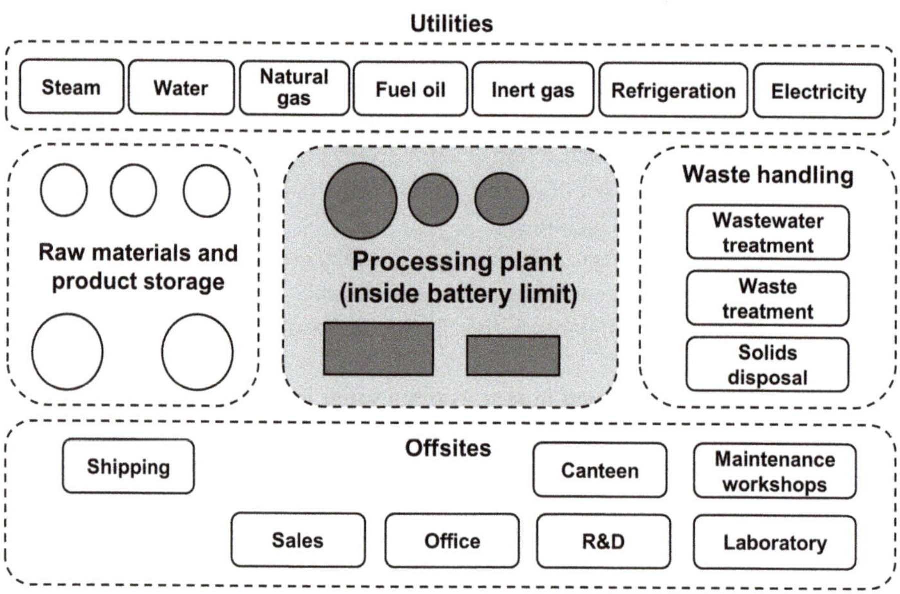

Fig. 10.6: Layout including auxiliary facilities (OSBL) of a chemical manufacturing plant.

Tab. 10.3: Approximate OSBL investment as percentage of ISBL investment [1].

Process complexity	Existing site	New site
Typical bulk chemical	30%	40%
Specialty chemical	30%	50%
Solids handling process	40%	100%

- Procurement of main plant items and bulks.
- Construction supervision and services.
- Administrative charges, including engineering supervision, project management, expediting, inspection, travel and living expenses and home office overheads.
- Contractor's profit.

In most cases, operating companies bring in one or more of the major engineering contracting firms to carry out all these activities. Initially engineering and construction costs can be estimated as 30% of the total plant costs (ISBL plus OSBL) for smaller projects and 10% of the total plant costs for larger projects.

As all cost estimates are uncertain, *contingency charges* are added into the project budget to allow for variation from the cost estimate. In addition, contingency costs also help to cover minor project scope changes, price changes (e.g., steel, copper, catalyst, etc.), currency fluctuations and other unexpected problems. A contingency charge can be thought of as an additional fee charged by the contractor to reduce the likelihood of losing money on a fixed price bid. A minimum contingency charge of 10% of ISBL plus OSBL cost should be used on all projects. For uncertain technologies higher contingency charges (up to 50%) can be used.

As illustrated by Figure 10.7, the DPC, OSBL investment, engineering and construction and contingency make up the TDC. The cost of land, typically about 1–3% of the

Abbreviation	Definition
PEC	Purchased equipment cost + delivery/installation/instrumentation/piping/electrical (ISBL)➔
C_{DPC}	Direct plant cost + buildings/site development/service facilities (OSBL) ➔
C_{TPC}	Total plant cost + engineering and construction➔
C_{DPI}	Direct permanent investment + contractor's fee, legal, contingency➔
C_{TDC}	Total depreciable capital + cost of land/plant startup➔
C_{TPI}	Total permanent investment + working capital (inventory, stock, reserves,...)➔
C_{TCI}	Total capital investment

Fig. 10.7: Connection between capital investment components.

TDC, is not included because its value usually does not decrease with time. Additionally, capital for plant startup/modification expenditures (materials, equipment, maintenance and repairs) is commonly required before the constructed plant can operate at maximum design conditions. It should be part of any capital estimate because it is essential to the success of the venture. Normally an allowance of 8–10% of the TDC is satisfactory, which is commonly included as a one time only expenditure during the first year of plant operation and therefore also not included. Adding the cost of land and plant startup yields the TPI. As already discussed before, the owner needs to invest some capital in maintaining plant operations, in addition to the fixed capital investment that was used to design and construct the plant. The capital that is tied up in maintaining inventories of feeds, products, and spare parts, together with cash on hand and the difference between money owed by customers (accounts receivable) and money owed to suppliers (accounts payable), is termed the *WC* of the plant. By adding the WC, the final TCI of a project is obtained.

10.4.2 Evolution and purpose of capital cost estimates

The accuracy of an estimate depends on the available amount of design detail, the cost data accuracy, and the time spent on estimate preparation. As illustrated by Figure 10.8, the available information only justifies approximate estimates in the early project stages and the accuracy of the estimate improves as the project progresses.

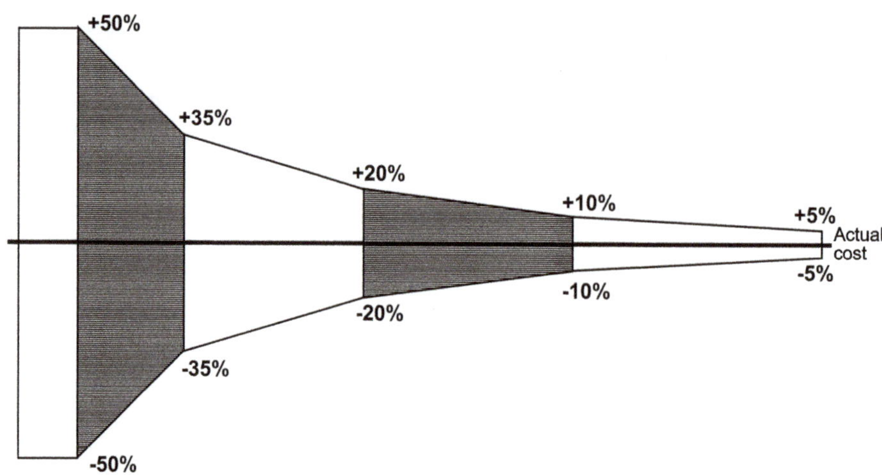

D-Phase (weeks) order of magnitude	D⁺-Phase (1–3 months) study estimate	C-Phase (3–6 months) preliminary investment estimate	B-Phase (1–2 years) definitive investment estimate	EPC-Phase (1–3 years) detailed contractor's investment estimate

Fig. 10.8: Evolution of investment estimates.

Commonly, according to their accuracy and purpose, the following capital cost estimate classification is used:

- *Order of magnitude estimates*, ±50%, usually applied for initial feasibility/screening studies, based on the costs of similar processes or functional units and requiring minimal to no design information.
- *Study estimates*, ±35%, used to make coarse choices between design alternatives based on limited cost data and design detail.
- *Preliminary estimates*, ±20%, used to make detailed choices between design alternatives and first optimizations based on more detailed design and more accurate cost data, including vendor quotes for critical equipment.
- *Definitive estimates*, ±10%, used to authorize funds to bring the design to the point where a more detailed and thus accurate estimate can be made. With experience and cost data from similar projects, acceptable accuracy estimates can be based on the project flowsheet, a rough P&I diagram and the approximate sizes of the major equipment items.
- *Detailed estimates*, accuracy ±5%, which are used for project cost control and estimates for fixed price contracts. This quality estimates require a completed front-end engineering design (FEED), including a (near) complete process design, firm equipment quotes, and a detailed construction cost breakdown and estimation.

To go beyond the preliminary estimate accuracy requires a fairly detailed design of the plant. The cost of estimate preparation thus becomes the cost of process design and sizing the main equipment items. To improve the accuracy further the contractor needs to establish the plot plan and plant layout to make accurate estimates of actual installation costs (piping, wiring, structural steel) that will be needed. The cost of estimate preparation can thus range from 0.1% of the total project cost for ±35% accuracy up to approximately 5% for a detailed ±5% estimate.

10.4.3 Order of magnitude estimates

During early design stages or preliminary marketing studies it is often desired to make a quick (order of magnitude) capital cost estimate without having to complete a plant design. Several short cut methods using existing plant data or step counting have been developed to enable total plant investment estimates within ±50% accuracy for preliminary studies. These methods are also suitable to perform a rough check on more detailed estimates based on process equipment costs later in the process design.

10.4.3.1 Existing plant data

The quickest way to make an order of magnitude estimate of plant costs is to scale it from the known cost of an earlier plant that used the same technology or from pub-

Tab. 10.4: Plant capacity and capital investment for some chemicals (year=2000) [4].

Product	Process	Reference capacity (10^3 tons/yr)	C_{REF} (10^6 \$)	n
Acetic acid	Methanol + CO	10	8	0.68
Acetone	Propylene	100	33	0.45
Ammonia	Natural gas	100	29	0.53
Ethanol	Ethylene hydration			
Ethylene	Refinery gas	50	16	0.83
Ethylene oxide	Ethylene oxidation	50	59	0.78
Methanol	Natural gas	60	15	0.60
Nitric acid	Ammonia	100	8	0.60
Sulfuric acid	Contact, sulfur	100	4	0.65
Urea	Ammonia and CO_2	60	10	0.70

lished data. This requires no design information other than the production rate, year the plant was built and plant location. The capital cost of a plant can be related to capacity by the following equation:

$$C_{\mathrm{TDC}} = C_{\mathrm{REF}} \left(\frac{\text{Desired Capacity}}{\text{Reference Capacity}} \right)^n \tag{10.15}$$

Here, C_{TDC} = total depreciable capital of the plant with the desired capacity and C_{REF} = total depreciable capital for the reference capacity. As illustrated by Table 10.4, the capacity exponent n can range from 0.4 for small scale processes (pharmaceuticals and specialty chemicals, for example) up to 0.9 for processes that use a lot of mechanical work or gas compression (solids or gas processing). As the exponent n is always less than 1.0, constructing larger plants tends to cost less per unit production capacity, which is known as *economy of scale*. Averaged across the whole chemical industry, n is about 0.6, and hence Equation 10.15 is commonly referred to as the *'six-tenths rule'* and can be used to get a rough estimate of the TDC.

As all cost estimating methods use historical data, and material/labor costs are subject to inflation, the old cost data have to be updated for estimating at the design stage, and to forecast the future construction cost of the plant. This is commonly done by using published composite cost indices published for various industries in the trade journals to relate present to past costs. These are weighted average indices combining data for equipment categories, construction labor, buildings plus engineering and supervision in proportions considered typical for the particular industry. For the chemical process industry, the *Chemical Engineering Plant Cost Index (CEPCI)* is the most widely used composite index. It is based on price developments in the United States process industry, and published monthly in the journal, Chemical Engineering. The estimated TDC can now be updated to the desired year by multiplying with the CEPCI index ratio between the desired and reference year:

$$C_{\mathrm{TDC}} = C_{\mathrm{REF}}\left(\frac{\text{Desired capacity}}{\text{Reference capacity}}\right)^{n}\left(\frac{\text{CEPCI}}{\text{CEPCI}_{\mathrm{REF}}}\right) \quad (10.16)$$

Figure 10.9 depicts the development of the annual CEPCI indices since 1957–1959, when the plant costs were relatively stable and the index defined as 100.

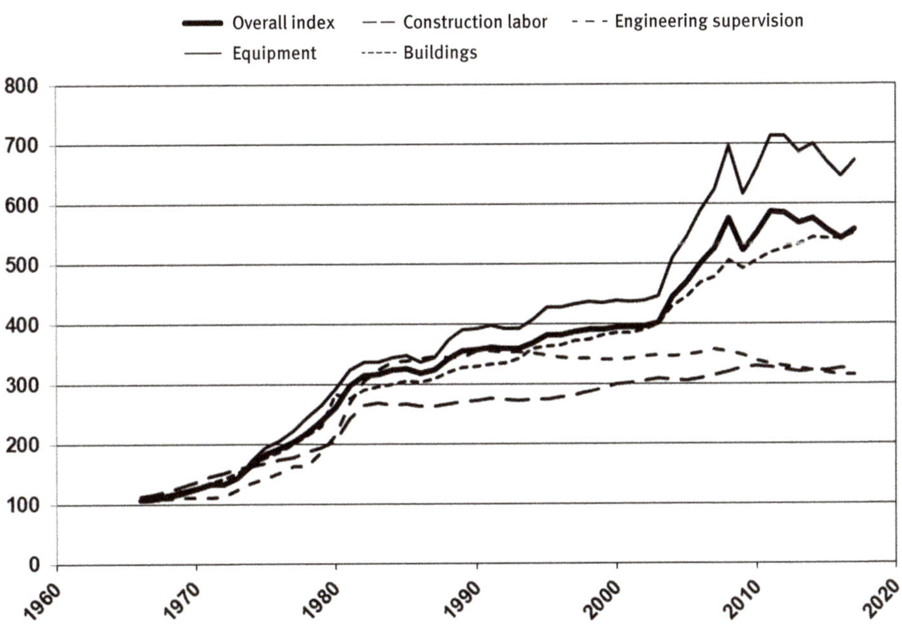

Fig. 10.9: Chemical Engineering Plant Cost IndexCEPCI indices between 1957–1959.

Most plant and equipment cost data are available on a U.S. Gulf Coast (USGC) or North West Europe (NWE) basis, as these have historically been the main centers of the chemical industry. IIowever, for other locations the cost of building a plant will depend on local costs for labor, fabrication, construction, shipping, transporting and importing (taxes) as well as currency exchange rates. In early stage cost estimates these differences are typically captured by using a *location factor* *(LF)* relative to the USGC or NWE basis to correct the TDC:

$$C_{\mathrm{TDC}}(\text{location}) = \mathrm{LF} \times C_{\mathrm{TDC}} = \mathrm{LF} \times C_{\mathrm{REF}}\left(\frac{\text{Desired Capacity}}{\text{Reference Capacity}}\right)^{n}\left(\frac{\text{CEPCI}}{\text{CEPCI}_{\mathrm{REF}}}\right) \quad (10.17)$$

LF for international locations are a strong function of currency exchange rates, and hence fluctuate in time. Table 10.5 provides example LF in U.S. dollars based on 2003 data.

Tab. 10.5: Location factors (LF) in U.S. dollars based on 2003 data [1].

Country	Region	LF
U.S.	Gulf Coast	1.00
Canada		1.00
Western Europe	France	1.13
	Germany	1.11
	Netherlands	1.19
	United Kingdom	1.02
Mexico		1.03
Brazil		1.14
China	imported	1.12
	indigenous	0.61
S.E. Asia		1.12
Japan		1.26
India		1.02
Middle East		1.07
Russia		1.53
Australia		1.21

10.4.3.2 Step counting methods

In case no process cost data are available, an order of magnitude estimate can be made by adding the contributions of the different plant sections, which are called *functional units*. This approach has been developed by *Bridgewater*, correlating plant costs against the number of processing steps for liquids and solids processing plants [1.7]:

$$Q \geq 60{,}000: \quad C_{\text{DPC}} = 4320\,N\left(\frac{Q}{s}\right)^{0.675} \tag{10.18}$$

$$Q < 60{,}000: \quad C_{\text{DPC}} = 380{,}000\,N\left(\frac{Q}{s}\right)^{0.30} \tag{10.19}$$

C_{DPC} represents the direct (ISBL) plant costs in U.S. dollars (U.S.Gulf Coast, Jan.2010 basis, CEPCI = 533), Q = plant capacity (metric tons/year), s = reactor 'conversion' (kg desired reactor product/kg total reactor input) and N = number of functional units. A functional unit is defined as:

- A significant process step, such as reaction section, separation section, or major unit operation such as main stream compressors.
- A multistream operation that is taken as one functional unit.
- Individual heat exchange and pumping are normally ignored as they are part of a functional unit unless substantial cost are involved (i.e., main stream compressors, refrigeration systems or process furnaces).
- Standard 'in process' and raw material storage are ignored because they tend to be a constant process function, unless mechanical handling for solids are involved.

Tab. 10.6: Hill method factors [1].

F_{PI}	**Process type**
1.85	Solids handling
2.00	Solids-fluids handling
2.15	Fluids handling
F_C	Construction factor
0.15	Outdoor construction
0.40	Mixed indoor and outdoor construction
0.80	Indoor construction
F_{OSBL}	OSBL factor
0.10	Minor addition existing facilities
0.30	Major addition existing facilities
0.80	Grass roots plant

- Large raw material, intermediate or product storages are normally treated separately from 'the process' in the estimate.
- Simple 'mechanical' separation without moving parts is also ignored (cyclone, gravity settler) as the cost is usually relatively insignificant.

The direct permanent investment, C_{DPI}, is now obtained by adding the OSBL factor, F_{OSBL}, as given in Table 10.6, to the Bridgewater direct plant cost that already includes engineering and construction for the ISBL part of the plant:

$$C_{DPI} = (1 + F_{OSBL}) \times C_{DPC} \tag{10.20}$$

Finally the total depreciable capital, C_{TDC}, total permanent investment, C_{TPI}, and the total capital investment, C_{TCI}, (see Figure 10.7) are obtained by including a contingency of 40%, additional 10% for costs of land, royalties and plant startup and taking 15% of the TPI as WC:

$$C_{TDC} = 1.40 \times C_{DPI} \tag{10.21}$$

$$C_{TPI} = 1.10 \times C_{TDC} \tag{10.22}$$

$$C_{TCI} = 1.15 \times C_{TPI} \tag{10.23}$$

An alternative as presented by Seider, Seader and Lewin [3], based on the method of Hill, which is based on the year 2006 (CEPCI = 500), a base production rate of 4536 metric tons/year, carbon steel construction and a design pressure less than seven bar located at the U.S. Gulf Coast. As a first step, the production rate factor, F_{PR}, is calculated using the six-tenths rule:

$$F_{PR} = \left(\frac{\text{main product flow rate (ton/year)}}{4536}\right)^{0.6} \tag{10.24}$$

Tab. 10.7: Material factors (FM) [1].

Material	F_M
Carbon steel	1.0
Aluminum	1.07
Cast steel	1.1
Stainless steel 304/316	1.3
Stainless steel 321	1.5
Hastelloy C	1.55
Monel	1.65
Inconel and nickel	1.7
Titanium clad	2.0

The F_{PR} is used to calculate the module cost, MC, for purchasing, delivering and installing each piece of equipment needed to perform/operate a significant process step, similar to the previously described functional units.

$$\mathrm{MC}_i = F_{PR}\, F_M \left(\frac{P_{\mathrm{design}}(\text{bar, if} > 7\,\text{bar})}{7} \right)^{0.25} \times \$160{,}000 \tag{10.25}$$

Here, F_M is a material factor accounting for the differences in construction materials as given in Table 10.7 and \$160,000 is the reference price for a base functional unit with reference capacity of 4536 metric tons/year, carbon steel construction and design pressure <7 bar located at the U.S. Gulf Coast in the year 2006.

The individual MCs are summed, updated to the current CEPCI cost index, and multiplied by the process type factor, F_{PI}, given in Table 10.6 to account for piping, instrumentation, controls and indirect costs to obtain the DPC, C_{DPC}:

$$C_{DPC} = F_{PI} \left(\frac{\mathrm{CEPCI}}{500} \right) \sum_{i=1}^{N} \mathrm{MC}_i \tag{10.26}$$

By multiplying the DPC with the engineering and construction factor F_C and OSBL factor F_{OSBL}, both given in Table 10.6, the direct permanent investment, C_{DPI}, is obtained:

$$C_{DPI} = (1 + F_C + F_{OSBL}) \times C_{DPC} \tag{10.27}$$

Finally the total capital investment is obtained from equation (10.23), through equations (10.21) and (10.22).

10.4.4 Study and preliminary estimates (factorial methods)

The more accurate study (±35%) and preliminary (±20%) estimates are based on estimating the purchase cost for all the major equipment items and adding the other costs as factor(s) of these purchased equipment costs. Clearly the accuracy of this approach will highly depend on the reached stage of design when the estimate is made,

Tab. 10.8: Examples of typical size exponents for equipment cost [4].

Equipment	n
Crystallizer	0.37
Compressor	0.69
Evaporator	0.54
Heat exchanger	0.44
Reactor	0.55
Tower	0.62
Tank	0.57

and on the reliability of the used equipment costs data. When detailed equipment specifications are available and firm vendor quotes have been obtained, more accurate preliminary estimates can be obtained, but in general this approach leads to the study estimate accuracy of ±35%.

10.4.4.1 Estimation of purchased equipment cost

The major effort in estimating the depreciable capital cost is estimating the cost of equipment. There are three main sources of equipment cost data: current vendor quotations, past vendor quotations and literature estimates. Vendor quotations are the most accurate, but the effort required to prepare detailed specifications and quotations are not usually warranted in the early stages of a project. Thus, for quick estimates, one has to rely on literature estimates and past quotations. Equipment costs reported in the literature are either free on board (FOB), delivered or installed cost, all usually given at some time in the past. Shipping and delivery cost will typically amount about 5–15% of the purchased cost. Before adding equipment costs they must all be on the same basis; either FOB, delivered or installed.

Usually, the literature contains equipment costs for other capacities than required. To scale the purchased equipment cost to the required capacity, we usually assume that its cost varies to some power of its capacity:

$$\mathrm{PEC}_{\mathrm{Base}} = \mathrm{PEC}_{\mathrm{REF}} \left(\frac{\text{Desired capacity}}{\text{Reference capacity}} \right)^{n} \tag{10.28}$$

In this equation, n is the *size-exponent* that can vary from less than 0.3 to larger than 1, depending on the type of equipment. Some typical values are given in Table 10.8.

Often, the cost literature contains equipment cost at base conditions, being at low temperature and pressure, carbon steel construction and a specific design. To update for the actual equipment cost at other conditions, the base is multiplied by correction factors:

$$\mathrm{PEC} = \mathrm{PEC}_{\mathrm{Base}}\, F_T\, F_{\mathrm{P}}\, F_{\mathrm{M}} \tag{10.29}$$

The material factor, F_{M}, has already been given in Table 10.7. Table 10.9 contains values for the temperature, F_{T}, and pressure, F_{P}, correction factors. These factors depend on

Tab. 10.9: Pressure and temperature factors (FT) [2].

Design pressure (atm)	F_P	Design temperature (°C)	F_T
0.005	1.3	-80	1.3
0.014	1.2	0	1.0
0.048	1.1	100	1.05
0.54 to 6.8	1.0	600	1.1
48	1.1	5,000	1.2
204	1.2	10,000	1.4
408	1.3		

the type of equipment. Therefore, using the same correction factors for all equipment is an approximation. For shell and tube heat exchangers these factors should be used with care as the shell material may be different than the tube material. Also, it is good practice to place the high pressure on the tube side to reduce material cost.

Finally the PEC need to be updated to the desired year by multiplying with the CEPCI index ratio between the desired and reference year:

$$\mathrm{PEC}_i^{\mathrm{year}} = \mathrm{PEC}_{\mathrm{REF}}\, F_{\mathrm{T}}\, F_{\mathrm{P}}\, F_{\mathrm{M}} \left(\frac{\text{Desired Capacity}}{\text{Reference Capacity}}\right)^n \left(\frac{\mathrm{CEPCI}}{\mathrm{CEPCI}_{\mathrm{REF}}}\right) \tag{10.30}$$

Equipment costs information can be found in the professional cost engineering literature, engineering textbooks and on various websites. Useful textbooks are Towler and Sinnot [1], Seider, Seader and Lewin [3], Peters, Timmerhaus, and West [4], Woods [6], Ulrich and Vasudevan [9], Cooper, Penney, Fair and Walas [11]. In these textbooks, correlations given for preliminary estimates are typically of the form:

$$\mathrm{PEC} = c_1 + c_2 S^n \tag{10.31}$$

Here, c_1 and c_2 are cost constants, S a size/capacity parameter and n the scale exponent for that type of equipment. A useful web based costing tool is available at www.matche.com. Other sources to obtain first estimates are resale websites such as www.equipnet.com and www.ippe.com, or websites such as www.alibaba.com. Alternatively, when available, cost estimating programs like Aspen Process Economic Analyzer can be used. It is highly recommended to thoroughly crosscheck the estimates obtained from multiple sources to arrive at a reasonable reliability.

10.4.4.2 Overall installation (Lang) factors

To achieve study estimate accuracy (±35%), it is typically sufficient to consider overall installation factors that relate the TDC (without WC) and TCI (including WC) of a plant

Tab. 10.10: Breakdown (percent of purchased equipment cost) of Lang factors [4].

Item	Process type Fluid	Fluid-solid	Solid
Purchase cost major equipment	100	100	100
– Installation	47	39	45
– Instrumentation and control	36	26	18
– Piping	68	31	16
– Electrical	11	10	10
– Buildings and structures	18	29	25
– Yard improvements	10	12	15
– Service facilities	70	55	40
Total direct plant cost	*360*	*302*	*269*
– Engineering and supervision	33	32	33
– Construction expenses	41	34	39
Total and indirect plant cost	*434*	*368*	*341*
– Contractor's fee and legal expenses	26	23	21
– Contingency	44	37	35
TDC	*504*	*428*	*397*
$F_{L(TDC)}$	5.0	4.3	4.0
– WC and land	89	75	70
TCI	*593*	*503*	*467*
$F_{L(TCI)}$	6.0	5.0	4.7

to the total purchased equipment cost by the equations:

$$C_{TDC} = F_{L(TDC)} \sum_i \mathrm{PEC}_i \tag{10.32}$$

$$C_{TCI} = F_{L(TCI)} \sum_i \mathrm{PEC}_i \tag{10.33}$$

In these equations, $\sum \mathrm{PEC}_i$ is the total delivered cost of all the major equipment items and F_L the installation factor, widely known as the *Lang factor*. These factors depend on the type of plant being considered and include process equipment, installation, instrumentation and control, piping, electrical, engineering, etc. The values and breakdown of the Lang factors are given in Table 10.10.

10.4.4.3 Individual factors

Further improvement in accuracy to achieve a preliminary estimate (±20%) can be achieved by using individual installation factors for different types of equipment. This requires the availability of much more detailed design information and is therefore only justified if reliable cost data are available and the design has been taken to the point where all the cost items can be identified and included. In Guthrie's [8] detailed cost estimating method, the costs for installation, piping, and instrumentation are taken separately for each piece of equipment. As an illustration, some of his factors are

Tab. 10.11: Guthrie's bare-module factors [8].

Equipment	Bare module factor
Shell and tube heat exchanger	2.19
Vertical pressure vessel	4.16
Horizontal pressure vessel	3.05
Gas compressor	2.15
Centrifuges	2.03
Crystallizers	2.06
Evaporators	2.45

given in Table 10.11. For more information about these individual factor methods, the reader is referred to Cooper [5], Gerrard [7], Guthrie [8] and Ulrich and Vasudevan [9], as this is beyond the scope of this book.

Nomenclature

BOM	Bill of materials	
CF	Cash flow	valuta/year
CCF	Cumulative cash flow	valuta
CDCF	Cumulative discounted cash flow	valuta
CEPCI	Chemical engineering plant cost index	–
C_{Land}	Cost of land	valuta
C_{DPC}	Direct plant cost	valuta
C_{DPI}	Direct permanent investment	valuta
C_{REF}	Total depreciable capital reference capacity	valuta
C_{TCI}	Total capital investment	valuta
C_{TDC}	Total depreciable capital	valuta
C_{TPC}	Total plant cost	valuta
C_{TPI}	Total permanent investment	valuta
c_1	Cost constant	valuta
c_2	Cost constant	valuta
D	Depreciation	valuta/year
DCF	Discounted cash flow	valuta/year
DPC	Direct plant costs	valuta
EP	Economic potential	valuta/year
FEED	Front-end engineering design	
FC	Annual fixed costs	valuta/year

F_C	Construction factor	–
F_M	Material factor	–
F_L	Lang factor	–
F_{OSBL}	OSBL factor	–
F_P	Pressure factor	–
F_{PI}	Process type factor	–
F_{PR}	Production rate factor	–
F_T	Temperature factor	–
FOB	Free on board	–
GP	Gross profit	valuta/year
IRR	Internal rate of return	–
ISBL	Inside battery limits	–
LF	Location factor	–
MC	Module cost	valuta
n	Year	–
n	Capacity exponent, size exponent	–
N	Number of years	–
N	Number of functional units	–
NP	Net profit	valuta/year
NPV	Net present value	valuta
OI	Operating income	valuta/year
OSBL	Outside battery limits	–
P	Pressure	bar
PBT	Pay back time	years
PEC	Purchases equipment costs	valuta
PFD	Process flow diagram	–
Q	Plant capacity metric	tons/year
r	Discount rate	–
RMC	Annual raw material costs	valuta/year
ROI	Return on investment	–
s	reactor 'conversion'	–
S	Sales	valuta/year
S	Size/capacity parameter	m^3, m^2, ton/hr
t	Taxaction rate	–
TCI	Total capital investment	valuta
TDC	Total deprecial capital	valuta
TPI	Total permanent investment	valuta
VC	Annual variable costs	valuta/year
WC	Working capital	valuta
WACC	Weighted average cos of capital	–
Φ	Fraction of total depreciable capital (TDC)	–

References and further reading

[1] G. Towler, R. Sinnot. Chemical Engineering Design: Principles, Practice and Economics of Plant and Process Design, 2nd edn, Elsevier, Amsterdam, 2013.

[2] H. Silla. Chemical Process Engineering, Design and Economics, Marcel Dekker, New York, 2003.

[3] W.D. Seider, J.D. Seader, D.R. Lewin, S. Widagdo. Product and Process Design Principles, 3rd edn, John Wiley & Sons, New York, 2010.

[4] M.S. Peters, K.D. Timmerhaus, R.E. West. Plant Design and Economics for Chemical Engineers, 5th edn, McGraw-Hill, New York, 2004.

[5] J.R. Cooper. Process Engineering Economics, Marcel Dekker, New York, 2003.

[6] D.R. Woods. Rules of Thumb in Engineering Practice, Wiley-VCH, Weinheim, 2007.

[7] A.M. Gerrard. Guide to Capital Cost Estimation, 4th edn, IChemE, London, 2000.

[8] K.M. Guthrie. Process Plant Estimating, Evaluation and Control, Craftsman, Solano Beach, California, 1974.

[9] G.D. Ulrich, P.T. Vasudevan. Chemical Engineering Process Design and Economics: A Practical Guide, Process Publishing, 2004.

[10] Dutch Association Cost Engineers. Price Booklet, 32nd edn, DACE, Nijkerk, 2017.

[11] J.R. Cooper, W.R. Penney, J.R. Fair, S.M. Walas. Chemical Process Equipment, 3rd edn, Butterworth-Heinemann, Oxford, 2012.

11 Evaluating for safety and health

11.1 Introduction

Safety concerns human safety for events of short durations, such as explosions. Health concern lifelong human health. The latter is concerned with workers exposed for working hours per day for their working life and for people exposed to emissions and radiations their life long. This chapter treats both types of evaluation and often the time of exposure duration is mentioned but the distinction between safety and health is in most cases not explicitly mentioned.

Assessing Health and Safety aspects during the design of an industrial facility or product is something that nowadays seems common practice for most major companies and engineering contractors. However, this has not always been the case. The wisdom of systematically assessing hazards and risks and the associated methodologies that have been developed to serve this purpose only came about after various major accidents had occurred. Some of the major accidents revealed blind spots and gaps in certain aspects of industry safety awareness at the time. The practical lessons learned and regulatory fallout from these accidents resulted in improving safety and risk management in the design and operation of oil, gas and chemical installations and the design of products [1]. Hazards and risk assessments should be performed in a systematic way and at the appropriate time during a design project. The goal of this chapter is to provide readers an introduction to basic safety concepts, and provide them with the tools to get started on performing risk assessments during early design stages. References are provided at the end of this chapter for further reading on this subject.

11.2 Risk reduction during design

Before going into details about the risk assessment and evaluation process, one needs to understand what is meant by 'risk' and 'risk management'.

Risk is a measure of the potential for human injury, environmental damage or economic loss in terms of both the probability of the loss, damage or injury occurring, and the magnitude of the damage, loss or injury if it does occur [2].

$$\text{risk} = f\,((\text{probability of occurance}) \cdot (\text{consequences}))$$

Risk management is the systematic application of management policies, procedures and practices to the tasks of analyzing, assessing, and controlling risk to protect employees, the public and the environment, as well as company assets while avoiding business interruptions [3].

https://doi.org/10.1515/9783110467741-011

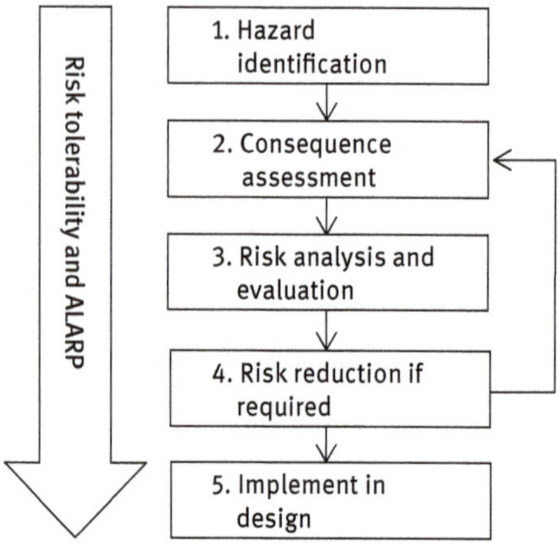

ALARP: risks As Low as Reasonably Practicable

Fig. 11.1: Risk reduction steps during design.

There are various models available that describe how risks are managed during design but the basic principle is that it is an iterative process in which hazards are identified and the risks evaluated until deemed acceptable. The following steps, illustrated in Figure 11.1, are general for risk reduction strategies:
(1) identify the hazards
(2) assess the consequences
(3) analyze and evaluate if the risks are acceptable
(4) decide on (more) mitigating measures based on risk evaluation
(5) implement in design

The principles applied to the risk management of process engineering can also be applicable to product engineering. The difference is more in where the focus of the assessment lies. When assessing the hazards and risks during product engineering, there is a shift of focus towards the end user. The production of toys is such an example. During the risk assessments of the process, the focus is on risk exposure to the workers present on the facility, surrounding population, local environment, etc.

The starting principle of a design should be to approach it from an inherent safety principle. This will be discussed in the next section.

11.3 Inherent safer design principle

The concept of inherent safer design was introduced by Trevor Klotz after the Flixborough disaster and his expression sums up the basic principle: "What you don't have cannot leak or burn" [4]. There are four main methods for achieving inherent safer designs:

(1) Avoidance – take measures in the design to remove hazardous materials, procedures or actions completely. For example, avoiding the use of toxic, flammable or corroding chemicals.
(2) Prevention – take measures to reduce the chance or likelihood of the hazard arising. For example, reducing the number of flanges to lower the number of potential leak sources in a design.
(3) Control – take measures to reduce the consequences (severity) before the hazard is realized. For example, segmenting the process in smaller sections with the help of emergency shutdown valves to limit the amount released during a loss of hydrocarbon containment.
(4) Mitigation – take measures to limit the effect or spread of the incident once the hazard has been realized such as emergency and evacuation arrangements. For example, fire walls, deluge systems and evacuation arrangements.

The principle of inherent safety in design is that first measures should be taken to avoid the hazard altogether and that mitigation measures should only be taken when the other options are not reasonably practical. It is for this reason that identification of hazards and the assessment of mitigating measures should be done early during design stages when possible. The true inherent safer design approach focusses on avoidance and prevention, thus choosing different technologies or materials to avoid hazards all together. This is as opposed to the safety in design approach, where additional safety systems are introduced to control and mitigate hazards.

There are some tools and techniques available that can help compare design options from an inherent safety perspective. An example of such a tool is the Dow indexes [5, 6] which, based on the properties of the chemicals intended for use as part of the design, can rank design options based on their fire, explosion and toxic hazards. Processes can also be compared based on the consequences during incidences. The principle has been adopted in many codes, standards, design engineering practices and regulations and is an important one to keep in mind during all design stages.

Having adopted the safety in design principle, and keeping this in mind during the risk management process, the first step is to start identifying the hazards of the proposed design.

11.4 Identification studies

The first step of systematic evaluation requires that the hazards be identified. But what is meant by a hazard? A hazard, in the context of this chapter, is defined as an inherent chemical or physical characteristic that has the potential to cause damage to people, property or the environment.

This section provides an overview of some common hazard identification methodologies used during engineering projects. It is not intended as a complete overview of all identification techniques or detailed treatise about the methodologies, but more as an introduction into this field. For a detailed description of the methodologies the reader should refer to books such as Lees' Loss Prevention in the Process Industries [7] and the international standard IEC/ISO 31010 Risk management – risk assessment techniques.

Chemical reactivity hazards

The first item to evaluate in a product risk assessment is the physical and chemical properties of the product. Following the inherent safer design principle, one should strive to avoid or minimize the use of chemicals and/or materials that have hazardous properties. Substitution or phasing out of chemicals that are known to be hazardous is now becoming more apart of international legislation. EU Regulation No. 1907/2006, Registration, Evaluation, Authorization and Restriction of Chemicals (REACH), came into force in 2007. The goal of REACH is to improve the protection of human health and the environment by identifying the intrinsic properties of chemical substances and phasing out chemicals of very high concern. When developing a product in the EU or for use in the EU, this regulation is important to be aware of and its requirements should be taken into consideration from the feasibility stage.

But how does one discover what the hazardous properties of a product are? There are various sources of information available that contain data with respect to toxic, fire and explosion hazards of chemicals such as but not limited to:

- material safety datasheets (MSDS)
- information contained in standards such as NFPA 49, 325, 432, 491
- Bretherick's Handbook of Reactive Chemical Hazards [8]
- incident reports

Having obtained information from a literature survey, the next step is to look at the chemical and thermodynamic properties since they can provide information on potential hazards. Properties that can provide information are:

- thermodynamic properties
- chemical composition, structure and bonds

Thermodynamic properties

The heat of reaction gives an indication of the potential temperature rise in a mixture. The higher the heat of reaction, the higher the potential for energetic, uncontrolled reactions. Using the heat of reaction and the heat capacity of the reaction mixture, one can calculate the adiabatic temperature rise, the maximum temperature rise that can theoretically occur without heat loss:

$$\Delta T_{\mathrm{ad}} = \frac{\Delta H_{\mathrm{r}}}{C_{\mathrm{p}}}$$

Where:
ΔT_{ad} = adiabatic temperature rise (K)
ΔH_{r} = heat of reaction (kJ/kg)
C_{p} = heat capacity of reaction mixture (kJ/(kg K))

The sum of the maximum process temperature and adiabatic temperature rise can give important information about potential hazards. When looking at the result, one should wonder if this temperature is below a temperature where:
- additional chemistry (such as decomposition) can occur
- phase transition (such as boiling or gas generation) can occur
- overpressurization from increased vaporization can occur

Another important thermodynamic property is the heat of formation. Compounds having a high endothermic heat of formation (positive heat of formation) contain a lot of energy and tend to be self-reactive. Correspondingly, they have a high exothermic heat of reaction.

Chemical composition, structure and bonds

Chemical structure and bonds provide a second clue. Be suspicious of [9]:
- compounds that have high bond strains
- carbon–carbon double bonds not in benzene rings (e.g., ethylene, styrene)
- carbon–carbon triple bonds (e.g., acetylene)
- nitrogen containing compounds (NO_2 groups, adjacent N atoms, etc.)
- oxygen–oxygen bonds (peroxides, hydroperoxides, ozonizes)
- ring compounds with only three or four atoms (e.g., ethylene oxide)
- metal- and halogen-containing complexes (metal fulminates; halites, halates; etc.)

Some chemicals have the tendency to produce strong oxidation reactions which can be explosive in nature. The oxygen required for this oxidation can be stored in the compound itself or be supplied from outside during the reaction. One estimating parameter for the potential hazard and instability of substances or reaction mixtures containing oxygen is the oxygen balance (OB). The oxygen balance is the amount of oxygen, expressed as weight percent, liberated because of complete conversion of the material to CO_2, H_2O, SO_2, Al_2O_3, N_2, and other relatively simple oxidized molecules [10]. If the oxygen balance (OB) is positive, there is an excess of amount of oxygen available in the compound for a conversion to simple components. If the OB is zero, there is just enough oxygen available for the conversion to simple components. If the OB is negative, there is insufficient oxygen for a complete oxidation of the compound.

The following decomposition reaction forms the basis of the OB:

$$C_xH_yO_zN_n \rightarrow xCO_2 + \frac{y}{2}H_2O + \frac{n}{2}N_2 + \left(z - 2x - \frac{y}{2}\right)\frac{1}{2}O_2$$

In case of perfect oxidation, no excess oxygen is left after the reaction and the carbon and hydrogen atoms have fully oxidized to carbon dioxide and water. The heat of reaction tends to reach a maximum in this case. The parameter $z - 2x - (y/2)$ is then equal to zero under such ideal conditions. The oxygen balance is defined as:

$$OB = \frac{1600\left[z - 2x - \frac{y}{2}\right]}{MW}$$

MW is the molecular weight of the compound. The closer the OB is to zero, the higher the hazards since the explosion sensitivity and severity will tend to reach a maximum at an OB of zero. The OB value can thus be used to assess the degree of hazard [10] as follows:

Hazard potential	Value of OB
Low	$-160 < OB < -240$
Medium	$-240 < OB < -120$
	$80 < OB < 160$
High	$-120 < OB < 80$

It should be stressed that the equation above is not applicable to molecules containing other atoms and that the OB only serves as an indicator. The way oxygen is bonded should be reviewed in relation to the OB. For an explosive such as nitroglycerine, the oxygen balance is +3.5, which corresponds well to the high hazard potential of the table above. Acetic acid on the other hand has an OB of 106.7 but is not an explosion hazard since the oxygen present cannot be used efficiently for the oxidation of carbon.

When chemical and/or thermodynamic properties are not known, then the next step in the hazards assessment is to perform tests. Guidance for performing such tests is provided in [10, 11] and [12].

HAZard IDentification (HAZID) and ENVironmental Identification (ENVID)

Plot plans preliminary layouts are developed early during the design, before process and instrumentation diagrams have fully matured. A relatively simple method to identify hazards during this stage is by using qualitative techniques that use key words to think of causes, consequences and safeguards in a design. One of the more common techniques that can be used during the feasibility stage up to detailed stage design is the HAZID study. The HAZID is a study technique for early identification of potential hazards and threats.

The major benefit of a HAZID is that the early identification and assessment of the critical HSE hazards provides essential input to project development decisions. This will lead to safer and more cost-effective design options being adopted with a minimum cost of change penalty. The deliverable of the HAZID study is a set of detailed minutes, showing the summary of the discussion with action recommendations directed to the separate involved parties who were present in the meeting.

The HAZID study has been developed specifically to reflect the importance of HSE issues on the fundamental decisions that are made at the inception of all development projects. The HAZID study is usually the first opportunity to assemble experienced operational, engineering and HSE staff together to address, in a short time frame, the issues surrounding the project.

During a HAZID study review session, a guideword list is used and the session itself facilitated by a chairman and supported by a scribe when required. An example guideword list and part of a HAZID table are given in Table 11.1 and 11.2 for a HAZID performed for a conceptual design of a new offshore facility. The guidewords should be adapted to reflect the specific circumstances or features of the design if possible.

Tab. 11.1: Example guidewords for a HAZID for an offshore installation.

HAZID category	Risk element, system or event
Process event	Unignited gas release Jet fire Flash fire Explosion Blowout Unignited liquid release Pool fire Chemical/toxic hazard Nonprocess fire Vent/exhaust/air intake Hazardous area classification
External event	Ship collision Dropped load Helicopter impact Environmental hazards
Safety system	Escape routes Muster station Lifeboat Life raft Active firefighting system Firewalls/blast wall/shielding/PFP Fire and gas detection Area alarm systems ESD/EBD HVAC

Tab. 11.1: (continued)

HAZID category	Risk element, system or event
Working environment, operations & maintenance	Access to maintainable equipment Limited workspace Interaction with rig LSA/Radiation Noise Handling of chemicals/explosives Mechanical handling Control room operations Well intervention operations Helicopter operations Crane operations Supply vessel operations
Other, e.g.,	Drawing comments Environmental aspects (air, soil, landscape, noise, energy, radiation, water, waste, light)

Tab. 11.2: Example table for a HAZID.

Risk element/ system or event	Cause/ guideword	Conse- quences	Evaluation/ safeguards	No.	Action required	Action party	Due date
Process event							
Unignited HC release							
Jet fire							
Etc.							

For each guideword, expanders are provided to help the brainstorm session on identifying possible hazards. The HAZID team brainstorms to find all potential hazards and its causes.

Each hazard is examined to see what:
- consequences may arise from the hazard
- controls are in place
- what action recommendations, if the team decides what controls are not sufficient or can be improved, action recommendations are formulated and assigned to the responsible party.

An ENVID uses the same principle as the HAZID, except that the purpose is to identify environmental hazards and thus the keywords are different. Below is an example list of keywords that can be used during an ENVID:

ENVID category	Guideword
Impact – air quality	Greenhouse gas emissions
	VOCs and fugitive emissions
	NO_x and SO
	Ozone depleting substances
	Particulate matter
	Toxic metals
	Odorous compounds
Impact – waste	Waste production
	Hazardous waste
	Domestic waste
Impact – resource use	Land use
	Land quality
	Loss of habitat
	Degradation of land quality
	Alteration of ecological/landscape processes
Impact – oil and chemical spills	Chemical use and storage
	Hydrocarbon use and storage
	Bunkering and equipment failure
	Human error and procedures
	Pipework, valves and seals
	Bunding and drainage
Impact – chemical use and discharge	Chemical use and storage
	Hydrocarbon use and storage
	Bunkering and equipment failure
	Human error and procedures
	Pipework, valves and seals
	Bunding and drainage
Impact – aqueous discharges	Drains
	Hydrocarbon and chemical contamination
	Sewage discharges
	Storm water
	Hydro test water
	Slurry
Impact – wildlife disturbance	Noise/vibration and light pollution
	Air pollution
	Water pollution
	Conservation of protected areas
	Conservation of protected species
	Ecosystem effects/biodiversity
	Displacement/loss of access

It is important that the actions are monitored after the session. A common pitfall is that action closeout is delayed due to demanding engineering schedules and not having assigned a person with the proper authority to follow up and expedite the actions with engineers. Agreeing at the start of the project on whom this responsibility will fall will help prevent costly rework during later stages of the design.

The HAZID & ENVID methodology is a tool that can be applied to process and product engineering assessments and is easy to understand.

HAZOP

H.G. Lawley of Imperial Chemical Industries (ICI) developed the HAZOP methodology in the seventies, to identify causes and consequences of disturbances in process installations. HAZOP stands for Hazard and operability analysis and is a structured hazard identification tool using a multidisciplined team. The methodology differs from the HAZID in the sense that a HAZOP focusses on hazards that originate from the process itself (inside out approach), while a HAZID assesses a broader range of potential hazards from within but also from outside the process (outside in approach). The HAZOP is accepted as the main technique for the identification of process hazards in the design and operation of a facility. A detailed description of the HAZOP methodology is available in the IEC code 61882 [13].

The HAZOP process is the systematic application of combinations of parameters (e.g., flow, pressure, temperature) and guide words (e.g., no, more, less) to produce deviations (no flow, less pressure) from the design intent or intended operational mode of the installation. The purpose of the HAZOP study is to:

- Review the design and consider whether any of the conditions which may occur, from operation, malfunction or maloperation, cause a hazard to people or cause damage to plant and equipment.
- Review whether the precautions incorporated in the design are sufficient to either prevent the hazard occurring or reduce any consequence to an acceptable level.
- Consider whether the plant remains safely operable, goes to a safe hold state, partially shuts down, or shuts down under the circumstances in the above given examples.

The deliverable of the HAZOP study is a set of detailed minutes, presented on worksheets for each deviation. These show the summary of the discussion with action recommendations directed to the separate involved parties which were present in the meeting.

A chairman will guide the meeting through the HAZOP and can be assisted by a scribe if required. Possible deviations/deficiencies from design will be addressed and consequences are identified during the session.

The installation being studied is divided in logical items (equipment and piping) that can be studied according the structured brainstorm method of the HAZOP study.

Tab. 11.3: Example parameters and guidewords for a HAZOP.

Parameters		Guidewords	
Flow	Maintenance	No	–
Pressure	Utilities	More/high	
Temperature	Startup	Less/low	
Level	Shutdown	Reverse	
Composition/phase	General	As well as	
Operations		Other	

These items are called nodes and are selected by the HAZOP chairman. HAZOP nodes are usually divided by phase flow (liquid, gas) and isolatable sections (e.g., between emergency shut down valves) or installation parts. The node or installation part is reviewed by the application of:

- parameters (e.g., flow, pressure, temperature, level), in combination with
- guidewords (e.g., no, more, less, reverse, other than)

The guideword/parameter combination is known as a deviation.

Example parameters and guidewords are available in Table 11.3.

The HAZOP team brainstorms to find all potential causes of the deviation. All potential causes are discussed. Each *cause* is *separately* examined to see what:

- consequences may arise with a certain deviation
- safeguards are in place (i.e., design criteria, indications, alarms, procedures, etc.)
- the team decides about whether the safeguards are not sufficient or can be improved, action recommendations are formulated and assigned to the responsible party

An example of a HAZOP minute is provided in Table 11.4.

HAZOPs are used primarily in process risk assessment and are less commonly used in product hazard identifications. The failure mode effect and analysis (FMEA) is a technique that resembles the HAZOP in some aspects and is much more commonly used for product hazards assessments.

Tab. 11.4: Example of a HAZOP minute.

Node	Deviation no.	Guideword	Cause	Consequence	Safeguard	Action no.	Action	Action responsible	Due date
2	50	Elevated temperature	Failure of heater control causing too elevated temperature	Exothermic runaway, over pressurization of reactor	Bursting disc on reactor prevents over pressurization	12	Ensure venting of bursting disc is to safe location	Jane Doe (Process engineer)	1st of December 2017

11.5 FMEA

FMEA is a well-known hazard identification technique. This method reviews failure on a component level and assesses the effect on the system as a whole. The review results in recommendations to improve the product or system design with respect to reliability and safety. Originally developed by the US military after the Second World War, the first formal application of the FMEA was in the aerospace industry in the 1960s for the NASA Apollo missions. It is a technique that is, nowadays, used in the product manufacturing and services industries. Like the other hazard identification tools discussed previously in this Chapter, the FMEA is performed as a team session facilitated by a chairman and sometimes supported by a scribe.

The objectives of a FMEA include [7]:

(1) Identification of each failure mode, the sequence of events associated with it and of its causes and effects.
(2) Classification of each failure mode by relevant characteristics, including detectability, diagonality, testability, item replaceability, compensating and operating provisions.
(3) Formulate recommendation for improvement if required.

One of the first steps during the preparation of a FMEA is to identify the systems boundaries. A FMEA can be performed on a system level, down to individual component level. For example, one could look at the failure of an ESD valve as a whole or break this system down to individual components (e.g., valve body, actuator, solenoid) and review failure of the individual components. The second step is choosing the appropriate failure or error modes. Failure modes for safety systems are provided in sources such as ISO 14224 [14]. During the team session, the failure causes, local and systems effects, detection, etc., will be identified by the team members.

The advantage of the FMEA is that it is a relatively easy method to apply and that it is suitable for hazard assessment on an equipment level. The FMEA approach also has a well deserved reputation for efficiently analyzing the hazards associated with electronic and computer systems, whereas the HAZOP approach may not work as well for these types of systems [15]. The disadvantage of the method is that, depending on the system boundaries and level of review, the study can become a lot of work to perform. The methodology is also less suitable for systems where complex logic is required to describe system failure [7]. Table 11.5 provides an example of an FMEA table.

Tab. 11.5: Example list of FMEA table.

System	Component (sub system)	Failure or error mode	Failure cause	Effect on other components	Effect on entire system	Detection method	Compensating provision	Actions	Action party	Remarks
Production export	Hydraulic actuated ESD Valve XXX	External leakage process medium	Leakage of valve body	Hydrocarbon release, potential for fire and explosions	Loss pf production Inability to fully isolate system during ESD	Large leakage detected by pressure drop PZTXXX Gas detection at ESD deck	Shutdown upstream production facilities to stop flow from ESD valve	None	None	None
		External leakage utility medium	Leakage of valve body	Hydraulic spill on ESD deck, loss of hydraulic pressure, valve closes	Loss of production	Hydraulic pressure detection PZTXXX	Platform will shut down on loss of hydraulic pressure	None	None	None
		Failure to close on demand	Jammed gate valve	None identified	Failure to close export, in case of ESD, possible escalation of fires on platform	Limit switches, timeout alarms	Shutdown upstream and downstream facilities Yearly functional tests	None	None	None

11.6 Dow index methods

The Dow Fire and Explosion Index (Dow F&E) [5] and Chemical Exposure Index (Dow CEI) [6] are internationally used ranking systems to evaluate the hazard category of a process plant and expected losses due to potential fires and explosions or chemical exposure. The Dow F&E uses the NFPA material factor which is a measure of the hazards of the substance which falls between 0 and 40 in a matrix of instability/reactivity and flammability [16]. The material factor is then modified by penalties and credit factors which depend on the operating conditions. If the results of the assessment are an index value higher than 100, the risk is deemed to be unacceptable and risk reducing measures need to be applied. Other outcomes of the assessment are an estimate of the potential area of damage, value of the equipment in the area and maximum probable property damage and business interruption. The Dow F&I is a useful tool during early engineering stages since it helps to assess multiple process options and identifies which process unit is the most hazardous. Modification or additional protection of mitigating measures can then be considered. The Dow CEI assesses the relative acute toxicity risks due to liquid spill, evaporation and dispersion of a vapor cloud.

11.7 Bowtie assessments

The Bowtie method is a risk assessment method that can be used to analyze and communicate how high risk scenarios develop. The essence of the Bowtie method consists of plausible risk scenarios around a certain hazard, and ways in which the organization stops those scenarios from happening. The method takes its name from the shape of the diagram that you create, which looks like a man's bowtie [17]. The methodology in this paragraph is based, for the most part, on the work described in the CGE Bowtie methodology manual [17], in which more detail is provided on creating Bowties and how to facilitate Bowtie sessions.

The Bowtie method has several goals:

- Provide a structure to systematically analyze a hazard.
- Help decide whether the current level of control is sufficient (or, for those who are familiar with the concept, whether risks are as low as reasonably practicable, or ALARP).
- Help identify where and how investing resources would have the greatest impact.
- Increase risk communication and awareness.
- Provide a framework to plot/collate safety related information from audits, incidents, maintenance backlogs, etc.

The advantage of using a Bowtie is that it creates an easy picture to understand and communicate on multiple levels of the organization. A complete Bowtie diagram, linked to the management system, is like a graphical table of contents, like a map, showing everything an organization does to control its major risks. During engineering projects, it can be used in various stages of design and is useful in helping to identify safety critical systems. It contains operational hardware barriers, behavioral barriers and organizational management systems, which makes it a useful tool to holistically look at where investing resources would have the greatest impact.

Bowties are best created during a team brainstorm session using input from various engineers and operations personnel. Constructing a Bowtie can be done by applying the following steps:

(1) Specify the hazard which will be assessed (anything that has the potential to cause harm e.g., stored flammable hydrocarbons, toxic compounds and object at height).
(2) Specify the top event, which is the event that occurs at the point in time when control over the hazard is lost (e.g., loss of containment or object falling).
(3) Identify threats, such as a possible cause that can release the hazard by producing the top event. (e.g., corrosion, overflow or breaking rope).
(4) Assess consequences after the top event has occurred, which are unwanted events caused by the release of the hazard (e.g., fire, explosion, toxic exposure to personnel or fatality from dropped object falling on personnel).
(5) Identify control barriers against the threats which can prevent or reduce the likelihood of the top event occurring (e.g., proper material selection and proper rope certification).
(6) Identify mitigation barriers against the consequences (e.g., fire walls, evacuation procedures and dropped object protection frame).
(7) Identify escalating factors, which are specific conditions or circumstances which could cause a barrier to fail or be less effective.
(8) Identify escalating factor barriers, which are barriers that can help to prevent or manage escalating factors.

A generic overview of an example Bowtie is presented in Figure 11.2, created using the CGE software Bowtie XP.

Unfortunately, barriers are seldom 100% effective. Escalation factors take this into account by stating a condition that reduces the effectiveness of a barrier. The following escalation factor categories are typically considered:

- Human factors: anything a person does to make a barrier less effective.
- Abnormal conditions: anything in the environment that causes a barrier to be put under strain.
- Loss of critical services: if a barrier relies on an outside service, losing that service might cause it to lose effectiveness.

When identifying escalation factors, the guideline is to keep in mind that they should not cause a top event or consequence, need to be credible, that lessons learned from other incidents are used and to focus on the critical barriers. Questions that a team

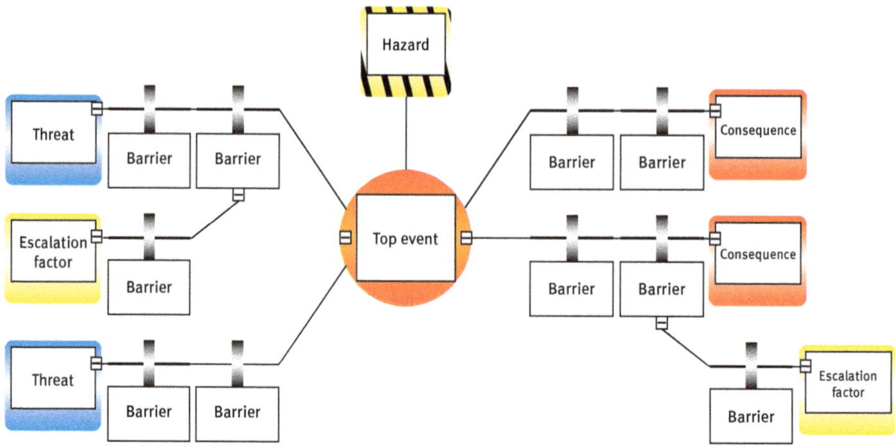

Fig. 11.2: Generic Bowtie.

should ask themselves are: if there are any circumstances under which the barrier will not work, how the barrier can be 'destroyed' and if the barrier has failed in the past.

Once the escalation factors have been established, the barriers for the escalation factors can be identified. Escalation factor barriers can be divided in the categories of human factors (e.g., training, supervision, etc.), abnormal conditions (e.g., maintenance, spares, etc.) and loss of critical services (backup systems, shutdown, etc.).

Bowties can be created for process related hazards but also for product related hazards. The challenge is to identify the proper hazard and top event. Some products are not hazardous by themselves and one should look more towards the intended use of the product to identify consequences.

11.8 Consequences assessment

Having identified the hazards, the next step in the risk management process is to assess the consequences. From a process or product safety perspective, the consequences of toxic exposure, fire and explosions will have to be reviewed. Depending on the engineering stage this can be a high level assessment for a risk screening purpose or a detailed one as input to a full QRA. The models discussed in this chapter provide the reader with relatively simple means to quantify consequences which can be used as input for high level assessments. However, modeling performed as input for quantitative risk assessments or as input to engineering decisions such as safety distances or specifications should consult sources where consequence assessment is discussed in more detail such as Lees' Loss Prevention [7], Cameron and Raman [18], the Yellow Book [19] and Casal [20]. When performing consequence modeling, it is very important to know what the limitations and underlying assumptions in the models are. This is to prevent using models that are not suitable for the case assessed, leading to an underestimation or overestimation of the consequences and therefore the risk.

Fire models

Fire consequences assessments allow the calculation of heat radiation transfer to a target with heat radiation expressed in kW/m^2. Different models are used depending on the type of fire. The following industrial fires are commonly considered during process risk assessments:

- pool fires
- jet fires
- fireballs resulting from boiling expanding liquid vapor explosions (BLEVE)
- flash fires

Simplified models are presented in the following section for pool fires, jet fires and fireballs. Flash fires are discussed only briefly. CFD models such as Fire Dynamic Simulator and KFX are used for more advanced assessments but are not discussed in this chapter.

To provide some feeling of what the effect of different levels of exposures are, some common tolerability levels used in the process industry are provided in Table 11.6.

Tab. 11.6: Heat radiation exposure effects [21].

Heat radiation	Effects
35 kW/m^2	Significant chance of fatality for people exposed instantaneously (can assume 100% lethality).
10 kW/m^2	1% lethality, structural failure of equipment can start to occur at these radiation levels and above.
6 to 9 kW/m^2	Pain within approximately 10 seconds, rapid escape only is possible.
3 to 5 kW/m^2	Will cause pain in 15–20 seconds and injury after 30 seconds exposure, special protected clothing of first responders provide just enough protection at 3 kW/m^2.
1 kW/m^2	Heat felt comparable to exposure to the sun during a hot summer day.

Pool fire

Pool fires result from the leakage and subsequent ignition of liquid substances. In an industrial facility they might be caused by mechanical impact, over pressurization, leaking flanges or valves, small bore fittings, etc. Fire models can consider the flame as a point source, multiple point sources and solid flame or modeled dynamically using computational fluid dynamics. The point source model is discussed in this section.

The point source radiation model models a flame as a single point radiator. The heat flux Q at a distance x from the flame center to the observer can be calculated by [18]:

$$Q = \frac{M \cdot \Delta H_c \cdot f \cdot \tau}{4 \cdot \pi \cdot x^2}$$

Where:

Q = heat flux (kW/m^2)
M = rate of combustion (kg/s)
ΔH_c = heat of combustion (kJ/kg)
f = fraction of thermal energy radiated (–)
x = distance from flame center to observer (m)
τ = atmospheric transmissivity

The above expression assumes that the position of the observer towards the flame is such that the observer is at maximum exposure to the flame radiation.

The radiation fraction will be dependent on the type of fuel, pool diameter, flame temperature, etc. Larger pool diameters will lead to more soot formation in some type of fuels and thereby reduce the fraction of thermal energy radiated. The fraction is a crucial factor and, in practice, difficult to estimate. The following expression has been proposed to calculate the fraction of thermal energy radiated for hydrocarbon pools with a pool diameter D_p (m) [20]:

$$f = 0.35 \cdot e^{-0.05 D_p}$$

In practice, the value of f ranges between 0.15–0.35 for various hydrocarbon pool sizes [18].

The atmospheric transmissivity τ accounts for the absorption of thermal radiation by the atmosphere. An approximate relation is [18]:

$$\tau = \log 10[14.1\,(\mathrm{RH})^{-0.108} \cdot r^{-0.13}]$$

Where:

RH = relative humidity (%)
r = distance (m)

The maximum rate of combustion M for a liquid pool fire on land can be estimated by the burning rate m_b and the pool surface area A (m^2) [22]:

$$M = m_b \cdot A$$

Where:

M = maximum rate of combustion (kg/s)
m_b = burning rate (kg/(s m^2))
A = pool surface area (m^2)

The burning rate m_b can be calculated by:

$$m_b = 1 \cdot 10^{-3} \frac{\Delta H_c}{\Delta H^*}$$

Where:
ΔH_c = heat of combustion (kJ/kg)
ΔH^* = modified heat of vaporization at the boiling point of the liquid (kJ/kg)

The modified heat of vaporization ΔH^* can be calculated by:

$$\Delta H^* = \Delta H_v + \int_{T_a}^{T_b} C_p \mathrm{d}T$$

Where:
ΔH_v = heat of vaporization of the liquid at the ambient temperature (kJ/kg)
C_p = heat capacity of the liquid (kJ/kg K)
T_a = ambient temperature (K)
T_b = boiling point temperature of the liquid (K)

A limitation of the point source model is that it overestimates the intensity of the thermal radiation near the fire, since it does not consider the flame geometry [20]. It is recommended that the model is only used for far field prediction at distances greater than 5 pool diameters from the center of the fire, thus using it to perform conservative calculations of danger to personnel as opposed to establish distances between adjacent equipment [20].

Jet fire

Jet fires are high momentum ignited releases. Their size and flame intensity are a function of the release rate and fuel properties. Jet fires can occur in facilities where pressurized flammable gasses and/or liquids are stored. Unlike pool fires which can, in principle, be extinguished using deluge and foam systems, jet fires are difficult, and in most cases impossible, to extinguish by active fire protection systems.

The best validated jet fire models are the Chamberlain model for vertical and angled jet fires [23] and the Johnson model for horizontal jet fires [24]. Calculation of thermal radiation effects on a target is an involved process and the reader should again refer to sources such as Lees' Loss prevention [7] and the Yellow Book [19] for details. The models treated here can best be used when an approximation of the flame dimensions is required or as input for a flame radiation model [25].

A simple correlation for the length L of a jet fire is [25]:

$$L = 18.5 \cdot m_r^{0.41}$$

Where:
L = flame length (m)
m_r = mass release rate (kg/s)

A generalized formula for different fuel types is [25]:

$$L = 0.00326 \cdot (m_r \cdot \Delta H_c)^{0.478}$$

Where:
ΔH_c = heat of combustion (J/kg)

Fire balls/BLEVE's

A boiling liquid expanding vapor explosion (BLEVE) occurs when a container holding a superheated liquid or liquefied gas experiences a sudden loss of containment. If the medium is flammable, a fireball can occur. The most well known disaster involving BLEVE's occurred in San Juanico, Mexico on 19 November 1984, in which the town of San Juan Ixhuatepec housing 40,000 people located adjacent to a facility owned by the PEMEX State Oil Company was almost destroyed due to a series of BLEVE's. Approximately 500–600 people died and between 5,000 and 7,000 suffered injuries.

The models presented here assume a BLEVE of a flammable material. The fireball diameter, duration and height can be calculated using the following correlations based on the TNO Yellow Book model [19]:

$$D_{fb} = 6.48\,(\cdot m_f)^{0.325} \quad t = 0.852 \cdot (\cdot m_f)^{0.260} \quad H = 0.75 \cdot D_{fb}$$

Where:
D_{fb} = fireball diameter (m)
m_f = mass of fuel (kg)
t = fireball duration (s)
H = height of fireball center from ground (m)

The radiation received by the individual, assuming that the fireball is in full sight of the receiver, can be calculated using the emitted flux E of the fireball and the view factor F, as illustrated in Figure 11.3 [18]:

$$I = \tau \cdot E \cdot F$$

Where:
I = received flux (kW/m^2)
F = view factor = $(D_{fb})^2 \cos\theta / 4x^2$ for $\theta \le 90° - \phi$
E = emitted flux $\cong \varepsilon \cdot \sigma \cdot T_s^4$
Where:
ε =emissivity (~ 1.0)
σ = stefan-boltzman constant ($5.67 \cdot 10^{-11}$ kW/(m^2 K^4))
T_s = flame surface temperature (K)
τ = atmospheric transmissivity (–)

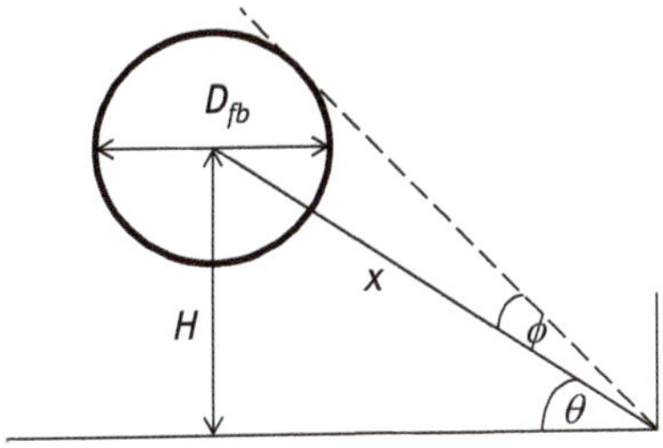

Fig. 11.3: Simplified model of fireball.

Flash fires

When a cloud of flammable gas forms over an area and finds an ignition source, the mass between the flammability limits will burn rapidly. People in the gas cloud will be exposed to very high heat fluxes for a short duration, but this is enough to assume that anybody inside will suffer fatal injuries. Under some conditions, the flash fire can transform into an explosion. This is treated in the following section.

Explosion models

The difference between a fire and an explosion is that the latter occurs in a premixed combination of the fuel and oxidant. The power of the explosion is dependent on the proximity of the fuel and the oxidant. Explosions common to the chemical industry can be divided into two types: deflagrations and detonations. Deflagrations are explosions in which the reaction front travels through the unreacted fuel in subsonic speeds and the propagation method is by heat transfer [26]. A detonation is an explosion with an extremely high, supersonic reaction speed. The propagation method is not heat transfer but an extremely rapid and sharp compression occurring in a shock wave. Under certain conditions, a deflagration could transition to a detonation. The deflagration type occurs most commonly in vapor cloud explosions and the detonation in dense phase explosions in cases where TNT or ammonium nitrates are involved. Thus various models have been developed over the years to model the diverse types of explosion behaviors and the most recent ones can be divided into two classes: blast curve models and computational fluid dynamic (CFD) models. The most common models are provided below:

- blast curve models
  - TNT equivalency
  - multi energy (ME)
  - Baker–Strehlow–Tang (BST)
- computation fluid dynamics models
  - AutoReaGas (not licensed anymore)
  - FLACS

This section will discuss briefly the TNT model which, in specific cases, is still used in the process industry and ME model which is used for modeling vapor cloud explosions. The reader should refer to other sources such as Lees [7] for more information on the other types of explosion models. Dust explosions are another important type of explosion which are not treated in this section. The reader should again refer to sources such as Lees [7] for further information. Damage levels from explosion effects are also discussed by Lees.

TNT equivalency

The TNT equivalent method is based on the relation between the flammable material and TNT factored by an explosion efficiency term. Since a lot of information is available with respect to TNT explosion characteristics and damage potential, this model is still widely used but the user should keep in mind that it is a poor model for vapor cloud explosions. TNT explosions are dense phase explosions producing a shockwave from the initial blast of the explosion. Vapor cloud explosions behave differently. The initial flame front is laminar and its speed is slow but, as it propagates into congested areas, the flame front will become more turbulent and the peak pressure will rise. Thus the explosion overpressure in a vapor cloud explosion is more dependent on the plant geometry. The blast shockwave has a different pressure time profile than that of a TNT explosion. In the near field it is higher and, in the far field, lower compared to a vapor cloud explosion. Because of this, the TNT model underestimates overpressures in the far field of a vapor cloud explosion. When the results of this consequence assessment are used for building design against vapor cloud explosions, the TNT method is not a suitable model and the ME method and Baker–Strelow–Tang or CFD models are better suited for that purpose. The TNT method is better suited for condensed phase explosions such as with ammonium nitrate. Even then, it is important to know that there is at least an order of magnitude margin of error in the TNT model due to large scatter in the results of the underlying experiments [7].

There is experimental evidence that if a spherical explosive charge of diameter D_c produces a peak overpressure ΔP at a distance R from its center, as well as a positive-phase duration t^* and an impulse i (the integral of the pressure time history), then a charge of diameter kD_c of a similar explosive in the same atmosphere will produce an overpressure wave with a similar form and the same peak overpressure p_0 (as well as a positive phase duration kt^* and an impulse ki at a distance kR from the charge center. This is called the Hopkinson or 'cube root' scaling law and illustrated in Figure 11.4 [20].

Typical blast curves for TNT have been obtained by military field experiments and these curves apply the cube root scaling law to provide blast characteristics versus scaled parameters. The first step is to calculate the equivalent TNT mass based on the heat of combustion, the explosion efficiency and the mass of the fuel [22]:

$$m_{\mathrm{TNT}} = \frac{\eta \cdot m_{\mathrm{fg}} \cdot \Delta H_{\mathrm{c}}}{\Delta H_{\mathrm{TNT}}}$$

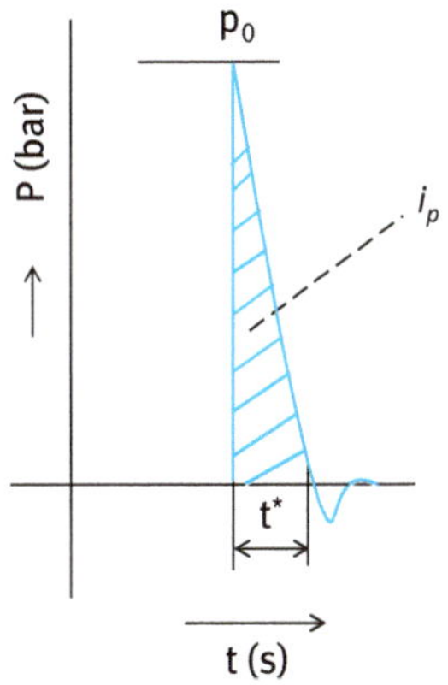

Fig. 11.4: Blast curve characteristics of a shock wave showing the side on peak overpressure p_0, the positive phase duration t^* and positive impulse i_p.

Where:
m_{TNT} = equivalent mass of TNT (kg)
η = empirical explosion efficiency (–)
m_{fg} = mass of hydrocarbon gas kg)
ΔH_c = heat of combustion (kJ/kg)
ΔH_{TNT}= heat of combustion TNT (4437–4765 kJ/kg)

The explosion efficiency depends on the substance. Cameron and Raman [18] propose the following classes based on various sources of information:
- class I: $\eta = 0.05$ (propane, butane, flammable liquids)
- class II: $\eta = 0.10$ (ethylene, ethers)
- class III: $\eta = 0.15$ (acetylene)

With the TNT equivalent mass calculated, the next step is to calculate the scaled overpressure properties [7]:

$$z = \frac{R}{(m_{TNT})^{\frac{1}{3}}}$$

Where:
R = real distance from the center of the explosion origin (m)
z = scaled distance from the explosion origin (m/kg$^{(1/3)}$)

$$p_s = \frac{p_0}{P_a}$$

Where:
p_0 = real peak overpressure (Pa)
p_s = scaled peak overpressure (–)
P_a = atmospheric pressure (Pa)

$$t_s = \frac{t^*}{(m_{TNT})^{\frac{1}{3}}}$$

Where:
t^* = positive phase duration (s)
t_s = scaled positive phase duration (s/kg$^{(1/3)}$)

$$i_s = \frac{i_p}{(m_{TNT})^{\frac{1}{3}}}$$

Where:
i_p = real impulse (s)
i_s = scaled impulse (Pa s/kg$^{(1/3)}$)

$$\tau_s = \frac{t_a}{(m_{TNT})^{\frac{1}{3}}}$$

Where:
t_a = real arrival time (s)
τ_s = scaled arrival time (s/kg$^{(1/3)}$)

Kingery and Bulmash [27] created blast parameter diagrams where, using the scaled distance z, the scaled blast parameters can be read from the diagram and the real explosion parameters calculated from these using the above equations. The diagram for a surface burst (hemispherical symmetry) is provided in Figure 11.5, which is the most representative type of geometry in case an explosion occurs at a process facility.

Vapor cloud explosions: ME method

The most common type of explosion in the process industry is the vapor cloud explosion. A vapor cloud explosion occurs when a flammable gas, between the concentration of the lower explosion limit (LEL) and upper explosion limit (UEL), ignites. The LEL is the minimum concentration of a flammable gas in the air that is capable of igniting and causing an explosion. Below LEL the gas concentration is too lean for ignition. The UEL is the maximum gas concentration in the air at which the gas can be ignited. Above the UEL the gas is too rich to ignite. The flammable mass for a vapor cloud explosion is the gas that is within the LEL and UEL concentration range. The explosion overpressure from a vapor cloud explosion will be more severe the closer the mixture is to the stoichiometric combustion concentration in air. But analysis from vapor cloud explosion incidents have shown that the fuel mass of the vapor cloud alone is not the only parameter that defines the effect of the explosion. Depending on the confinement and congestion in different areas where the gas is located, the explosion characteristics can differ widely. Figure 11.6, shows its principle.

A vapor cloud explosion in an area of no congestion and confinement will have a low laminar burning speed, which is too low to cause any significant blast overpres-

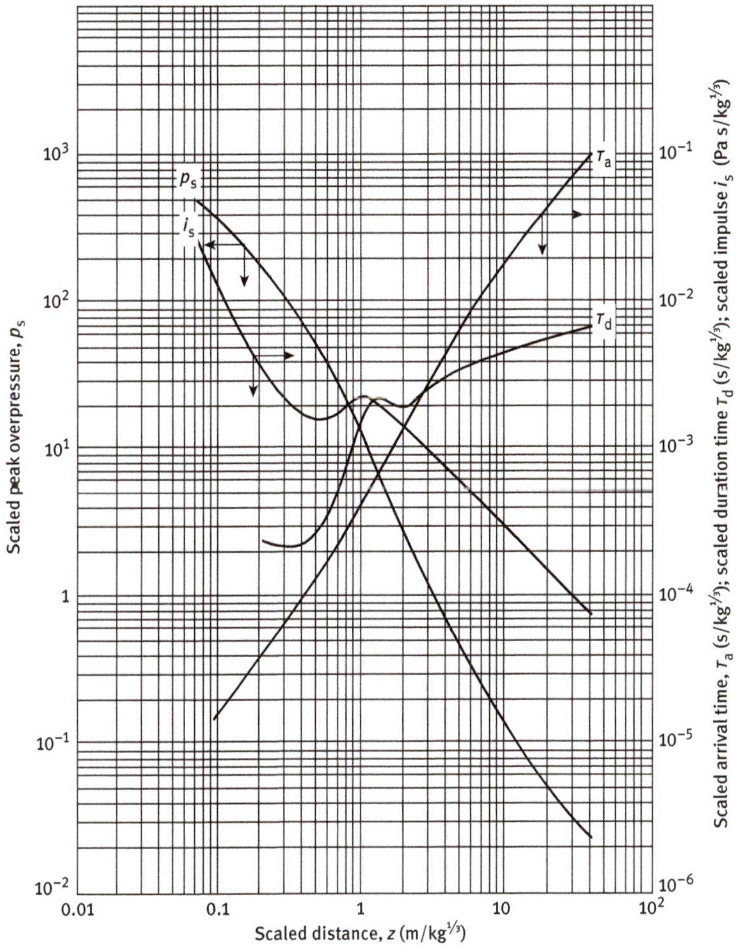

Fig. 11.5: Some side on blast parameters for a TNT explosion for a surface burst, hemispherical symmetry explosion (*reused with permission from Elsevier*) [28].

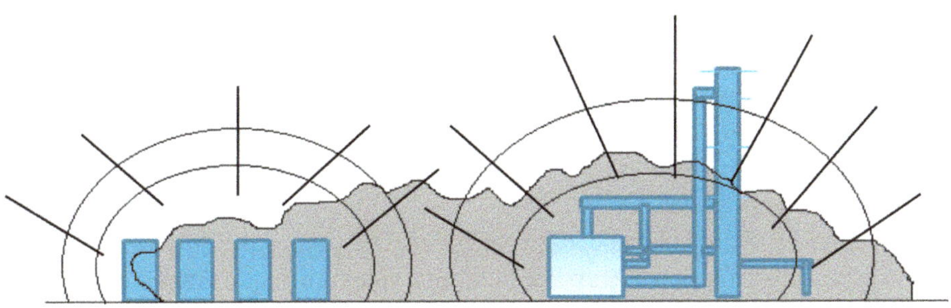

Fig. 11.6: Principle of the ME method showing that the blast characteristics of a vapor cloud explosion is assessed individually for each region on the facility. Adapted from [19].

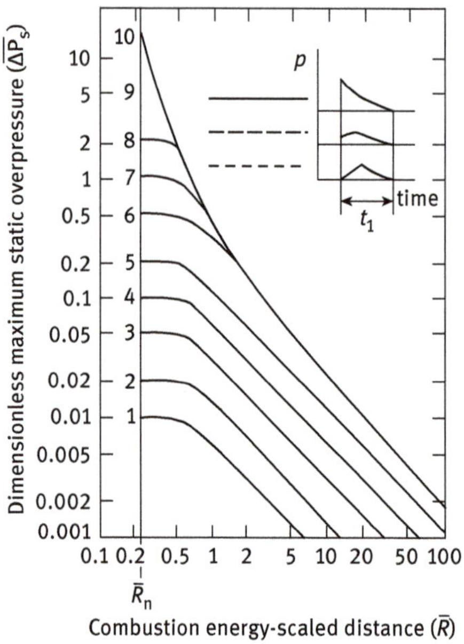

Fig. 11.7: Blast chart of ME method showing the scaled peak side on overpressure and combustion energy-scaled distance for ten different blast source strengths. Similar charts are available for blast characteristics such as impulse and dynamic overpressure. *(Reused with permission from Elsevier) [28].*

sures. The vapor cloud explosion will behave like a large flash fire in this case. On the other hand, a vapor cloud explosion in a heavily confined and/or congested environment can create high blast overpressures approaching the characteristics of dense phase explosions. To consider this difference between dense phase explosion behaviors and vapor cloud explosions, specific models have been developed for the modeling of vapor cloud explosions. One of the most accepted models is the ME method developed by TNO [30]. The method is applicable to unconfined vapor cloud explosions and not to vented vapor cloud explosions or internal explosions in piping or vessel [30].

The ME model is based on numerical simulations of blasts which produced scaled blast parameter characteristics versus a combustion energy scaled distance (Figure 11.7). The ME method is applied by identifying each blast source, characterizing the congestions and confinement in each source and finally determining their contribution to blast generation.

The following steps are followed to determine the explosion characteristics of a vapor cloud explosion using the ME method [19]:

(1) determine cloud size
(2) identify the congested areas
(3) determine the free volume of each congested area (unobstructed areas)
(4) estimate the explosion energy E of each area: volume of cloud times the explosion energy of the fuel (3.5 MJ/m^3 combustion energy of most stoichiometric hydrocarbon mixtures in air assumed in the ME model)
(5) determine the free volume of each congested area

(6) determine the source strength of each area (1 to 10)
(7) determine the location of the center of each area
(8) calculate the combustion energy scaled distance from the center of each area
(9) calculate the blast characteristics at the real distance by using the ME blast charts

The combustion energy scaled distance r' is calculated as follows:

$$r' = \frac{r}{\left(\frac{E_{ex}}{P_a}\right)^{\frac{1}{3}}}$$

With:

$$E_{ex} = V_{cloud} \cdot E_v$$

Where:
r = real distance from center of area (m)
r' = scaled distance ($m^{(2/3)}$)
E_{ex} = explosion energy (J)
V_{cloud} = cloud volume in area (m^3)
E_v = combustion energy of hydrocarbon mixture = $3.5 \cdot 10^6$ J/m^3
P_a = atmospheric pressure (Pa)

The scaled side on overpressure can then be calculated as follows:

$$P'_S = \frac{\Delta P_S}{P_a}$$

Where:
ΔP_S = real peak side on overpressure (Pa)
P'_S = scaled peak side on overpressure (–)

Critical things to consider during the use of the ME model are the validity limits of the model, potential blast interaction between blast sources and the selection of blast source strength.

The ME is best suited for vapor cloud explosions where overpressure effects need to be considered at relatively large distances from the blast source. The ME model is not suitable for assessing overpressure effects close to the explosion source and in complex geometries such as offshore production facilities, drilling rigs and floating production storage and offloading vessels (FPSO's). CFD models need to be used in such cases.

A point of consideration is up to what distance individual blast sources can still be independent. If two blast source areas have sufficient separation distance, the explosion will start to build up pressure in the area where ignition occurred, then reach the free field between the sources and lose flame speed and overpressure before reaching the second area. If the two sources are close to each other, the blast wave will lose some flame speed but then continue in the second area. In this case, the blast sources cannot be considered independent and behave as one large explosion source. Thus, the potential for interaction between blast sources also needs to be considered. The Yellow Book [19] defines the critical separation distance as the parameter to help make this distinction. The project 'RIGOS' [31] concluded that the guidance provided in the Yellow Book [19], with respect to critical separation distance, is not conservative and could lead to underestimating the explosion overpressures. The research showed that critical separation distance is dependent on the overpressure of the donor explosion. For further guidance, the reader should refer to the research report [31].

11.9 Toxic exposure

Another hazard that is important to assess is toxic substances. Toxic substances can enter a biological organism by various routes [26]:
- inhalation – uptake via the lungs
- dermal contact – penetration through the skin
- ingestion (oral) – uptake via gastrointestinal tract
- in utero – uptake by the fetus via the placenta

In process risk assessments, the inhalation and dermal contact routes are the most common. In product engineering, other routes can also become relevant. An example is in the assessment of pharmaceutical products. In pharmaceutical practice, additional routes of exposure are also possible such as intravenous, intraperitoneal, subcutaneous and intramuscular [26].

From a consequence assessment perspective, the exposure time and concentration of the substance are the parameters to consider. Threshold values have been developed for different circumstances such as working environment concentration limits and emergency concentration limits. Some examples are:
- TLV-TWA – time weighed average exposure over 8 hours or 40 hours
- TLV-STEL – short term exposure limit, not longer than 15 minutes
- TLV-C – threshold limit value, ceiling limit which may not be exceeded at any time
- IDHL – immediate danger to life and health, exposure up to 30 minutes
- ERPG – emergency response planning guidelines, exposure up to 1 hour assumed

A threshold limit value is selected based on the goal of the assessment and, in some cases, regulatory requirements. Calculations then need to be performed to assess the concentration and exposure time and these are compared to the threshold limit value to estimate the effects.

To assess the consequences of toxic exposure to people, dose response relationships are sometimes used. A 'dose' is the amount of substance that is taken up by an organism. The dose is expressed as a concentration c, raised to a chemical specific parameter n, multiplied by the exposure time t. Since the amount taken up is time dependent [20]:

$$D_{\text{toxic}} = c^n \cdot t$$

Where:
T_{toxic} = the received dose
c = concentration of toxic substance exposure
t = exposure time

It should be noted that the toxic concentration in limit values or dispersion models are sometimes expressed in mg/m^3 or in ppmv. To transform the units from ppmv to mg/m^3 and vice versa, one can use the following equation assuming ideal gas behavior [20]:

$$c_{\text{ppmv}} = \frac{22.4}{M} \cdot \frac{T}{273} \cdot \frac{1.013}{P_{\text{a\_bar}}} \cdot c_{\text{mg/m}^3}$$

Where:
M = molecular weight (kg/kmol)
T = ambient temperature (K)
$P_{\text{a\_bar}}$ = atmospheric pressure (bar)

Since the absorbed dose is time dependent due to a varying concentration exposure during an emergency, the expression becomes:

$$D_{\text{toxic}} = \int_0^t [c(t)]^n \, \mathrm{d}t$$

The dose is used in so called 'probity relations', which will be explained briefly in this section. Probit stands for 'probability unit' and probit functions are empirically derived mathematical expressions linking risk (effect and probability) to exposure (intensity, time, and duration) [26]. It is assumed in the probit function that the log-exposure-risk relation follows a Gaussian normal distribution. By using the probit relation, one can calculate the probability expressed as a percentage of an exposed population which will exhibit a given adverse effect as related to a given exposure. Probit functions are expressed in the form:

$$\text{Pr} = a + b \cdot \ln(D_{\text{toxic}})$$

Where Pr is the probit value, a and b are regression constants and D_{toxic} is the toxic dose. The regression constants depend on the toxic effect that is assessed and are available for a limited number of substances. Transformation tables for probits to response fractions (%) have been developed by Finney [32] and adopted in the TNO Green Book [33]. Analytical expressions are also available for calculations, which are useful in cases when computer models are used for the calculations [34].

The reader should refer to the Green Book [33] in which the subject is treated in more detail. Note that toxic exposure relations provided in 1992 might have changed so be advised to mainly use the source to get familiar with the methodology and check more recent literature sources before using the dose response relations mention in the Green Book.

Note that the application of probit relations is not limited to toxic effects. There are probit relations available in literature that can be used to calculate the consequence for fire and explosions, such as those provided in the Green Book [33] and Lees [7].

11.10 Codes, standards and designing ALARP

Having identified the hazards and assessed the consequences, how does one choose the relevant engineering design cases? What is, and when is it 'safe enough'. First, there are several information sources, approaches and methods that can assist engineers making these choices, such as:
- regulations, codes and standards
- company specific design engineering standards
- ALARP and risk analysis

These will be elaborated on in the following sections.

Regulations, codes and engineer standards

Engineering regulations, codes and standards all serve to ensure that products and processes are safe and of superior quality. Each has their own characteristics and there are differences between them, as follows:
- *Technical regulations:* mandatory requirements set by a regulatory authority, defining the characteristics and/or performance requirements of a process, product or service.
- *Engineering codes:* laws and regulations specifying minimum requirements to protect safety, health and the environment. Codes will typically refer to standards for specific details on requirements that have not been specified in the code itself. Codes are enforced by one or more governmental entities and are important for establishing best engineering practices.
- *Engineering standards:* technical documents defining technical methods or characteristics of a product, process or service. They are not laws and are thus not enforced by regulatory authorities but are considered as engineering practice.

In other words: codes specify what needs to be done and standards detail how to accomplish this.

Regulations, codes and standards specify in some cases what the design cases should be. But in many instances, this is not the case. And even when they do, there are cases during engineering projects where there are conflicting design requirements between standards or they provide too little information on how to deal with a specific design challenge. In such cases, further analysis or reviews are required to make the proper design choices. This is where the ALARP principle and risk analysis comes into the picture when reviewing design aspects that could have an impact on safety, health or the environment.

ALARP principle

ALARP has its origin in the UK Health and Safety at Work, etc. Act from 1974 [35]. The principle's goal is to avoid making sacrifices grossly disproportionate to the benefits gained. An extreme example might be to spend $1m to prevent five staff suffering bruised knees, which is obviously grossly disproportionate. But to spend $1m to prevent a major explosion capable of killing 150 people is obviously proportionate [36].

The principle is outlined in Figure 11.8:

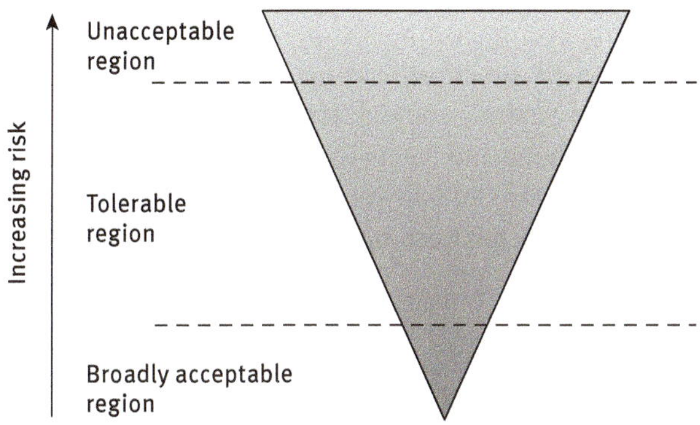

Fig. 11.8: ALARP triangle.

The risk is divided into three regions, shown in Figure 11.8 [37]:

- *Unacceptable region:* risks are too high and therefore risk reduction measures are required, unless there are exceptional reasons for the practice or activity to be retained.
- *Tolerable region:* risk reduction should be applied if the sacrifices are not disproportionate to the benefits gained.
- *Broadly acceptable region:* risks falling into this region are generally regarded as insignificant and adequately controlled. They are typical of the risk from activities that are inherently not very hazardous or from hazardous activities that can be, and are, readily controlled to produce very low risks.

The challenge of the ALARP principle lies in the practical application of the principle. What is considered 'reasonably practical' can be highly subjective. The UK's Health and Safety Executive website provides some guidance to inspectors to assess if duty

holders have reduced risks to as low as reasonably practical. Camaron and Raman [18] state factors that come into consideration when assessing practicability under the ALARP principle:
(1) Can the consequences be eliminated?
(2) If elimination is not possible, can they be mitigated?
(3) What are the possible mitigation measures?
(4) Are the mitigation measures within engineering practicability?
(5) If it is not practicable to design the mitigation measure due to engineering constraints, then can the likelihood of an occurrence resulting in the consequence be reduced?
(6) What are the possible likelihood reduction measures?
(7) Are the likelihood reduction measures within engineering practicability?
(8) What are the benefits of the risk reduction measures? This may not be a reduction in public risk alone, but should also address risk to employees, benefits gained in minimizing interruptions to operations, etc.

After having established what is deemed 'practical' it is required to assess if these measures are 'reasonable'. Cost benefit analysis and risk assessment are some of the important tools to assess this. The UK Health and Safety Executive provides guidance on how to perform the cost benefit analysis [36] but warns that it should not be the sole criteria to determine if the ALARP level has been reached.

11.11 Risk analysis

Risk analysis is a tool that can be used as part of an ALARP assessment and help in making design decisions. There are two types of assessments:
- Qualitative, where risk is expressed in a nonquantified way (e.g., low, medium and high).
- Quantitative, where the result of the assessment is a number and compared to a risk acceptance criterion.

The following sections provide a brief overview of the methods and their advantages and disadvantages.

Qualitative assessments

A commonly used qualitative tool in the chemical and process industry is the risk matrix. The risk matrix is a graphical representation of the risk as a function of the consequences and frequency or probability of occurrence. Consequence categories can be defined for people, environment, assets and reputation. An example is provided in Figure 11.9. It has three risk categories (low, medium and high) but more could be

Consequences		Frequency of occurrence				
Severity level	People	1	2	3	4	5
		Very unlikely	Unlikely	Possible	Likely	Very likely
		$< 10^{-4}$ /yr	10^{-4}/yr -10^{-3}/yr	10^{-3} - 10^{-2} /yr	10^{-2} - 10^{-1} /yr	$>10^{-1}$ /yr
5	More than 5 fatalities					
4	Between 1 and 5 fatalities				**High**	
3	Major injury			**Medium**		
2	Minor injury		**Low**			
1	Slight injury					
0	No injury					

Fig. 11.9: Example risk acceptance matrix defining consequences for people.

allocated. The categorized regions on the matrix correspond with the regions of the ALARP triangle (Figure 11.8). When the outcome of the risk assessment is in the 'high' category, the risk is unacceptable and risk reducing measures are required. If a risk is in the 'low' category, the risk must be managed for continual improvement. If a risk is 'medium' category, risk reduction measures must be incorporated until ALARP.

The matrix in our example has a 5 × 5 dimension but other dimensions such as 3×3, 4×4, 6×6, 4×6, etc., are also possible with more or less allocated severity and frequency categories. Other examples are available in industry guidelines such as from the International Organization of Oil and Gas Producers [38] and ISO standards [39]. The same principles apply to all the matrixes, irrespective of their dimensions. Major chemical and oil and gas producing companies have a corporate defined risk matrix which is used as a basis for all the process safety related risk assessment studies. The advantage of having a corporate defined risk matrix is that it helps to achieve an increase in consistency between risk assessments in a company.

Qualitative risk assessments, such as the risk matrix, have the advantage that they are relatively easy and quick to use and can help to prioritize actions. Despite its apparent ease of use, there are in fact several areas where the risk could be incorrectly addressed leading to a wrong prioritization of actions and incorrect screening of high risks. A critical point to remember when using the risk matrix is its limited suitability for assessing complex and high risks. In those cases, decisions should be based on more quantified risk assessment methods to decide if the risks have been reduced to ALARP.

Quantitative assessments

In case of complex systems with the potential for major hazards, quantified risk assessment can be employed to assist in making engineering decisions. This section will briefly discuss some common techniques and their advantages and disadvantages.

Quantitative risk analysis (QRA) requires failure frequency of equipment and an assessment of the consequences as input for the assessment. Consequence assessment models have been discussed previously in this chapter. Frequency data can be obtained from online and offline information sources. Several databases containing failure frequencies are available free of charge from sources such as from the international association of oil and gas producers [40] and the UK Health and Safety Executive [41]. The quantitative assessments that will be discussed in the following sections are the fault tree analysis (FTA), event tree analysis (ETA) and the QRA.

Fault tree analysis

A FTA investigates how a top event (fault), such as an exothermic runaway, loss of containment, etc., originates from individual causes. It is a deductive technique meaning that one starts at the outcome. The goal of the FTA is to calculate the probability or frequency of the top event. It also provides insight into how a top event scenario develops from its individual causes. An example is provided in Figure 11.10. The name is derived from the fact that the top event is the apex that branches to individual causes, thus resembling a tree. A fault tree considers that the system components are in one of two states: a correct state or a fault state. Note that there is a difference between a failure and a fault in a component. A fault is an incorrect state which may be due to a failure of that component or may be induced by some outside influence. Thus, fault is a wider concept than failure. All failures are faults, but not all faults are failures [7]. A fault tree is constructed by using 'gates'. Gates are points in the diagram where branches meet. Two common gates are the 'AND' and 'OR' gates. At the 'AND' gate, all the components meeting at this gate must fail to progress further towards the top event. At the 'OR' gate, one of the components meeting at this point must fail to further progress towards the top event.

To effectively use the fault tree the following needs to be prepared and decided before the analysis:
- The goal of the analysis such as screening of scenarios, verification of codes, comparing design alternatives, etc.
- Scope of the system to be investigated.
- Frequency and probability data.

A fault tree diagram can then be constructed through the following steps:
(1) Define the top event.
(2) Decide where the analysis will start on the system diagrams such as Piping & Instrumentation Diagrams (P&ID) and Process Flow Diagrams (PFDs). For safety related assessments, this will typically be the point where the safeguarding system starts to act.

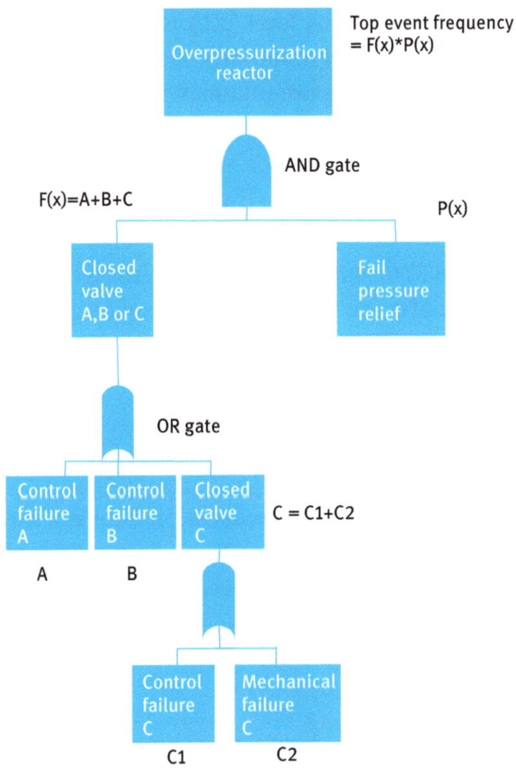

Fig. 11.10: Example fault tree.

(3) Move through the system against the flow of liquid, gas, energy, current, etc., and consider the faults of all relevant components.
(4) Continue modeling up to the point where the goal of the analysis has been met.

The FTA and has the advantage that risk contribution to the top event from individual components can be quantified. The limitation of a FTA is that only Boolean states are assessed (fail/correct) and thus no account is given to partial functioning, temporal effects and spare parts [16]. There are various references for the construction of fault tree diagrams such as Lees' Loss Prevention in the chemical industries [7].

The FTA also forms the basis of the layers of protection analysis (LOPA). The LOPA is becoming more common in the process industry as a tool to determine how reliable independent barriers will have to be to prevent the realization of a single process hazard. The steps of a LOPA consist of:

- calculate initial event frequency without taking barriers into consideration
- compare result to a quantitative criterion and determine the 'gap'
- calculate if current barriers reduce the risk sufficiently
- if not, redesign, increase reliability of barriers and/or introduce new independent barriers to ensure criteria are met

The technique is sometimes used during safety integrity level (SIL) assessment. During a SIL assessment, the level of required protection of an instrumented system is based on the IEC61508 and IEC61511 standards. For more information on the LOPA methodology, the reader should refer to the book 'Layer of Protection Analysis: Simplified Process Risk Assessment' [42].

ETA

The ETA is a complementary method to the FTA, employing an inductive method to assess branching consequences from an initial cause. Probabilities or frequencies can be assigned to each of the branches. An example for a hydrocarbon gas scenario is provided in Figure 11.11.

Construction of an event tree consists of the following steps:

(1) Define the initiating event.
(2) List the sequence of protections of the system.
(3) Start from the first layer and branch for a yes/no for the success or failure of the layer.
(4) Continue branching for each layer until the sequence has been completed.
(5) Describe the consequences at each end of the event tree. Note that it is possible that multiple branches have the same consequences, meaning that multiple pathways to these consequences are possible.

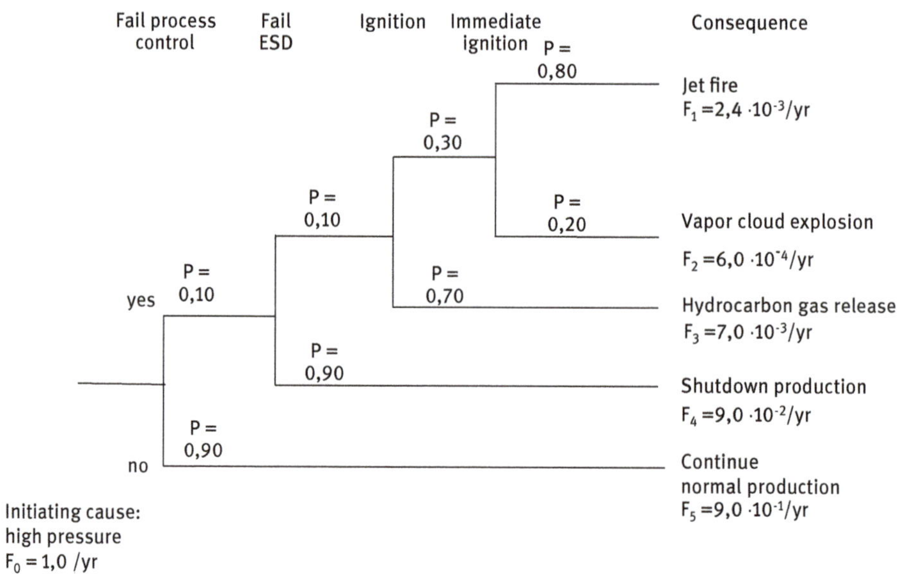

Fig. 11.11: Example event tree for a pressure peak in a gas production facility. Frequencies and probabilities are only for illustrative purposes.

(6) Assign numeric values to the branches and calculate the frequency of the consequences. Calculation of the consequence frequencies is performed by multiplication along the branches. One should check that each probability node on the event tree sums up to unity and that the consequence frequencies sum up to the initial event frequency.

The advantage of the event tree is that it is a good method to show many possible consequences from a single scenario. The event tree is also used as a precursor for QRAs.

Individual and group risk

The tools mentioned in this chapter can be used as input to a full QRA of a system. A QRA will assess the frequencies and consequences of a variety of scenarios and is therefore a more involved study compared to other assessments. The goal of the QRA can vary from land planning use to assessing if the risk to operators on the industry site is acceptable. In the process industry, it is common to investigate the risk of a fatality, since this is a severe consequence that is more easily definable compared to the risk of injury. General assumptions in QRA assessments are that contribution of all incidences on the facility are additive and that the risk receiver is exposed to the presence of the risk source over a certain period (usually one year). There are two categories of risks that can be investigated:

- Individual risk (IR): the frequency at which an individual may be expected to sustain a given level of harm from the realization of specified hazards [43].
- Group risk (GR): expresses the relationship between the frequency and the number of people suffering from a specified level of harm [18].

If consequence severity values are calculated for a meshed area around a risk source, points of equal risk can be connected to produce individual risk contours at various levels, for example, 10^{-4}, 10^{-5}, 10^{-6} or 10^{-7} [16]. If directional effects are absent, the contours are circular, and if a dominating wind direction exists, or directional blast or fragment projection occurs, contours become ellipsoidal or even irregular [16].

Group risk (also called societal risk) differs from IR in the fact that it expresses the relationship between the frequency and the number of people suffering from a specified level of harm as opposed to the individual approach [16].

References and further reading

[1] Pat-El IE, et.al, Major Accidents and Their Impact – The EU Directive for Offshore Safety, Society of Petroleum Engineers conference paper 172094, 2014.

[2] CCPS. Guidelines for Chemical Transport Analysis, American Institute of Chemical Engineers, New York, 1995.

[3] CCPS. Chemical Process Quantitative Risk Analysis, 1999
[4] Klotz TA. Inherently Safer Design – The Growth of an Idea, Process Safety Progress, 15, 1996, 5–8.
[5] AIChE. Dow's Fire & Explosion Index Hazard Classification Guide, 7th edn, June 1994.
[6] AIChE. Dow's Chemical Exposure Index Guide, 1st edn, 1994.
[7] Lees FP. Lees' Loss Prevention in the Process Industries, edited by Sam Mannan, Burlington, Butterworth-Heinemann, 3rd edn, 2005.
[8] Urben PG (ed). Bretherick's Handbook of Reactive Chemical Hazards, Oxford, Academic Press, 7th edn, 2007.
[9] CCPS. Essential Practices for Managing Chemical Reactivity Hazards, 1st edn, 2003.
[10] CCPS. Chemical Reactivity Evaluation and Applications to Process Design, 1995.
[11] CCPS. Guidelines for Safe Storage and Handling of Reactive Materials, 1995.
[12] Leggett, DJ. Chemical Reaction Hazard Identification and Evaluation: Taking the First Steps. AIChE Loss Prevention Symposium, New Orleans. March. 2002.
[13] IEC 61882, Hazard and Operability Studies – Application Guide, 2001.
[14] ISO 14224, Petroleum, Petrochemical and Natural Gas Industries – Collection and Exchange of Reliability and Maintenance Data for Equipment, 2016.
[15] CCPS. Guidelines for Hazard Evaluation Procedures, 2nd edn, 1992.
[16] Pasman HJ. Risk Analysis and Control for Industrial Processes – Gas, Oil and Chemicals, 1st edn, 2015.
[17] CGE. Bowtie XP Methodology Manual, Rev.15, March 2015.
[18] Camaron IT and Raman, Process System Risk Management, Volume 6, 1st edn, Elsevier, 2005.
[19] VROM. Methods for the Calculation of Physical Effects – 'Yellow Book' Publication Series on Dangerous Substances PGS 2, 3rd edn, 2005.
[20] Casal J. Evaluation of the Effects and Consequences of Major Accidents, Industrial Safety Series, Vol 8, Amsterdam, Elsevier, 2007.
[21] HSE. Methods of Approximation and Determination of Human Vulnerability for Offshore Major Accident Hazard Assessment, 2010. Sourced 26 January 2018 from: http://www.hse.gov.uk/foi/internalops/hid_circs{/\penalty\exhyphenpenalty}technical_osd/spc_tech_osd_30/spctecosd30.pdf.
[22] CCPS. Guidelines for Consequence Analysis of Chemical Releases, 2nd edn, Hoboken, Wiley, 1999.
[23] Chamberlain, GA. Developments in Design Methods for Predicting Thermal Radiation from Flares, Chem. Eng.Res.Des., 65, 1987.
[24] Johnson, AD, et.al. A Model for Predicting the Thermal Radiation Hazards from Large Scale Horizontally Released Natural Gas Jet Fires, Trans.IChemE, 72, Part B, 1994.
[25] OGP. Consequence Modeling, Report No. 434-7, March 2010. Sourced 26 January 2018 from: https://docslide.com.br/documents/ogp-434-7-consequence-modelling.html.
[26] Lemkowitz SM, and Pasman HJ. *Chemical Risk Management*, Faculty of Applied Sciences, Chemical Engineering, Delft University of Technology, 2002.
[27] Kingery CN, and Bulmash G. Air Blast Parameters from TNT Spherical Air Burst and Hemispherical Surface Burst, Tech. Rep. ARBRL-TR 02555, US Army, Ballistic Res. Lab., Aberdeen Proving Ground, MD, 1985.
[28] Lees FP, Lees' Loss Prevention in the Process Industries, edited by Sam Mannan, Burlington, Butterworth-Heinemann, 4th edn, 2012.
[29] Van den Berg, AC. The Multi-Energy method – A Framework for Vapor Cloud Explosion Blast Prediction, Journal of Hazardous Materials, 12, 1985.
[30] TNO. Application of Correlations to Quantify the Source Strength of vapor Cloud Explosions in Realistic Situations, Final Report for the Project 'Games', Report PML 1998-C53, October 1998.

[31] HSE. Research to Improve Guidance on Separation Distance for the Multi-Energy Method (RIGOS), Research Report 369, 2005.
[32] Finney DJ. Probit Analysis, 3rd edn, Cambridge, Cambridge Press 1971.
[33] Green Book, Methods for Determining Possible Damage, 1st edn, 1992, Sourced 26 January 2018 from: https://www.scribd.com/doc/61170131/Green-Book-Methods-for-the-Determination-of-Possible-Damage-CPR-16E.
[34] Vilchez JA, et.al. J. of Loss Prevent. Proc., 14, 2001.
[35] HSE. Guidance, Sourced from: http://www.hse.gov.uk/risk/theory/alarpglance.htm, accessed on 18-12-2016.
[36] HSE. Cost Benefit, Sourced from: http://www.hse.gov.uk/risk/theory/alarpcheck.htm, accessed on 18-12-2016.
[37] HSE. Reducing Risks, Protecting People – HSE's Decision Making Process, London, HMSO, 2001.
[38] OGP. HSE Management – Guidelines for Working in a Contract Environment, report 423, June 2010.
[39] ISO 31000, Risk Management, 2009.
[40] OGP. Risk Assessment Data Directory, Report 434, March 2010.
[41] HSE. Failure Rate and Event Data for use within Risk Assessments, Sourced from: http://www.hse.gov.uk/landuseplanning/failure-rates.pdf, accessed January 2017.
[42] CCPS. Layer of Protection Analysis: Simplified Process Risk Assessment. 1st edn, Hoboken, Wiley, 2001.
[43] Jones DA. *Nomenclature for Hazard and Risk Assessment in the Process Industries*, 2nd edn, Rugby, England, Institution of Chemical Engineers, 1992.

The following references are recommended for further reading: [9, 10, 15, 16, 18–20, 22, 28, 33, 42]

12 Evaluating for environmental, social and sustainable development aspects

12.1 Environmental evaluations

12.1.1 Introduction to environmental evaluations

The scope of environmental impact assessment has increased enormously in the last 60 years. In the 1960s, legislation was set by individual countries and was very limited. In the Netherland, for instance, laws were implemented to reduce low level smog, by limiting low level acid gas emissions. Very tall factory chimneys, of 200 meters and higher, were then constructed to make sure that acid flue gases no longer caused local low layer smog. In the seventies and eighties, these tall chimney emissions caused acid rain and thereby caused forests to die in other countries; notably in Sweden. This in turn caused stringent flue gas emission legislation in Western European countries, which were later adopted by the European Union. Other industrialized regions, such as North America, East Asia and South America followed suit.

For product design, such as for refrigerants, long range environmental effects, such as atmospheric ozone layer depletion and river water quality deterioration due to laundry washing products, were considered. In the nineties and beyond, enhanced global warming by burning fossil fuels caused worldwide concerns, as expressed by the United Nations and global companies.

Presently, the scope of environmental impact assessment is on a global scale and concerns environmental effects on the Earth's atmosphere, on water, soil, and space [1]. Environmental evaluation of designs is now an essential element all over the world. Companies not only comply to environmental laws, but go beyond laws as they anticipate that in future these laws will be more stringent. As such, it is more economical to make designs more robust to meet the requirements of these future legislations, rather than having to invest later in end of pipe solutions. Many companies have now have their own environmental strategy, policy and regulations regarding environmental impacts of novel product and process operations. Some do it to keep their license to operate, some to anticipate new, more stringent environmental laws and some as part of their contribution to sustainable development. The next sections provide methods and guidelines for environmental evaluation for each innovation stage gate.

12.1.2 Environmental evaluation: discovery stage

In the discover stage, very little environmental information is available or can be generated on the novel product and/or process idea. On the other hand, only little infor-

https://doi.org/10.1515/9783110467741-012

mation is needed as the only question to be answered is whether the idea is better than the commercial product and the process with which the idea competes.

Here are two simple metrics to evaluate the novel idea on environmental impact. It is based on an estimate of the maximum amount of product (hence, minimum amount of waste) that will be produced by the novel idea. For a product, this can be directly estimated from output/input mass ratio. For a simple chemical product, it can be derived from the stoichiometric reaction equation. The metric is the mass efficiency of the stoichiometric reaction (MEST) determined from the equation [2]:

$$\text{MEST} = \text{product mass}/(\text{total mass of all molecules produced})$$

The MEST value should be calculated for the novel product or process and be compared with the MEST value of the reference case.

An alternative metric often used is atom efficiency [3]. This also is defined as ratio of product mass and total mass of molecules generated by the reaction. If the stoichiometric equation is used to determine the atom efficiency, then the atom efficiency will have the same value as the MEST value.

It is also useful to calculate the actual mass efficiency (ME) of the reference case. The mass efficiency is the inverse of the E factor of [3]. The ME metric has the advantage that it can be directly compared to MEST values.

$$\text{ME} = \text{actual product mass}/(\text{total actual mass waste and product})$$

If the MEST value of the novel idea is lower than the MEST value of the reference case, than the novel idea scores poorly on the environmental criterion, even if the MEST value of the novel idea is higher than the ME value of the reference case.

The reasoning for this latter statement is that the ME value of the reference case will increase in time through improvements to the reference case. This may well happen within the time needed for the idea to become implemented.

In the discover stage the environmental impact of the novel product and process should be enormously lower than reference case, if the reference case has a large environmental impact. In the present production of fine chemicals and pharmaceuticals, the amount of waste produced far exceeds the amount of product due to stoichiometric consumption of additional chemicals [2, 3]. By applying catalysis this waste can be reduced enormously.

12.1.3 Environmental evaluation: concept stage

In the concept stage, the novel product and process design should be evaluated on its environmental impact in comparison with the reference case on all aspects. However, limited resources are available so the evaluation should still be simple. The best method available is the conceptual life cycle assessment method (LCA), which is proposed in the next section.

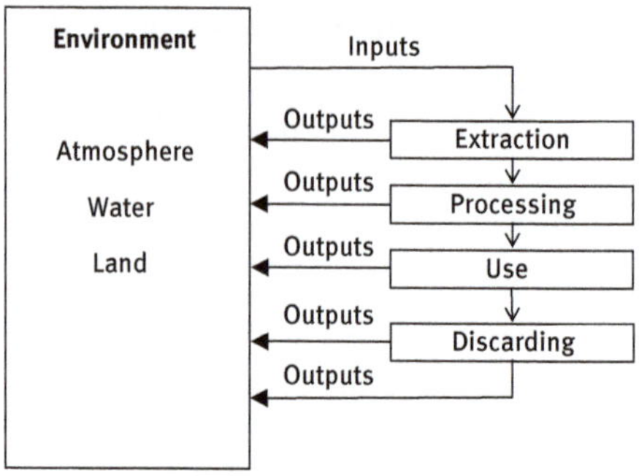

Fig. 12.1: LCA system elements.

LCA for concept stage

There are three levels of LCA: conceptual, simplified and detailed. For the concept stage, the conceptual LCA is most suitable. In my courses on sustainable design for products, processes and systems, I worked out a simple stepwise approach to conceptual LCA, called rapid LCA [4]. First a simplified systems description for LCA is provided in Figure 12.1.

System elements of LCA are the environment (with subsystems, atmosphere, water and land) and the steps (extraction, processing, use and discarding). The concept LCA considers all mass inputs from the environment and all mass outputs to the environment for these life cycle steps. In a complete LCA, far more steps will be involved, such as transport between the major steps and more process steps. Other feeds will also be involved, but these are neglected in this limited LCA. In most cases, the conclusion that the concept is better than the reference case will appear to be correct, when in latter stages a complete LCA is executed.

The inputs and outputs are at first instance defined as mass flows with their composition the same as streams in process design. Then inputs and outputs are further defined on their environmental impacts, such as eco toxicity.

The rapid LCA involves:

(1) defining the functional unit
(2) defining all life cycle steps from primary production to end of life and preferable recycle steps
(3) selecting one or two key environmental impact types
(4) identifying all major emission streams for each life cycle step
(5) quantifying the environmental impact
(6) comparing the rapid LCA of the novel design with the reference case
(7) concluding whether the novel design has a lower environmental impact or not

Ad 1: Functional unit. For a comparison, it is very important to have the same functional unit definition for the alternatives and the reference case. For a process design the functional unit will be the product produced by the process. The functional unit definition is then given by the amount of product to be produced and the product specifications. The definition of the amount of product produced can be simply design capacity expressed as the amount in tons produced per year.

A word of warning should be stated here; economist often express the amount in absolute mass dimension (ton), while they mean the amount per year (ton/year). Chemists also often express the amount in absolute mass (grams), but they really mean absolute mass. For a comparison between different processes with the same product, the functional unit definition should be the same. As all inputs and outputs will be linearly related to the amount of product, the selected dimension and size of the functional unit can be any if it is the same for all alternatives considered.

For novel product design, the definition of the functional unit is in general far more complicated as often it is an action, such as a single hair wash, rather than an absolute amount of a material. If the novel product design replaces several existing products, then the functional unit should be defined in such a way that it is clear how much of each product of the reference case makes up the functional unit.

Ad 2: All life cycle steps. For a novel product, most upstream life cycle steps can be found by a literature search on LCA about the input components. The downstream steps after product use may need some imagination. For a novel process for an existing product, the LCA of the existing product can be used and adapted to the input and output streams of the novel process compared to the reference process case.

Ad 3: Selection of the key environmental impact types will, in most cases, directly follow from the major environmental impact types (see Table 12.1) and is obtained from LCA data of the reference product(s).

Ad 4: Literature on feedstock stream manufacturing descriptions helps to find emission streams data.

Ad 5: Crude estimates of each stream for the selected impact types will be made and added up.

Ad 6: A comparison on environmental impacts of the novel design and the reference case will be made

Ad 7: From the comparison, a conclusion on the environmental impact of the novel design compared to the reference case for the key impact type will be made. A qualitative statement about the assumptions for the other environmental impact types will also be made.

A helpful method for a rapid evaluation in the concept stage is to fill in template Table 12.2 for all environmental impacts for which data is available. For impacts with no data, 'no data yet' is filled in. For some of those impacts it may be possible to reason that the impact for the modern design will be lower than the reference case. Then 'lower' can be filled in for the 4th column.

Tab. 12.1: Environmental impact types for LCA [5].

Atmosphere	Water	Biotic	Soil
Global warming	Eco toxicity	Eco toxicity	Eco toxicity
Ozone layer depletion	Nitrification	Habitat reduction	Physical degradation
Photochemical pollutants	Heat	Biotic resource reduction	Dehydration
Acidification		Physical ecosystem reduction	Resource depletion
Radiation			
Noise			
Smell			
Visual landscape			

Tab. 12.2: Template environmental impact ratio innovative design/reference case.

Impact type	Reference impact	Innovative design impact	Impact ratio new/reference

The numbers of the last column are used for the evaluation. If all ratios are less than one then the environmental evaluation has a positive outcome. If for some impacts the ratio is less than one, but for at least one impact the ratio is higher, than the outcome is negative. It has a resemblance to a Pareto improvement.

The advice for this choice of improvement is based on the experience that it is nearly impossible to convince managers, stakeholders and/or customers that the product is environmentally better if it has a lower environmental impact for some impact types but has a higher impact for at least one impact type. It also avoids lengthy discussions on how to use weight factors for the individual impact types to get to single environmental impact figures.

Process output evaluation using atom balances

The process concept design can be quickly evaluated on the completeness of all output streams by a simple atom balance over the process input and output streams. Any atom that enters the process will leave the process. Often when a crude input such as biomass is used as input to the process, the concept designer neglects or forgets that the biomass contains atoms, such as potassium and sulfur, which will leave the process in one of the outlet streams. A simple atom balance analysis then reveals that the output streams are not sufficiently defined and that the envisaged output destination of some output streams are infeasible.

Fugitive emissions evaluation process concept design

For the process concept itself, fugitive emissions via valves, etc., can also be estimated from simple process flow sheets, where major unit operations or process equipment is defined. Environment protecting agency (EPA) provides for most unit operations and process equipment directly the amount of fugitive emissions in kg/h [6].

12.1.4 Environmental evaluation: feasibility and development stages

A simplified LCA method can be chosen to assess the environmental impact of the novel design compared to the reference case.

The simplified method consists of:

- Screening for important inputs and outputs of the life cycle.
- Simplifying by selecting a few most important impact categories for the inputs and outputs.
- Assessing reliability of simplified LCA for comparison innovative design with reference case.

The simplified method still looks like the concept method. It takes more impact categories in consideration and uses more quantified literature data. It is, however, prone to uncertainties whether the novel design is better than the reference case. Because the purpose of the feasibility stage is to take a decision of a major investment in product prototype testing and/or process pilot plant purchase, these uncertainties are in general not acceptable. Therefore, a full detailed LCA is recommended.

In the follow up stages, a detailed LCA is the best method to fully assess environmental impacts associated with all the steps of a product's life from cradle to grave, i.e., from raw material extraction through materials processing, manufacture, distribution, use, repair and maintenance and disposal or recycling. The procedures of LCA are part of the ISO 14000 environmental management standards [7]. It is an elaborate method and only provides an accurate result if the ISO 14040 method is applied [8].

A complete LCA of the novel product should then be determined for the feasibility stage, using the certified well defined LCA method ISO 14040 and 14044 [9]. ISO 14040 and ISO 14044 are the leading standards for LCA. In addition to applying the LCA method properly, reliable LCA databases are needed in addition to get reliable LCA of the novel product and the reference case. These databases may be purchased from institutes specialized in LCA data gathering and processing.

The quality of the LCA result not only depends on the quality of the databases used but also on the choice of database type, and the origin and future developments. The database can be based on measured data for the case at hand (called foreground data) or based on averaged data, holding data for a certain country or region. As such, choosing the relevant country is very important. For breakthrough innovations, which will take several years to be implemented and then will be applied for a long time, the

LCA database will not be very relevant for the long term future. This is due to emissions and intakes from other steps in the LCA changing due to innovations. Given this uncertainty, the best approach is to make sure that the innovation scores are far better on environmental impacts.

The LCA results will reveal impacts for each impact type. There are ways to normalize these different outcomes to a single impact figure, but this is not recommended. A novel product that is far better on one or more impact types, but worse on one or more others, compared to the reference case, will lead to endless discussions inside and outside the company (NGO's). It is far better to ensure that the novel product and/or process is not worse on any of the impact types and considerably better on the major impact types. The product and process will be easily communicated as environmentally improved.

For a detailed description and application, the reader is referred to the LCA handbook [5]. For a nice overview of all LCA methods and ranking, the reader is referred to a booklet generated by the European Commission [10].

12.1.5 Environmental evaluation: detailed design stage

In addition to the full LCA, environmental laws and regulations of the countries where the product will be marketed, where the process will be located and the countries where the feedstock will be obtained will also be considered. Often a public environmental impact report must be made to facilitate local civilian's objections to certain emissions or risks of emissions.

In the detailed design stage, *fugitive emissions* for the process should also be determined. These are emissions from flanges and seals. The EPA method for this is easy to use [11].

12.2 Evaluating for social acceptance

12.2.1 Importance of social acceptance

Social acceptance of novel products and processes is enormously important. It can be the show stopper. Think, for instance, about nuclear power plants in Germany, which are no longer accepted by the government and will be phased out. Also consider the Brent Spar case of Shell. Because German petrol buyers did not accept the Shell stance, dumping the platform as waste in the ocean seemed the best solution to Shell. Petrol buyers bypassed Shell petrol pumps and bought petrol and diesel elsewhere. Within six weeks, Shell reviewed its position and invited parties to come up with better solutions for the platform which led to many solutions being presented. The best selected solution was moving the platform to a Fjord in Norway, dismantle it there and reuse all steel and concrete parts in other applications.

For new processes, a license to operate for decades is therefore essential. That license is not only governed by law but also by social acceptance. For consumer products in a free market with competition, social acceptance is the first criterion for success or failure. It cannot be separated from market acceptance. Social acceptance is related to the beliefs and cultures of societies in which the product and process will be located. When, for instance, using animal material for products while depending on the cultures in which the products will be marketed, religious criteria should be considered.

The next section shows a comprehensive list of items relevant for social acceptance. This will be of use in the concept stage as a checklist for the novel design. For the feasibility and development stage, social and market acceptance for novel product is of major importance. For these stages, market surveys and inquiries are the key methods for determining acceptance. In other sections of this book those surveys are described.

12.2.2 Social acceptance aspects

Aspects relevant for social and market acceptance are shown in Table 12.3 and briefly discussed thereafter. The word innovation used here stands for the novel product, process, value chain and life cycle.

Tab. 12.3: Checklist for social and market acceptance of novel product and process. Derived from [Verkerk [12], Potter [13] and de Soto [14].

Modal aspect	Essence of social and market acceptance
Beliefs	The innovation is acceptable to believers
Ethical	The innovation agrees with moral values
Juridical	The innovation agrees with laws
Aesthetic	The innovation is appealing
Economic	The innovation price fits its functions (value for money)
Social	The innovation fits in the social context
Linguistic	Symbols and language provide the correct perception of the innovation
Historic	The history of the innovation and company is attractive to customers
Logical	The logic of the innovation is clear
Sensitive	Senses (smell, taste, hearing, sight, feeling), and emotions associated to the product and process are appealing

The upper part of the modal aspects of artifacts of Verkerk [12] has been used for the first column of the checklist (Chapter 4, Table 4.8 contains the full list of modal aspects). The second column has been filled in using information from Potter [13] and de Soto [14] and my own observations. A brief description for each modal aspect related to novel product acceptance follows hereafter.

Beliefs of a society can have a strong influence on social acceptance of the novel product and/or process. This holds for societies with a named governing belief, such as Hinduism (no animal derived feeds), Islam (no ethanol) and Humanism (no human rights violations).

Ethical and moral aspects of a product over its value chain are often very important for social acceptance. Think, for instance, of child labor associated with production of an intermediate in the value chain.

Juridical aspects such as laws and regulations can make the product unacceptable, because the product (or its manufacturing process) violates laws related to safety, for instance. Also, the laws on property and land ownership regulations are very important [14].

Aesthetic aspects such as the shape of the product (packing) can make or break a consumer product.

Economics of a novel product are often very important for acceptance. The product price should fit the product function. It should also be at least like the reference product for the same function.

Societal aspects of the novel product should be such that it fits with the local society and preferably enhances its welfare. Contributing to socially sustainable development goals will help in this respect. Also, the market type (centrally governed, free market or regulated market, informal market and worker union presence) is relevant for introducing a novel product or process. Furthermore, the government structure, i.e., democracy/dictatorship, is of importance for introducing novel products and processes.

Linguistic aspects are particularly important for new consumer products. The product is often not seen as an artifact fulfilling a function, but as providing meaning to the person and its social relations. For example, a novel nail polish may send a message to the person who sees it on the nails of the lady.

Historic aspects can be very important. When the new product is intended to replace an existing product, while the history related to the reference product has such strong ties in society that the new product is not accepted.

Logical aspects of the new product and process should be applied. If the novelty makes no sense, the new idea will not survive.

Sensitivity is an enormously important aspect for consumer products. The impression it has on the human senses will make the product attractive or unacceptable.

12.3 Sustainable development assessment

12.3.1 Sustainable development history and consensus

Sustainable development was, for the first time, defined by Brundtland for the United Nations in 1987 [15]:

"Sustainable development is not a fixed state of harmony, but a process of change in which the exploitation of resources, the direction of investments, the orientation of technological development and institutional change are made consistent with future as well as present needs. Sustainable development is development that meets the needs of the present generation without compromising the ability of future generations to meet their own needs."

From then onwards in subsequent UN world conferences the definition has been further refined.

Since then, this definition has been applied and adapted several times in UN general assemblies. The UN general assembly of September 25th 2015, set up the Sustainable Development Agenda for 2030 with the aim to end poverty, protect the planet, and ensure prosperity for all as part of a new sustainable development agenda. This time the World Business Counsel for Sustainable Development (WBCSD) was also involved in setting this agenda and in defining the Sustainable Development Goals (SDGs) for 2030. These SDGs are summarized in the Table 12.4 [16].

For a design evaluation, these 17 goals can be used. The question is then does the design, when implemented, contribute to these SDGs. To make this practical for rapid evaluations, the following questions can be used.

Evaluating a novel product, process or service on sustainable development criteria means positively answering the questions:

- Does the novel product, process or service meet a need or requirement?
- Does it not harm the environment over the life cycle?
- Does it create prosperity over the value chain?
- Does it benefit societies over the value chain?

If the answer to any of these questions is negative, then that technology does not contribute to sustainable development. If the answers are all positive, a comparison with the commercial reference case should be made.

SDG number nine focusses on innovation. So academic research, contract research, technology innovators, startups and technology centers of companies all can place this SDG on their strategy.

Apart from this consensus on the definition of sustainable development there are numerous other descriptions of sustainability, often only addressing the environmental aspect. These descriptions are of little use for companies and for education as they lack global consensus and therefore have no authority.

Tab. 12.4: SDGs for 2030 [16].

Goal 1.	End poverty in all its forms everywhere.
Goal 2.	End hunger, achieve food security and improved nutrition and promote sustainable agriculture.
Goal 3.	Ensure healthy lives and promote wellbeing for all at all ages.
Goal 4.	Ensure inclusive and equitable quality education and promote lifelong learning opportunities for all.
Goal 5.	Achieve gender equality and empower all women and girls.
Goal 6.	Ensure availability and sustainable management of water and sanitation for all.
Goal 7.	Ensure access to affordable, reliable, sustainable and modern energy for all.
Goal 8.	Promote sustained, inclusive and sustainable economic growth, full and productive employment and decent work for all.
Goal 9.	Build resilient infrastructure, promote inclusive and sustainable industrialization and foster innovation.
Goal 10.	Reduce inequality within and among countries.
Goal 11.	Make cities and human settlements inclusive, safe, resilient and sustainable.
Goal 12.	Ensure sustainable consumption and production patterns.
Goal 13.	Take urgent action to combat climate change and its impacts.*
Goal 14.	Conserve and sustainably use the oceans, seas and marine resources for sustainable development.
Goal 15.	Protect, restore and promote sustainable use of terrestrial ecosystems, sustainably manage forests, combat desertification, and halt and reverse land degradation and halt biodiversity loss.
Goal 16.	Promote peaceful and inclusive societies for sustainable development, provide access to justice for all and build effective, accountable and inclusive institutions at all levels.
Goal 17.	Strengthen the means of implementation and revitalize the Global Partnership for Sustainable Development.

A word of warning regarding the loose use of the word sustainability. In industrialized countries, the meaning of the word is often reduced to living in harmony with the environment. What I also have noticed is that due to the many other definitions of sustainability, some people think that the word is meaningless or that they can make their own definition and then work with that definition. It will be clear that this approach will bear little fruit as that definition has no common ground and has not been reached by global consensus.

The only powerful term and definition of sustainable development, reached by global consensus, is the one reached within the United Nations and which has been accepted by the World Business Counsel of Sustainable Development (WBCSD).

Sustainable development hot spot scan

The sustainability hot spot scan is a comprehensive tool that provides insights in the most important potential 'hot spots', i.e., risks and opportunities in sustainability in a company's existing and future worldwide value chains.

It is based on many databases from all over the world on safety, health and social impacts and LCAs. The databases are from the UN, WBCSD, BASF social impact tool SEEBALANCE, and many others. The scan provides information about which of the 17 SDGs are addressed by the company and where are the major areas of concern (hot spots) are. It is an excel spreadsheet and can be downloaded for free from the MVO Netherlands website [17].

The tool can be used in many ways. For example, it can be used to do a sensitivity analysis of design parameters on the hot spots. By simply changing the country of origin of a feedstock, for instance, or by changing the choice of a solvent, sensitivity analyses can be carried out and directions for improvement can be identified. It can be used by the designer on its own, but also with others from marketing and legal departments. The choice of feedstock and the related choice of countries can be quickly evaluated. For example, it can give direct insight into the risk of feedstock produced by child labor in a certain country. By changing the country and or feedstock type, informed decisions regarding sustainable development can be made.

References and further reading

[1] Vogler J. The global commons, Chichester: J. Wiley & Sons, 2000.

[2] Sheldon RA. Atom efficiency and catalysis in organic synthesis, Pure Appl. Chem., 72, 2000, 1233–1246.

[3] Sheldon RA. The E factor: fifteen years on, Green Chem., 9, 2007, 1273–1283.

[4] Jonker G, Harmsen J. Engineering for sustainability: A practical guide for sustainable design, Amsterdam: Elsevier, 2012.

[5] Guinee JB, et.al. Handbook on life cycle assessment. Operational guide to the ISO standards. Dordrecht: Kluwer Academic Publishers, 2002.

[6] EPA. Protocol for Equipment Leak. Emission Estimates. Emission Standards, EPA-453/R-95-017. 1995. Sourced from: https://www3.epa.gov/ttnchie1/efdocs/equiplks.pdf, accessed on 25 January 2017.

[7] Madu C. Handbook of Environmentally Conscious Manufacturing, Berlin: Springer Science & Business Media, 2012.

[8] Niederl-Schmidinger A, Narodoslawsky M. Life Cycle Assessment as an engineer's tool? Journal of Cleaner Production 16(2), 2008, 245–252.

[9] ISO. Environmental management – Life cycle assessment – Principles and framework, sources 30-1, 2017, sourced 26 January 2018 from: http://www.pqm-online.com/assets/files/lib/std/iso_14040-2006.pdf.

[10] ILCD. General Guide to LCA, 2010. Sources 26 January 2018 from: http://publications.jrc.ec.europa.eu/repository/bitstream/JRC48157/ilcd_handbook-general_guide_for_lca-detailed_guidance_12march2010_isbn_fin.pdf.

[11] EPA. Methods for Estimating Air Emissions from chemical manufacturing plants, EIIP Vol. II: Chapter 16 2007. Sourced 26 January 2018 from: https://search.epa.gov/epasearch/epasearch?querytext=Methods+for+Estimating+Air+Emissions+from+chemical+manufacturing+plants&areaname=&areacontacts=&areasearchurl=&typeofsearch=epa&result_template=2col.ftl.

[12] Verkerk MJ, et.al. Philosophy of Technology, New York: Routledge, 2017.

[13] Potter D. Dimensions of society, Sevenoaks: Hodder and Stoughton, 2nd edn, 1978.

[14] De Soto H. The Mystery of Capital, London: Bantam Press, 2000.

[15] Brundtland GH. Our common future, report of the World Commission on Environment and Development (WCED), Oxford: Oxford University Press, 1987.

[16] UN. Transforming our world: the 2030 Agenda for Sustainable Development, Resolution adopted by the General Assembly on 25 September 2015. Sourced 26 January 2018 from: http://www.un.org/sustainabledevelopment/sustainable-development-goals/.

[17] MVO. Sustainability Hotspot Scan, 2017. Sourced 26 January 2018 from: https://mvonederland.nl/publicatie/sustainability-hotspot-scan.

13 Communicating

13.1 Communicating: project team and stakeholders

A design project comprises many activities carried out by the team members and in interaction with stakeholders. This was made clear in Chapter 4 and also becomes clear from all other chapters. In this Chapter, we want to focus not only on the reporting of the (final) design results, but also emphasize that 'communicating' is far more than completing the final project documentation.

In order to support the communication during the project with project team members and with the project's stakeholders, a number of communication tools and tips will be discussed. From experience, it is clear that in order to arrive at the best design solution within resources (time, tools and designers) made available, communicating about the design goal, approach, intermediate results, evaluation, decision making and argumentation in the team and with stakeholders are crucial activities. Only by sharing this information in an effective and efficient way, reflection and improvements by team members' and stakeholders' feedback makes improving the design possible. Based on our experience, project teams may spend up to 40% in communication activities.

In this Chapter, we will link a number of key communication tools to the Delft Design Map (DDM) (see Section 4.3).

13.2 Communicating using the Delft Design Map (DDM)

In the DDM (see Section 4.3) a 'reporting' step is included in all design levels (see Table 13.1). This is done for a special purpose, and links to this Chapter's introduction.

By introducing the 'reporting' step, we want to emphasize the need for capturing the goal, approach, results, decisions and next steps in a series of design level reports. It invites the project team to collect and report their findings at least at each design level.

These reports also form the basis for, e.g., Stage-Gate™ project review meetings and decisions about the project's next steps.

As explained in Section 8.1, 'designing' is 'taking decisions' about many different aspects, and at many different levels to arrive at the optimal design. Reporting these design results, design models (schematic, mathematical, verbal and sometimes even physical/iconic models), the design decisions that were made, and the argumentation why these were made is crucial.

Many design projects will not make it to full implementation, and will be stopped in the Stage-Gate™ process because of various business, technical and environmental reasons. However, the knowledge gained from these projects is extremely valuable for

https://doi.org/10.1515/9783110467741-013

Tab. 13.1: Delft Design Map (DDM) design levels, design steps, including the 'reporting' step (AR: Activity Report(s), DLR: Design Level Report).

Stage gate™:		Concept				Feasibility			Development			
AR: Activity report(s) **DLR: Design Level Report**	**Framing**	**Supply chain imbedding**				**Process technology**			**Process engineering**			**Final**
Design levels ⇒ **⇓ Design steps/ activities**	**(PF)**	**(CW-PQ)**	**(PQ-PF)**	**(I-O)**	**(SP)**	**(TN)**	**(UN)**	**(PI)**	**(ED)**	**(OI)**	**(FO)**	**(Final)**
1 Scope	AR	AR	AR	AR	AR	AR	AR	AR	AR	AR	AR	AR
2 Knowledge	AR	AR	AR	AR	AR	AR	AR	AR	AR	AR	AR	AR
3 Synthesize	AR	AR	AR	AR	AR	AR	AR	AR	AR	AR	AR	AR
4 Analyze	AR	AR	AR	AR	AR	AR	AR	AR	AR	AR	AR	AR
5 Evaluate	AR	AR	AR	AR	AR	AR	AR	AR	AR	AR	AR	AR
6 Select	AR	AR	AR	AR	AR	AR	AR	AR	AR	AR	AR	AR
7 Report	DLR	DLR	DLR	DLR	DLR	DLR	DLR	DLR	DLR	DLR	DLR	DLR

the company's decisions at that time and in the future. A well-documented design project that includes what and for what reasons the design decisions were made, and the learnings and future outlooks, is highly valuable. Business conditions can change so that these design results and teachings can be brought to life again. With many team members having moved on to other parts of the company (or to other companies), it is only the documentation that remains. With this documentation, a new design team can have a head start in tackling this project or, in absence of this historical documentation, can be outperformed by competitors.

It should be stressed that only reporting design activities at the end of each design level is far from sufficient in the communication between the team members, and with the stakeholders. These reports are a collection of information and decisions from many more activities. In fact, in each of the design steps design activities take place at each level that should be communicated.

A good working habit for a design team is to use a mix of reporting and communicating tools. These can be used by each team member and by the project team in order to support the communication.

In the next paragraphs the communication tools that are in use by MSc students and PDEng trainees in the TU Delft design courses will be discussed:

- activity reports
- meetings: agenda and minutes of meeting (MOM)
- models (including brain writing and drawings)
- presentations
- reports

13.3 Activity reports

Well organized and written project documentation is not only valuable at the end of each design level and at completion of the entire project. *During* the project, reports and other documentation are needed even more.

After project planning meetings (see Section 13.4), team members will start executing their (individual or team) design activities. Reporting design activities in 'activity reports' assists the designer(s) to first formulate the objective and deliverables of the/their activity. Then he/she/they can choose and report the approach and resources needed. Finally, results, discussions and teachings from the activity can be reported. Typically, activities that take a couple of hours to a day will be concluded with a concise activity report. A format for such a design activity report is shown in Section Appendix A13.1).

The 'activity reports' greatly assist the designer to focus on activities' targets in relation to the overall design. Reporting the results forces the designer to think over these, and gives time for alternative options or argumentation to surface. It also avoids executing too many and possibly unnecessary activities, and thereby losing valuable time. A proven guideline for designers to consider is: '*if a design activity is not worth reporting, it is not worth executing*'. The creation and use of 'schematic models' like figures, drawings and tables is highly recommended. The proper use of these models cannot be stressed enough. Schematics and drawings take time to prepare, but are far more effective in communicating design results and teachings than text. It is very tempting to create schematic models that are very close/similar to information found in other sources. It is much more productive to construct these schematic models to what is required for *your design*, always focusing on your target(s). Some examples can be mentioned here:

(1) House of Quality (HoQ) diagrams (see Section 4.3.3 and Section 8.3).
(2) Stream specifications table (streams crossing battery limit, see Table A13.1).
(3) Concepts/criteria decision matrix or Pugh matrix [1] (see Table A13.2):
   (a) *Do* list your requirements first, and indicate per design option or technology how each option fits in meeting these requirements.
   (b) *Do not* make a long list of available technologies, or of design options and compare their strengths/weaknesses (in general). Link it directly to your project's needs.
(4) Pure component properties table (see Table A13.3):
   (a) *Do* make a list of components properties that are essential/useful for your design and fill out what is easily retrievable and what is not (or not yet retrievable).
   (b) *Do not* make a list of pure components with only properties that are easily retrievable.

The activity reports are not only valuable for the team member executing the activity. They are equally valuable for the other team members, as they can keep abreast with other activities than their own and they can do this at their own convenience. Reading is much faster than listening to a team members' presentations in team meetings. Team members can come well prepared to team meetings, and these meetings can be used for taking decisions on the design direction and future planning. This greatly improves the information exchange and knowledge sharing.

The storage of the activity reports can be very well organized using the DDM's design level/design step coding: reports can be coded from PQ–PF/Scope, all the way to final report. At each design level the 'report' design step can be used to consolidate the various activity reports in e.g., the design level report: 'PQ–PF/report' (see Table 4.3).

Reporting/documentation when done with the 'target setting and knowledge delivering approach', as explained above, communicates a clear result oriented mindset in the project. In the DDM the target setting approach becomes already very apparent in the first design level, 'project framing'. Before starting the actual design work the overall project goals, objectives, deliverables, etc., are determined and 'framed'.

Reporting and documenting the results of this design level gives an 'intermediate result' that can always be consulted by the team members. From here on, all team members will adopt the approach of reporting while designing so that design activities results become available to the team in a continuous workflow.

Last but not least, it is very good practice to decide very early on in the project on the reporting tools/software to use. This ensures that all stakeholders (also the ones outside the organization) are able to open the files, read the report(s) and provide (digital) feedback. Of course, this is equally important for mathematical modeling and simulation software, spreadsheets and presentation software.

13.4 Meetings: agenda and minutes of meeting (MOM)

Standard agenda, MOM structure (internal and external)

During a design project a multitude of meetings are organized. The design team will schedule (internal) progress meetings for the team and (external) meetings with stakeholders. Reports of these meetings (MOM) are special versions of activity reports. In order to set the objective(s) of the meeting time planning is essential, as is the preparation of an agenda. Also, meeting rooms and other facilities (audio/video equipment for presentations and access for remote audiences) need to be arranged. The devil is in the detail as each hiccup in the availability or proper functioning of these items will interfere with the meeting's effectiveness. Meticulous care should be put in these preparations, especially for the stakeholder meetings, and nothing should be taken for granted.

The preparation and recording can be enhanced by using standard templates for agenda and MOM (see Section A13.5).

13.5 Models

Models play a very important role in design as explained in Section 8.1. The verbal, schematic, mathematical and physical models form the backbone of the design results and decisions. Preparation of the schematic, mathematical and physical models and communication thereof are important tasks. These models should be prepared and explained in such a way that they can be understood by stakeholders without any additional extensive text. The designers need to give this sufficient thought and attention, as misunderstanding about insufficient clear models will lead to confusion, mistakes, delays and frustration.

Many models have been discussed in previous chapters. Some further examples are:
- mindmap
- brain writing or brain drawing maps
- fishbone or Ishikawa diagrams
- process block scheme (PBS) (Figure 8.1)
- process flow sheet (PFS) (Figure 8.2)
- process stream summary (PSS) [2]

13.6 Presentations

During the course of the design project, the team not only prepares and delivers reports, but also presentations. These presentations will be held at different phases of the project, and for various audiences/stakeholders. The design team should be aware that their presentations should always be prepared with their audiences in mind. Also the limitations of presentations and their effectiveness in conveying information, knowledge and messages to the audience should be known to the design team members.

The goal of the presentation to the audience should be clear. During design projects the presentations at the key project milestone meetings are to inform the audience about project progress, design decisions made, proposals about how to continue the project and to seek stakeholders' approval and feedback.

The main 'storyline' can be supported (but not overshadowed) by supportive information. The relationship between design results, decisions and the overall project context and objectives should always stand out. Supportive information, mostly on design options analysis and evaluation, comes second: it is needed but not at the expense of omitting the main storyline.

The discussions and feedback of the stakeholders should be promoted, as this information is crucial for the next phases. This implies that the information presented and discussed should be easily understood by the audience, so the exchange of ideas is realized. Recording the discussions and feedback in MOM (Section 13.5) is the main deliverable of most presentations/meetings. The next section provides further guidelines for preparing and delivering presentations.

13.6.1 Quality checks: FOOFI list for presentations

At regular intervals, formal presentations are given to the design project's stakeholders. The preparation and delivery of these presentation should not be underestimated. The structure, content and layout of the slides need particular attention. Audiences have a limited concentration span. The presentation's content should be very clear, and the audience should be focused on the main 'storyline' and message. The stakeholders need to be taken from a helicopter view of the project's objective to the results, conclusions and recommendations. The detailed information is only supportive of this.

Over the years, a quality checklist for design project presentations has been composed by P. Swinkels while supervising and coaching MSc and PDEng trainee design project teams. This list has been named FOOFI, which translates to 'frequently occurring opportunities for improvements'. FOOFI is the phonetic equivalent of (the Amsterdam) slang word for the Dutch '*foefje*', meaning 'knack or agility'.

This FOOFI list for presentations is presented in Section Appendix A13.6. A similar list is also available for reporting quality checks (see Section 13.7.3 and Section Appendix A13.8).

13.7 Reporting in stage gate reviews

In the following sections the contents of the consolidated reports for the 'concept and feasibility stages' are presented.

13.7.1 Concept stage

For the 'concept' stage gate, a consolidated report is prepared for the concept 'stage gate' review. This consolidated report will be composed of selected contents from the consolidated reports at each of the design levels:
- framing
- product concept/consumer wants
- product concept/product function
- input/output, and sub processes

Often, this document is also called 'concept stage report', or 'basis of design' report. It is important that, at this stage, all key data for the design, as agreed by the design team and client/stakeholders, are tabulated and provided with background information.

As can been seen from the previous chapters, the first design levels concern the consumer needs, requirements and specifications, possible product compositions and structures and possible processing technologies using various feed streams, etc.

Not only the specifications of the product, but also of (saleable) byproducts, commercially available feed streams and utilities, and last but not least any environmental limits on gaseous, liquid and solid waste streams should be clearly specified (for an example, see Table A13.1). At TU Delft MSc and PDEng level design projects a report structure for this report is suggested, and standardized tables and formats are proposed (see Table 13.2 for product focused design and Table 13.3 for a process focused design). Further information on this topic can also be found in [3].

Tab. 13.2: Suggested concept stage report structure and contents: product design.

Section (DDM levels: Table 4.3)	Contents	Topics
1 PF	Description of the design; project incentive (opportunity assessment)	Project brief/charter: background; framing (BOSCARD – background, objectives, scope, constraints, assumptions, risks, and deliverables); stakeholder needs; design type and driver (market pull, technology push, innovation map). Market assessment (customers, markets, competitors, 'SHEETS' (safety, health, environment, economic, technically feasible, and social criteria) based opportunity definition); market segmentation; value chain analysis.
2 PF	Design approach	Design challenges, design methodology, creativity methods, group methods and tools (SWOT – strength, weakness, opportunities, and threats); project and time planning.
3 CW–PQ	Consumer needs and requirements; project opportunity assessment	Stakeholder/consumer needs (voice of customer); product quality requirements in relation to product application and entire life cycle.
4 PQ–PF	Product requirements	House of quality (HoQ) (composed by quality function deployment (QFD), relating consumer needs, product quality performance, product property function and manufacturing requirements.
5 PF; CW–PQ; PQ–PF	Database creation	Literature and patent search; product application process conditions and knowledge; chemical components; materials; pure component properties (physical, chemical, SHE (safety, health, and environmental criteria), price); formulation and physical/chemical models and data; thermodynamic models and data; stoichiometry.

Tab. 13.2: (continued)

Section (DDM levels: Table 4.3)	Contents	Topics
6 CW–PQ; PQ–PF	Preliminary product concepts synthesis	Based on the product application process' understanding and functional requirements, generate product concepts – both top down idea generation, and bottom up, based on molecular identity, product (nano) and microstructures, combined into a product/application process combination.
7 CW–PQ; PQ–PF; I-O; SP; TN; UN	Product concepts and preliminary process synthesis	Product concept/criteria evaluation/selection; chosen superior product concept (components, structure, physical/chemical product properties, in use performance). Based on product structuring/manufacturing technology choices, stoichiometry, yields and split factors, determine: plant capacity and location; battery limit; available utilities; other design determining assumptions; *quantitative* Input/output diagrams; (sub)process block diagrams; stream summaries.
8 I-O; SP	Preliminary business case (product and process design) 'SHEETS' analysis	Product: performance versus requirements in use; product and process: performance verses 'SHEETS': safety ('Bowtie'), health, environment, economy (gross margin), technology, social analysis using concept stage knowledge.
9 PF	Planning review and update	Review project and time planning, and use lessons learned to improve future project and time plan; work allocation between team members.
10 PF	Creativity and group process tools	Review the team's creativity and group performance. Plan any improvements for the remainder of the project.
11 PF to UN	Conclusions and recommendations	At completion of the concept phase and feed forward to the subsequent phases.
	List of symbols	
	Literature	
	Appendices	

Tab. 13.3: Suggested concept stage report structure and contents: process design.

Section (DDM levels: Table 4.3)	Contents	Topics
1 PF	Description of the design project incentive (opportunity assessment)	Project brief/charter: project background; framing (BOSCARD); stakeholder needs; customer requirements; technical requirements. Market assessment (customers, markets, competitors; 'SHEETS' based opportunity definition.
2 PF	Design approach	Design challenges, design methodology, creativity methods, group methods and tools (SWOT); project and time planning.
3 PF; CW–PQ; PQ–PF; I-O; SP	Database creation	Literature and patent search; chemical components; materials; pure component properties (physical, chemical, SHE, price) thermodynamics; stoichiometry.
4 I-O; SP; TN; UN	Preliminary process concepts synthesis	Based on technology choice, stoichiometry, reaction yields and separation split factors, determine: Plant capacity and location; battery limit; available utilities; other design determining assumptions; *quantitative* input/output diagrams; (sub)process block diagrams; stream summaries.
5 I-O; SP; TN; UN	Preliminary business case 'SHEETS' analysis	Business case analysis based on 'SHEETS': safety ('Bowtie'), health, environment, economy (gross margin), technology, social analysis using concept stage knowledge.
6 PF	Planning review and update	Review project and time planning, and use lessons learned to improve future project and time plan; work allocation between team members.
7 PF	Creativity and group process tools	Review the team's creativity and group performance. Plan any improvements for the remainder of the project.
8 PF to UN	Conclusions and recommendations	At completion of the concept phase and feed forward to the subsequent phases.
	List of symbols	
	Literature	
	Appendices	

13.7.2 Feasibility and development stages

A suggested report structure and contents for the feasibility and development stage reports can be found in Tables 13.4 (product focus) and 13.5 (process focus). The concept stage report is further extended. Parts of the concept stage report becomes part of the feasibility and development stage report.

Tab. 13.4: Suggested feasibility and development stages report structure and contents: product design focus.

Section (DDM levels: Table 4.3)	Contents	Topics
	Summary	In not more than one page the highlights of the project (objective, approach, results, including quantification of the key design results and performance criteria), conclusions and recommendations should be mentioned.
	Table of contents	Use descriptive chapter and section titles.
1 PF	Description of the design; project incentive (opportunity assessment).	Project brief/charter: background; framing (BOSCARD); stakeholder needs; design type and driver (market pull, technology push, innovation map). Market assessment (customers, markets, competitors, 'SHEETS' based opportunity definition); market segmentation; value chain analysis.
2 PF	Design approach	Design challenges, design methodology, creativity methods, group methods and tools (SWOT); project and time planning.
3 CW–PQ	Consumer needs and requirements; project opportunity assessment	Stakeholder/consumer needs (voice of customer); product quality requirements in relation to product application and entire life cycle.
4 PQ–PF	Product requirements	HoQ (composed by QFD), relating consumer needs, product quality performance, product property function and manufacturing requirements.
5 PF; CW–PQ; PQ–PF	Database creation	Literature and patent search; product application process conditions and knowledge; chemical components; materials; pure component properties (physical, chemical, SHE, price); formulation and physical/chemical models and data; thermodynamic models and data; stoichiometry.
6 CW–PQ; PQ–PF	Preliminary product concepts synthesis	Based on product application requirements and understanding of application process, generate product concepts – both top down idea generation, and bottom up, based on molecular identity, product (nano) and microstructures, combined into a product/application process combination.

Tab. 13.4: (continued)

Section (DDM levels: Table 4.3)	Contents	Topics
7 CW–PQ; PQ–PF; I-O; SP; TN; UN	Product concepts and preliminary process synthesis	Product concept/criteria evaluation/selection; chosen superior product concept (components, structure, physical/chemical product properties, in use performance). Based on product structuring/manufacturing technology choices, stoichiometry, yields and split factors, determine: plant capacity and location; battery limit; available utilities; other design determining assumptions; *quantitative* input/output diagrams; (sub)process block diagrams; stream summaries.
8 I-O; SP	Preliminary business case (product and process design) 'SHEETS' analysis	Product: performance versus requirements in use; product and process: performance versus. 'SHEETS': safety ('Bowtie'), health, environment, economy (gross margin), technology, social analysis using concept stage knowledge
9 PF	Planning review	Review project and time planning, and use lessons learned to improve future project and time plan; specify work allocation between team members.
10 PF	Creativity and group process tools	Review the team's creativity and group performance. Plan any improvements for the remainder of the project.
11 CW–PQ; PQ–PF; I-O; SP; TN; UN	Detailed product design and validation; product/process interaction	Superior product concepts (concept stage) are further designed in detail. Product performance validation (by product modeling, product testing and product prototyping) is carried out. The design of the manufacturing process, the processing equipment and their interactions with the product physical properties is further modeled, tested and validated.
12	Detailed business case evaluation	Product: performance verses requirements in use; Product and process: performance verses 'SHEETS': safety ('Bowtie'), health, environment, economy (gross margin), technology, social analysis using concept stage knowledge
13 PF to Final	Conclusions and recommendations	Based on the design knowledge gathered at this phase, present the strengths and weaknesses of the product/process design (in relation to the needs and requirements). Depict the major risks (technical, financial, etc.) and their consequences in failing to achieve the project objectives. Suggestions are given how to mitigate/eliminate these risks, and what should be the next steps.
	List of symbols	
	Literature	
	Appendices	

Tab. 13.5: Suggested feasibility and development stage report structure and contents: process design focus.

Section (DDM levels: Table 4.3)	Contents	Topics
	Summary	In not more than one page the highlights of the project (objective, approach, results (including quantification of the key design results and performance criteria), conclusions and recommendations should be mentioned.
	Table of contents	Use descriptive chapter and section titles.
1 PF	Description of the design project incentive (opportunity assessment)	Project brief/charter: project background; framing (BOSCARD); stakeholder needs; customer requirements; technical requirements. Market assessment (customers, markets, competitors; 'SHEETS' based opportunity definition.
2 PF	Design approach	Design challenges, design methodology, creativity methods, group methods and tools (SWOT); project and time planning.
3 PF; CW–PQ; PQ–PF; I-O; SP	Database creation	Literature and patent search; chemical components; materials; pure component properties (physical, chemical, SHE, price) thermodynamics; stoichiometry.
4 I-O; SP; TN; UN	Preliminary process concepts synthesis	Based on technology choice, stoichiometry, reaction yields and separation split factors, determine: plant capacity and location; battery limit; available utilities; other design determining assumptions; *quantitative* input/output diagrams; (sub)process block diagrams; stream summaries.
5 I-O; SP; TN; UN	Preliminary business case 'SHEETS' analysis	Business case analysis based on 'SHEETS': safety ('Bowtie'), health, environment, economy (gross margin), technology, social analysis using concept stage knowledge.
6 PF	Planning review	Review project and time planning, and use lessons learned to improve future project and time plan; specify work allocation between team members.
7 PF	Creativity and group process tools	Review the team's creativity and group performance. Plan any improvements for the remainder of the project.
8 I-O; SP; TN; UN; PI	Detailed and integrated process synthesis	Superior process concepts from the preliminary process concepts evaluation are designed, developed in further detail, and analyzed and evaluated (SHEETS) in the task network and unit network levels, leading to (a) superior process flow scheme(s) (PFS), process stream summary (PSS), and mass and energy balances. Mass and heat integration are applied in the process integration level. Utility requirements and product yields are discussed. SHEETS performances are reassessed

Tab. 13.5: (continued)

Section (DDM levels: Table 4.3)	Contents	Topics
		based on the more detailed analysis. SHEETS performances are reported in subsequent sections. The superior PFS is described in detail using the PFS and PSS. Design choices and possible alternatives are described in relation to the SHEETS performance.
9 ED	Equipment summary lists and equipment unit design specification sheets	Based on the functional specification of the equipment in the PFS, the equipment design variables are determined and sized to fulfill these specifications. It is especially important to identify the key equipment design variables (sizes, internals, material of construction, etc.), and how values are assigned to these variables for the best performance. See Section A13.7 for equipment summary lists and equipment specification sheets.
10 I-O; SP; TN; UN; PI;	SHE and S	Safety, health, environment and sustainable development detailed evaluation (HAZOP, DOW's Fire and Explosion Index (F&EI).
11	Economic evaluation	Detailed economic analysis: in addition to gross margin at this stage, also equipment costs, total capital investment and operating costs can be calculated and economic performance criteria – rate of return (ROR), payout time (POT), discounted cash flow rate of return (DCFROR), net present value (NPV) – can be determined (see Chapter 10).
12 PF to Final	Conclusions and recommendations	Based on the design knowledge gathered at this phase, present the strengths and weaknesses of the design (in relation to the needs and requirements). Depict the major risks (technical, financial, etc.) and their consequences in failing to achieve the project objectives. Suggestions on how to mitigate/eliminate these risks, and what should be the next steps are given.
	List of symbols	
	Literature	
	Appendices	

13.7.3 Quality checks: FOOFI list for reports

At major design project Stage Gate^{TM} milestones, design reports are prepared for the stakeholders by the design team. In the previous section the contents for the concept, feasibility and development milestones were presented. In this section guidelines are provided on ways to improve the quality of how the information is presented, taking into account the different roles of the readers of the report. What many (junior) designers do not appreciate is that very few readers will read the complete report (except university lecturers in design courses). Different stakeholders are interested in different parts of the report, and need to be able to navigate efficiently through the report and also to absorb and understand the information they are after with minimal effort. This means that, in order of priority, the following parts of the report need to be self-explanatory and understandable as a standalone text:

- summary
- table of contents
  - should contain descriptive text about content
- project objectives
- conclusions and recommendations
- figures and tables, especially of the design requirements, and design results:
  - stakeholder needs, project and product and process (performances) requirements
  - stream tables
  - process block scheme and PFS
  - etc.

Attention to figures and tables is compulsory, as (scanning) readers tend to focus on these report components first. Figure and table titles together with the content should be sufficient for the reader to understand without reading the main text. The text surrounding figures and tables also receives attention by the scanning reader, so it makes sense to put important information in these areas. The readers should be able to quickly retrieve the main storyline first, and then focus on the domain knowledge that is important for their role(s), e.g., related to the SHEETS performance criteria. These detailed chapters should be clear and make sense when read in isolation. One should include key information from other chapters that is crucial to understand the individual chapters.

As already mentioned in section 13.6.1, a quality checklist for design project reports has been composed (see Section Appendix A13.8).

Abbreviations

BOSCARD	Background, objectives, scope, constraints, assumptions, reporting, deliverables
CW–PQ	Consumer wants – product quality (performance function) (DDM)

DCFROR	Discounted cash flow rate of return
DDM	Delft Design Map
DOW's F&EI	DOW's Fire and Explosion Index
ED	Equipment design (DDM)
FO	Flow sheet sensitivity and optimization (DDM)
FOOFI	Frequently occurring opportunities for improvement
HAZOP	Hazard and operability study
HoQ	House of quality
I-O	Input-output structure (DDM)
MOM	Minutes of meeting
NPV	Net present value
OI	Operability integration (OI)
PBS	Process block scheme
PF	Project framing (DDM)
PFS	Process flow scheme
PI	Process integration (DDM)
P&ID	Piping and instrumentation diagram
POT	Payout time
PQ–PF	Product quality (performance function) – (product) property function (DDM)
QFD	Quality function deployment
ROR	Rate of return
SHE	Safety health environment
SHEETS	Safety health environment economy technology social
S	Sustainable development
SP	Sub processes (DDM)
SWOT	Strengths weaknesses opportunities threats
TN	Task network (DDM)
TOC	Table of contents
UN	Unit network (DDM)

References and further reading

[1] Pugh S. Creating innovative products using total design, New York, NY, USA, Addison-Wesley-Longman, 1996.

[2] De Haan AB, Swinkels PLJ, de Koning PJ. Instruction manual conceptual design project CH3843 – from idea to design. Department of Chemical Engineering, Delft University of Technology, Delft, Netherlands, 2016.

[3] Seider WD, Seader JD, Lewin DR, Widagdo S. Process and Process Design Principles: Synthesis, Analysis and Evaluation, 3rd edn Hoboken, NJ, USA, John Wiley & Sons Inc., 2010, 681–692.

Part D: **Education**

14 Education

In this Chapter, the authors give an overview of the BSc, and truly international MSc, PhD and PDEng programs at TU Delft in the field of (bio)chemical engineering, and with a focus on the interaction between education, research and design. The cooperation with industry and public/private partnerships and the valorization of knowledge in support of innovation is presented from a 'systems point of view'. We hope this forms an inspiration for other universities and countries to adopt a similar approach.

14.1 (Bio)chemical design education: a long history

The authors of this book have built their diverse product and process design and innovation skills through our education at the Universities of Technologies in Delft, Eindhoven and Twente, through our industrial careers at various companies (Corbion, Cosun, DSM, ICI, Shell, Unilever). We have nearly two decades of experience in teaching product and process design at Delft University of Technology (TU Delft), and also at the Technical University Eindhoven (TU/e), University of Twente (UT) and Rijksuniversiteit Groningen (RUG). Inspired by our predecessors, and also having worked closely with many of them in the last decades, we also would like to mention them in this chapter on education. At TU Delft, the process design courses, including a large open ended capstone design project, combined with the setup of a computer simulation course and facilities, have been in existence since the mid 1980s. As a special recognition for their drive to set up and build a world class design course including projects, we would like to mention, the predecessors and current colecturers in curricula of (bio)chemical product and process design at BSc, MSc, and Professional Doctorate in Engineering (PDEng) (see Table 14.1).

In this Chapter, the developments in product and process design education in the (bio)chemical engineering discipline, and the fit in the TU Delft BSc, MSc and PDEng programs are described.

14.2 Education programs

Since the year 2000, higher education at the Technical Universities in The Netherlands has quickly converted to the Bologna 1st/2nd and 3rd cycles of education (Figure 14.1):
- 1st cycle: BSc: three years (180 European Credit Transfer System (ECTS) credits
- 2nd cycle: MSc: two years (120 ECTS credits)
- 3rd cycle: PhD: four years, research orientation
- 3rd cycle: PDEng: two years (120 ECTS credits), design orientation

https://doi.org/10.1515/9783110467741-014

Tab. 14.1: TU Delft lecturers in (bio)chemical product and process design since the mid 1980s – predecessors and current lecturers.

Predecessors in lecturing (bio)chemical product and process design at TU Delft:	Current (co)lecturers in (bio)chemical product and process design:
Em.Prof.ir. A.G. Montfoort	Ir.drs. G. Bierman PDEng
Dr. F.A. Meijer	Dr.ir. G.M.H. Meesters
Em.Prof.dr.ir. J. de Graauw	Dr.ir. H.W. Nugteren
Em.Prof.ir. J. Grievink	Dr. P.J. Hamersma
Dr.ir. P.J. Verheijen	Dr. A.J. Houtepen
Prof.ir. K.Ch.A.M. Luyben	Dr.ir. J.L.B. van Reisen PDEng
Dr.ir. C. van Leeuwen	Dr.ir. P.J. Daudey
Dr.ir. J.C. Goebel	Dr. A.J. Groenendijk
Ing. M. Krijnen	Prof.dr.ir. L.A.M. van der Wielen
Ing. M. Valentijn	Dr.ir. A.J.J. Straathof
Em.Prof.dr. G. Frens	Dr.ir. M. Ottens PDEng
Em.Prof.dr.ir. P.W. Appel	Prof.dr.ir. G.J. Witkamp
Prof.dr.-ing. J. Gross	Dr.ir. M.C. Cuellar Soares PDEng
Prof.dr.ir. P.J. Jansens	Prof.dr.ir. H.J. Noorman
Ir. C.P. Luteijn	Prof.dr.ir. M.C.M. van Loosdrecht
Ir. J. Dijk	A.M. Echavarria Alvarez PDEng
Em.Prof.dr.ir. K. van 't Riet	Dr.ir. L.D.M. van den Brekel
Dr.ir. J.F. Jacobs	Prof.ir. G.J. Harmsen
Prof.dr.ir. J.J. Heijnen	Prof.dr.ir. A.B. de Haan
Ir. M.W. Lambrichts	Ir. P.L.J. Swinkels

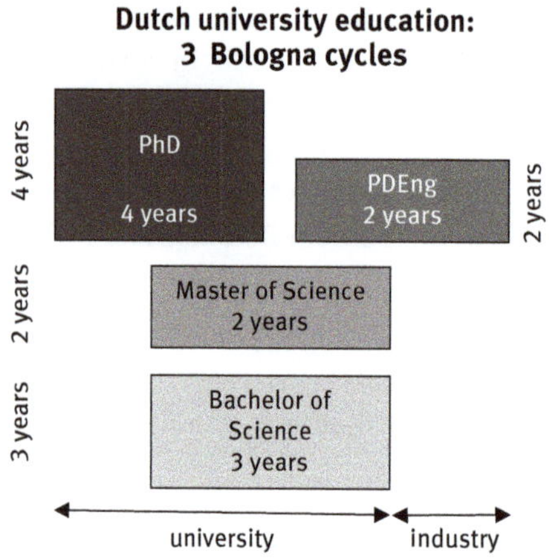

Fig. 14.1: Three cycles of higher education at the Dutch universities of technology (1st (BSc), 2nd (MSc) and 3rd (PhD or PDEng: Professional Doctorate in Engineering) cycle, based on the Bologna declaration.

The three cycles, based on the Bologna declaration, provide the opportunity to students to complete different parts of their studies at different universities within different countries. In addition to the European Erasmus program (European Region Action Scheme for Mobility of University Students), this is stimulating the international mobility of students between European countries. It has been named the largest success of the European Union.

The connectivity and curricula of the BSc, MSc and the third cycle programs (PhD or PDEng) will be discussed with a special focus on the design content.

14.2.1 BSc programs: TU Delft

The TU Delft BSc programs 'Molecular Science & Technology (MST)' and 'Life Science & Technology (LST)' are offered in cooperation with Leiden University. Students attend courses and take exams at both universities and, at graduation, earn a joint degree from TU Delft and Leiden University. This cooperation was started over ten years ago to attract more students to these education fields. It allowed high school graduates to delay their choice to the 2nd year for the more science related chemistry or life science direction in Leiden, or the more technology oriented direction at TU Delft. This has been a very large success. Student numbers rose quickly to larger values than the separate bachelor programs had before. TU Delft saw even a substantial increase after this cooperation. Being exposed to technology oriented lectures and research group research activities made more students choose to continue their BSc education at TU Delft. The same holds for the biotechnology focused BSc program, LST.

The MST and LST BSc curricula have a study load of 180 ECTS credits, are full time and take nominally three years to complete. The program is held in Dutch, but textbooks are mostly in English. The MST program combines chemistry and (chemical) technology. The LST program combines life sciences (cell biology and chemical processes) with biotechnology. Students are introduced to the specializations within chemistry and chemical technology (MST) and life sciences and biotechnology (LST) listed below. Students also gain ample laboratory experience in the very well equipped laboratories in Delft and Leiden. The exposure to research work and to the staff in the research sections starts early in the first year. The experience shows that this boosts the students' understanding of, and interest in, the field. The application of knowledge and research questions inspires the students. It greatly helps their decision making process about whether to continue in the chemistry (MST) and life sciences (LST) field at Leiden University or in the technology directions at TU Delft.

The BSc-MST chemistry track focuses in more depth at disciplines like inorganic, theoretical physical, organic and molecular chemistry. The BSc-MST technology track covers a wider range of mathematics, numerical techniques, thermodynamics, physical transport phenomena, chemical reaction engineering, separation technology and a (chemical) product design project [1]. As mentioned before, building competencies in experimentation skills and participating in scientific research projects are important in both tracks. In the 3rd BSc program year a more in depth or a broad minor can be selected and the BSc final research project completes the curriculum.

The BSc-LST curriculum [2] is also run at Leiden and Delft, but all students follow the same program in the first two years. The more fundamental topics and laboratory projects (cell biology, biochemistry, molecular cell biology and immunology, microbial physiology and molecular genetics) are covered in Leiden, and the more biotechnology topics (mathematics, statistics, thermodynamics, transport phenomena and biotechnology process design project) are taught in Delft. An introductory course on the design of sustainable biotechnology processes is a key design project in year two. The choice of the third year direction (minor and BSc final research project) depends on the MSc choice. MSc-LST in Delft requires the third year BSc to be completed in more technology related topics at TU Delft. The MSc-LST in Leiden requires the third year BSc to be completed at Leiden University.

14.2.2 MSc programs: TU Delft

The MSc programs, Chemical Engineering (CE) and Life Science & Technology (LST), are truly international programs and fully taught in English. The fraction of international students joining this program is currently about 30%. This number is rising, as is the total number of MSc students. The very high rankings of the TU Delft (bio)chemical engineering field in various indices like QS World University Rankings, and Shanghai Ranking, supported by the excellent research and teaching facilities, have a large attraction on international talent. Both MSc programs cover 120 ECTS credits and take two years to complete.

The MSc Chemical Engineering has three tracks: chemical product engineering, process engineering, and nuclear science and engineering. The process engineering track is traditionally the largest, but the other two tracks are steadily growing. Focus of the product engineering track is on the design and synthesis of products, ranging from new construction materials, materials for energy conversion & storage to pharmaceuticals. The nuclear science and engineering track focuses on nuclear engineering and its applications making use of radioisotopes. The TU Delft nuclear reactor is a key facility for these radioisotopes for, e.g., medical applications, and the field is developing rapidly into other health and industrial applications.

The MSc LST has three tracks: biocatalysis, biochemical engineering, and cell factory. These two MSc programs comprise of deepening domain knowledge, executing a large design project, an internship placement and an individual research project.

Top talent students in these MSc programs can apply for the Bologna third cycle of education: PhD (Section 14.2.3) or PDEng positions (Section 14.2.4). MSc students with exceptional academic performance are invited after the first two quarters of the MSc program to participate in the 'honors' program. During this program the students select an (additional) 20 ECTS credits. This program should be completed within the two year MSc program. 'Honors' specializations can have a 'academic research focus' (preparing for a PhD position), a 'design focus' (preparing for a PDEng position) or an 'industry focus' with a research project with an industrial partner.

14.2.3 PhD programs: TU Delft

Research in the area of (bio)chemical engineering at TU Delft is clustered in the 'Delft Process Technology Institute' (DPTI) [3]. DPTI is a partnership and encompasses 19 research groups from the departments Biotechnology, Chemical Engineering, Process and Energy and Energy Technology (Figure 14.2). DPTI's research, education and design activities are focused on three main areas:

- biochemical process engineering
- process intensification and information intensification
- technologies for advanced materials

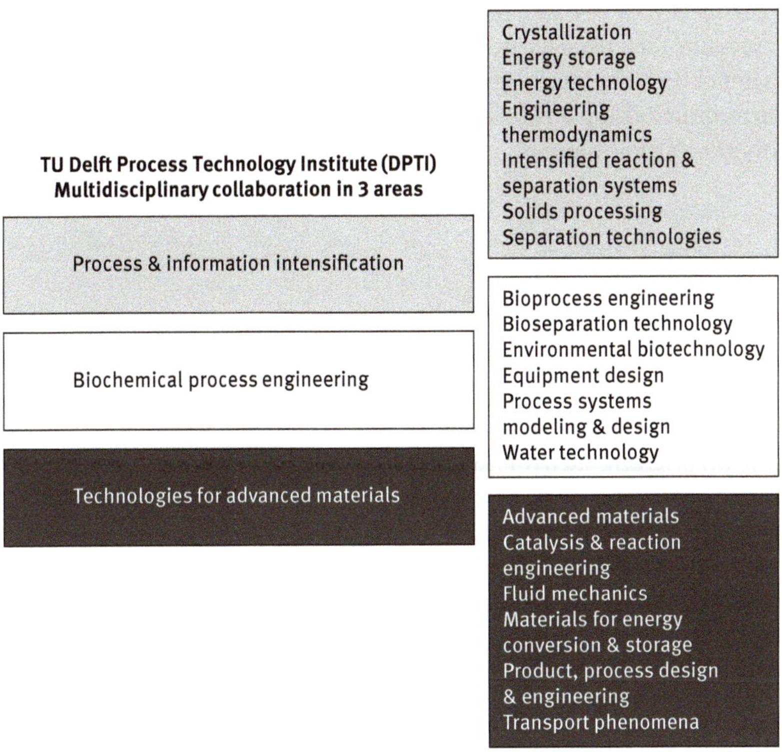

Fig. 14.2: Overview of the three key research areas of the TU Delft Process Technology Institute (DPTI), based on [3].

DPTI aims for aligning the activities in the research groups, and intensifying collaboration with other (inter)national universities and industries. DPTI offers industry and knowledge institutions a wide variety of partnership opportunities. It ranges from BSc or MSc internships to participation in large international consortia for PhD and PDEng projects.

DPTI also has a strong impact on the visibility and impact of its research and education. Over the last few years, the TU Delft in the Chemical Engineering subject always reached top fifteen ranks in the QS World University Ranking [4].

In general, the PhD research tracks are four year projects, funded by national, European and worldwide initiatives. PhD tracks are mostly part of larger projects with many partners in public/private partnerships (PPP). Top performing MSc graduates are hired as PhD students after a careful selection process. The applicants' educational performance, research, and personal competences and motivation are tested in interviews. Once hired, the PhD student will soon setup a personal development plan in addition to their research project plan. PhD students select several personal development courses organized by the TU Delft's Graduate School. After one year, a 'go/no go' presentation and assessment is held. After the 'go ahead' decision, PhD students are expected to complete their research project with minimum delay. The PhD students' progress in executing their personal development plan, research project and realizing publications is monitored by their scientific supervisors. The Graduate School organization aids in recording said plans, execution progress, and in advising PhD students and supervisors during this process.

14.2.4 PDEng programs: TU Delft

During more than thirty years, PDEng programs are a very successful cooperation between the Dutch technical universities and industry. Because of its unique setup, and the fact that these programs hardly exist in other countries, a rather detailed overview is presented here.

The text below is largely cited from Swinkels [5], who has given a concise description on the history, nature and current situation of the over thirty year old post Master PDEng traineeships at the Dutch technical universities. In some places, the text has been adapted to the current (2018) situation. At the end of this section, the contents of the TU Delft PDEng traineeship programs in the (bio)chemical product/process design fields are presented.

"The three universities in The Netherlands – Delft University of Technology, Eindhoven University of Technology and University of Twente – offer two year post MSc technological design programs (traineeships) to selected top level MSc graduates recruited from all over the globe. The Dutch Government and the Dutch industries jointly initiated these traineeships in 1986, with the aim to train top talent in design competencies, and in business and personal skills, to prepare them as key 'knowledge workers' who strongly contribute to the design of new products and processes, and can boost innovation. The PDEng traineeships are designed to train and educate top MSc graduates to become qualified excellent designers capable of designing 'fit for purpose' and 'first of its kind' products, processes and systems. The PDEng trainees are encouraged to actively look beyond the perimeters of their own disciplines and recognize the challenges and restrictions imposed by product chain management, time and money."

"Over a period of thirty years many of these programs are still going strong – supported by long lasting and multinational industrial partners who actively and continuously pursuing (open) innovation and cooperation with the technical universities. Other programs were terminated, and yet new ones matching with emerging innovation areas were started. This lifelike dynamic very closely follows industrial demand and opportunities for innovation, the demand for technical innovation talent and the universities' research, design and education capabilities and strengths. Contribution from, and participation by these stakeholders in innovation, is a prerequisite for every successful PDEng program. Regular contacts and active cooperation projects between the industrial partners, universities and PDEng trainees prove to be essential in this process."

"In the late 1990s the Dutch Universities started to internationalize their MSc curricula, and the PDEng (and PhD) recruitment followed suit. Currently over 90% of the PDEng trainees originate from outside the Netherlands, and both genders are equally represented. The PDEng traineeships have turned into truly global and gender balanced doctoral training centers. This has increased the attractiveness of the traineeships for the industry, university and last but not least for the PDEng trainees themselves."

"The position of the PDEng programs in the Dutch technical university education system is shown in Figure 14.3. It builds on the first (BSc) and second (MSc) cycles of the Bologna agreement, and forms together with the research oriented PhD track, the third tier of academic education. The design oriented PDEng education reaches out into the industry domain, as after one year of competence building and knowledge acquisition, the full second year is spent on an individual design project initiated by and executed (mostly) at the industrial partner's site."

"On successfully completing the PDEng program, the graduates receive a certified diploma and are entitled to use the academic degree 'Professional Doctorate in Engineering (PDEng)'. All diplomas are registered in a central register kept by the Royal Dutch Society of Engineers (KIVI: 'Koninlijk Instituut Van Ingenieurs')."

"Since 2006 the PDEng designer programs have been coordinated under the flag of the 3TU.School for Technological Design – Stan Ackermans Institute (3TU.SAI), a cooperation between the three technical universities in the Netherlands (Delft, Eindhoven, Twente). Currently also *Wageningen University & Research (WUR)* has joined the now called 4TU.SAI cooperation."

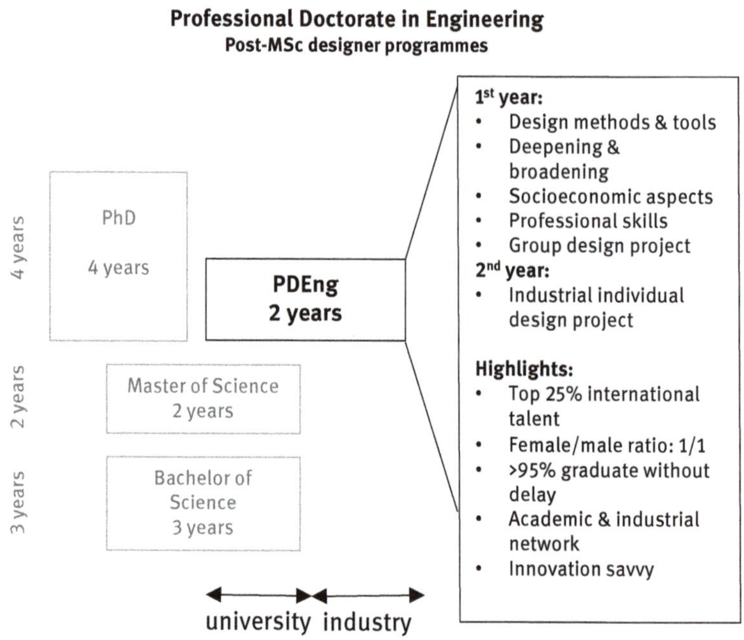

Fig. 14.3: Position of PDEng traineeships at TU Delft (bio)chemical engineering fields and specifics regarding graduation efficiency and gender balance in the Dutch technical universities' education system.

"Nowadays between 200–250 trainees join – after a stringent recruitment and selection process – one of the 20 PDEng programs in the various disciplines every year. They work as a salaried employee (PDEng trainee) for one of the three technical universities, on strengthening their skills through workshops and design cases, and on solving a real design case for (and at) an industrial partner during the full second year of their employment."

"On October 1st 2014 TU Eindhoven Rector Magnificus Prof.dr.ir. C.J. van Duijn awarded the 3500th PDEng (Professional Doctorate in Engineering)." On October 6th 2017, the 4000th PDEng graduate was celebrated.

"It should be noted that over 80% of the 3500 PDEng graduates originate from countries outside the Netherlands, and that the female/male ratio has steadily increased to over 50%. *More than 75%* (and rising) of the PDEng graduates find *technological challenging* and permanent jobs in the Netherlands. From a recent alumni overview from the (then) 23 year old program, Process & Equipment Design (TU Delft), it was also evident that the large majority of these alumni progress their industrial careers in the technology or technology management career track. This is a very positive observation, as traditionally many (MSc) graduates from Dutch universities move into operation management, financial or general management roles rather soon in their careers, and stop contributing directly to technological innovations. As their business and financial counterparts don't make the reverse career switch, this loss of technical talent could easily become a threat. PDEng graduates seem to very naturally and effectively occupy these positions, and keep pursuing a technical career track for a very significant part of their careers. Their technological design talent and ambitions leads and keeps them in this high in demand career track."

"Training PDEng designers parallel to educating PhD researchers is an approach that has catalyzed and delivered many innovation benefits to the industries based in the Netherlands. With vast worldwide challenges ahead of us in the areas of energy, food, water, scarce resources and health, it is an example that could be adopted worldwide by more countries. The startup and implementation of such PDEng programs is not an easy task. To initiate these programs it requires joint initiative and conversion on program content, but also financial and intellectual property aspects between universities, companies and government. To sustain these programs a continuous dialogue between these stakeholders is absolutely necessary. These are conditions that are not quickly and easily met. Dutch industry took the initiative in 1986 and universities and government converged quickly. Over twenty-eight years these partners managed to keep aligned in the interest of the Dutch innovation agenda and of the many (inter)national technological design talents that found their career for life (and are living it)."

"The PDEng programs thrive on the industry's interest in cooperating, and supporting (both in supervising the PDEng trainee, and in financial contribution) the one year individual design assignments as topic for the PDEng design thesis. Many of DPTI's multinational industrial partners in large multinational consortia, also regularly provide design problems for the PDEng traineeships. Also, small and medium enterprises and startup companies are able to find their way via the TU Delft Valorization Center to this cooperation with the PDEng traineeships. All industrial partners (small or big) not only benefit from a novel design solution for their design assignment (by the inputs of fresh minds, a sound design methodology, project management approach, and the latest academic research results), but also build and maintain a strong and lasting network with the university and with top level designer trainees who mostly start their industrial career in the Netherlands."

"In order to prepare the PDEng trainee for the challenging one year individual design project (IDP) for an industrial partner, over the years a technical training program has been developed to strengthen selected knowledge and competence areas. The structure of the programs as described by Swinkels [6] hasn't undergone major changes over the last decade. Strengthening the design methodology tools, and fine tuning and extending the advanced domain knowledge into the quickly developing research and education strengths are the major developments. The specific example

listed below is taken from the PDEng traineeships in the (bio)chemical product and process engineering at TU Delft) [7]:
- Personal skills (personal effectiveness, technical writing and editing, presentation skills).
- Business skills (project management, technoeconomic evaluation).
- Advanced methods for product and process design (design methodology for sustainable product and process design and computer tools).
- Advanced domain subjects (process intensification, catalysis and reactor engineering, nano engineered materials and products) linked to the research focal areas of the TU DPTI and completed with a design assignment.
- A real design case from industry, executed as a group design project (GDP) for three full months, and intensively coached by industrial principals and TU Delft scientists and experienced design coaches."

"The PDEng trainee's knowledge and competencies continue to be trained and coached during the full year Individual Design Project executed at the industrial partner."

The contents of the key PDEng design courses are presented in paragraphs 14.3.3 and 14.3.4.

14.3 Design oriented courses at TU Delft's BSc, MSc, PhD and PDEng level

14.3.1 BSc molecular science and technology (MST) and BSc life science and technology (LST)

14.3.1.1 Chemical product design (4052TLEOY, 6 ECTS)

The course provides insight into the aspects that play a role in developing chemical products. Teams of five to six students need to effectively use their creativity and acquired chemical technology knowledge to develop an innovative, but realistic, new product. Course topics comprise: performing as a team, product design and development, relation between product properties and physical chemical composition, developing physical chemical models to describe product properties, identifying customer needs and idea generation and selection. The course setting mimics the company situation by assigning a product development task to a development team of engineers (i.e., students) by their manager (i.e., a staff member of the Department of Chemical Engineering). The teams are graded on their report (75%) and presentations (25%). All students will fill in a peer review form on their fellow teammates. This peer review is used to adjust the average group grade and to determine the grade of the individual students.

14.3.2 MSc chemical engineering

14.3.2.1 Product and process design (CH3804, 5 ECTS)

This course is a preparatory course for the Conceptual Design Project (CH3843). Topics covered include:

- product and process design methodology
- material supply chains
- opportunity identification
- relation product performance and composition
- process integration
- process flow sheet modeling (Aspen Plus®)
- energy and mass integration
- process and product evaluation and optimization
- health, safety, environmental and sustainability aspects of design
- economic performance estimation and evaluation
- creativity methods and team performance

Lectures are combined with multiple graded product and process design assignments in which teams of students directly apply the obtained knowledge. The students are graded through a combination of individual tests (50%, product, and flow sheet modeling) and team assignments (50%). An individual mark for the team assignments is established by conducting a peer review in which all students evaluate the contribution of their fellow teammates.

14.3.2.2 Conceptual design project (CH3843, 12 ECTS)

The design project involves a conceptual design generation for a chemical product, device or process. To offer students a realistic experience, all assignments are of high practical relevance and mostly delivered and cosupervised by experts in the industry acting as 'clients'. The projects are carried out in teams (between five and six students from mixed educational backgrounds and nationalities) with a stage gate based intermediate reporting and presentations to train teamwork and communication skills. Each project is supervised by a university team coach. Furthermore, the team is supported by multiple university scientific staff members who are selected based on the project content and act as technical advisors. Creativity is stimulated through a creativity workshop and active application of the various creativity tools during the project execution. Final grading is based on:

- Design skills, theoretical knowledge and understanding (context, problem analysis, alternatives generation and design quality)
- Design project organization, planning and execution (realization, motivation, creativity and open mindedness)
- Communication and social skills (reporting, presentations and team performance)

With input collected from all supervisors and advisors. Individual marks for the team members are established by conducting a peer review in which all evaluate the contributions of their fellow teammates.

14.3.3 Process and product design PDEng courses

14.3.3.1 Advanced principles of product and process design (ST6064, 6 ECTS)

The 'Advanced Principles in Product and Process Design' (ST6064) course is divided in three parts:

- Part 1: online individual product design assignment
- Part 2: the core of the course on product design and process design methodologies
- Part 3: comentor and critically review and evaluate a MSc level conceptual product or process design (draft final report)

The first part of the course provides the PDEng trainees with the basics of design work and is executed as an individual self study course with assignments. This first part of the ST6064 course starts immediately after starting the PDEng traineeship.

The core of this course consists of four parts. Course participants:

- Read the course materials on product and process design principles and methodologies discussed in this book and based on the Delft Design Map (DDM) for product and process design and other referenced text books. They engage in online discussions on the material. This preparation is followed up in six half day workshops. During these workshops the material is summarized by the lecturer, examples are provided and a question and answer session is held.
- Apply the design methodology, project planning, team working skills, and communication tools in an open ended team project. Teams consist of three to four PDEng trainees who are coached by the lecturer(s). The team members each spend 80 hours on this project. In three workshop meetings, the teams present and discuss progress and receive feedback on the design approach, results and team cooperation. The final results are reported, and formally presented and discussed with a PDEng trainee peer group.
- Attend and contribute to a one day national conference or workshop organized by, for example, Netherlands Network Process Systems Engineering – NL (PSE-NL) [8], Process Intensification Network (PIN-NL) [9] and/or others.

14.3.3.2 Process simulation laboratory (ASPEN Plus®) (ST6063A, 2 ECTS)

This course's objective is to train students to build a flow sheet in *Aspen Plus®* using a few example cases, to understand an existing flow sheet simulation, to simulate the process with Aspen Plus®, and to solve problems related to the modeling and conversion of these simulations. Also, practical skills in modeling and analysis of complex systems is acquired, and insight into the simulation of several processes is gained.

The course covers mass and energy balances, sequential modular flow sheeting, application of various thermodynamic models, use of simulation convergence methods, and finally a two week process design and simulation project on the simulation of a complete (bio)chemical process.

The course is completed with presenting and discussing the project results at the Fluor Haarlem office for a professional audience and with the assessment of the project report.

14.3.3.3 Advanced process energy analysis and optimization (ST6101, 2 ECTS)

Industrial processes often have large utility requirements for heating and cooling process streams to be able to convert, separate and transport raw materials to products. Efficient use of utilities reduces operating costs and CO_2 emission per unit product. Since the 1970s, analysis and optimization techniques have been developed to assess heat requirements and integration opportunities within process systems and industrial sites. The basic principles are generally available in chemical engineering textbooks. This course gives an advanced guide to applying basic and advanced energy analysis techniques to common industrial cases. It uses these techniques to optimize industrial heat exchanger networks, both in grassroots and retrofit designs. Additionally, attention is given to the design of common utility systems for process heating and cooling. Cases are elaborated on using *SuperTarget™ Software*. The course covers both individual and team assignments. The team assignments are presented at the CB&I The Hague office to an audience of experienced designers.

14.3.3.4 Technology management, economical evaluation in the process industry (ST6612, 6 ECTS)

This course develops an understanding of a broad range of economic, financial, strategic and marketing issues. It also examines the characteristics and working of the process industry business in general and of the development, design and construction funnel in particular. The course covers best practice in capital and operating cost estimating, and assessing project profitability in lifecycle terms. Furthermore, it offers an introduction to financial methods for nonaccountants, and discusses a company's financial performance in the light of its financial statements.

The course has been developed specifically for postgraduate engineers and researchers wishing to pursue a career in process development and design in a manufacturing or engineering/construction environment. At the same time, it is also geared towards the needs of those already working in the industry for some years.

Being able to determine and evaluate investment costs is of vital importance for the process industry. Participants in the course will learn to think in nontechnical terms as costs and market share. The course consists of three parts. In the first part of the course, the structure and characteristics of the process industry are analyzed. Next, attention is paid to organization and layout of projects and to the role of economics. The second part aims at methods for determining investment costs, annual costs and profitability. The third part deals with evaluating the financial performance of a company as a whole. Also, attention is paid to strategic aspects of investment decisions. In concordance with the course participants, other subjects can be dealt with (for example, the 'Porter industry analysis' and 'mergers and acquisitions').

The course starts with interactive lectures combined with workshops. The workshops comprise of exercises and case studies. The course is concluded with an assignment in which the economic feasibility of an actual project is assessed.

Course outline

- Part 1
  - introduction industry/economy general
  - process (chemical) industry business
  - case study
  - project organization
  - economic evaluation/investment appraisal
- Part 2
  - capital cost estimating
  - course project
  - case studies
  - manufacturing cost estimating
  - case studies
  - evaluation profitability/discounted cash flow (DCF) analysis
- Part 3
  - cost of capital, risk and uncertainties
  - finance and accounting fundamentals
  - case study
  - finance and accounting fundamentals (continued)
  - management accounting
  - case study/project

14.3.3.5 (Personal and) project management (ST6111, 2 ECTS)

The (Personal and) Project Management training is set up to learn how to combine personal, project and process management theory with daily (project) work. In addition, it demonstrates how to apply these methods into practice during the various projects in the PDEng traineeship. 'Hard' project management skills like project structuring, project planning and scheduling, are covered. However, an analysis of personal competencies and preparing and working on a personal improvement plan are also key parts of this course. The course builds on multicultural differences, and peer sharing and learning. Individual coaching sessions are held with all participants in the first year and the personal improvement plan is further monitored during the design projects.

14.3.3.6 Sustainable design of processes, products and systems PDEng course (ST6792, 4 ECTS)

Educating design in general and educating sustainable design in particular to students is paved with difficulties. On the one hand the students should be free to use their creativity to come up with breakthrough novel designs that meet a need and are a factor four to ten better in environmental, economic and social impacts than conventional solutions. On the other hand, because they are still inexperienced in design, they can become stifled due to this difficult task and their little experience in designing in general. Moreover, students learn designing in various ways. Dorst [10] distinguishes six different forms of intelligence relevant for design: linguistic, logical, spatial, musical, bodily and personal. He also observed different characters such as pragmatic problem solvers, autonomous designers and visionary artists. It is therefore important to have the teaching and coaching adapted to these differences in personal make up. To teach the students, a method has been developed that facilitates different learning methods and that provides an atmosphere of trust between students and teachers. The design course is set up in six steps, each taking half a day, with two to three weeks in between for assignments. Each step consists of:

- A plenary teaching lecture of a course element group design work starting in the classroom with the teacher being available for coaching on request
- Assignment task work: a course book by Jonker and Harmsen [11] is used which allows students to look up information
- Presentation of intermediate results
- Feedback by the teacher on these results

In this way, the design effort is cut up in smaller steps and allows for a large amount of coaching specifically geared towards each group.

Course program schedule:

- Day 1: Theory of *triple P sustainable development* and relation to product design
  - Team task 1: Define triple P sustainable design scope, goal and constraints for given case
- Day 2: Presentation group results team task 1
  - Theory of industrial symbiosis
  - Team task 2: Redefine design scope, with all inputs and outputs and destinations using industrial symbiosis theory
- Day 3: Presentation of results team task 2
  - Theory precaution principle. Theory lifecycle assessment (LCA) and interactive workshop. Theory closing material cycles and atom balances
  - Team task 3: Define LCA functional unit and all LCA steps for a case and make an LCA block flow diagram with all input and output streams. After day three: PDEng trainees elaborate at home the problem definition and LCA for the reference case

- Day 4: Presentations of results team task 3
  - Theory on people dimension: social fit with culture and acceptance. Theory scenario set building for future robustness test
  - Team task 4: Describe the social and economic context relevant to the design and redesign the solution to fit with the social context.
  - Team task 5: Make a set of two scenarios, then analyze the design solution for robustness using scenario set.
- Day 5: Presentation of results team tasks 4 and 5
  - Question and answer session
  - Team task 6: Finalize design case. After day five, the PDEng trainees will finalize their design and design report/presentation in teams. The final design report is a substantial amount of work to earn the 4 ECTS credits (which totals 112 hours). Six half days of coursework (24 hours) leaves 88 hours per student for the home problem and final project assignment work. The end deliverable is a written report to be sent to the course lecturer four days before the final oral presentation.
- Day 6: Presentation of the triple P design result to the PDEng peer groups and lecturer

14.3.3.7 Group design project (GDP) (ST6802, 21 ECTS, ST6814, 17 ECTS, or ST6815, 17 ECTS)

The objective of the group design project (GDP) for all PDEng programs (Process & Equipment Design – ST6802, Designer in Bioprocess Engineering – ST6814 and Chemical Product Design – ST6815) is to enable the participants to develop their design skills in a real life design case in teams consisting of three to five members. The group as a whole is responsible for the final outcome of the project. In turn, each member of the group will be assigned the group leadership. The participants are forced to use their deepened skills of product and process fundamentals and acquired design skills in an integrated way. They are made aware of other engineering disciplines in the areas between conceptual product and process design and an operational plant. The theory and learnings from the prerequisite courses Advanced principles of product and process design (ST6064) and (personal and) project management (ST6111) are applied as follows:

- Each participant spends 600 hours on the assignment.
- The total amount of (elapsed) time for the assignment is limited to half a year, so each participant spends at least 24 hours a week (60%) on the assignment on average.
- The hours not spent on the project (400 hours) are used for other PDEng program subjects.
- The implementation of the assignment is the responsibility of the team as a whole.
- Generally, three reports are written during the project and two intermediate and one final presentation(s) is held at the following milestones meetings:

  - Scope and basis of design
  - Results of conceptual design (including product composition and structure, process flow diagrams and balances)
  - Final review meeting: evaluation of the design, final conclusions and recommendations
- These meetings take place at TU Delft or at the location of the (industrial) principal and for a wider audience of the principal's employees.
- The team members manage their own planning and project execution. During the project the supervisors act solely as consultants. After project completion, the supervisors review and assess the results and the team performance.

Progress and consultation meetings are scheduled on a regular basis and are attended by the participants and supervisors (at least every three weeks although often more frequently). The project is carried out under supervision of TU Delft staff members, who are in an advisory role of an experienced process engineer/designer from the industry. The industrial supervisors join in the progress meetings (at TU Delft or at their location). In between meetings, communication takes place by telephone, online video communication or email. The supervision by (an) experienced designer(s) from industry is a directive of the Netherlands Committee for Certification of (post MSc) Technological Designer Programs (CCTO).

14.3.3.8 Loss prevention in process design (ST6042, 5 ECTS)

The main issue of the course Loss Prevention in Process Design, is the design of safe chemical processes. Important subjects are: inherent safe process design, equipment design, selection of material, piping systems, heat transfer systems, thermal isolation, process monitoring and process control, High pressure systems, electrical systems, control of dangerous emissions (for instances flares) and documentation are also included.

The goal of the course is to familiarize PDEng trainees with advanced knowledge of risks of fire, explosion and toxicity of chemical products and processes. Students should be able to design a safe and environmentally friendly chemical product and plant using modern loss prevention design methods upon completion of the course. The course consists of the following parts:
- lectures/exercises on chemical risks (fire, explosion, toxicity)
- student presentation on chosen safety topic
- preparation of written conceptual design for hazardous chemical process, with detailed analysis of and solution to safety, health and environmental risks, making use of various approaches: layer of protection analysis (LOPA), inherently safer design (ISD), green chemistry and green chemical engineering
- presentation of project assignment results. Students will work in groups of two to four.

The objective of the assignment is to create a conceptual design of a chemical process for safe and environmentally friendly production of a hazardous substance (where 'hazardous' refers to toxic, flammable and explosive substances), using a modern design philosophy. The scope of the assignment comprises of: synthesis route and reaction stoichiometry, fire/explosion/toxicity data on all substances involved (reactants, products and byproducts), basic thermodynamics, basic kinetics, process flow sheets including most the important unit operations, description of process including process conditions, with special emphasis on the design of the most hazardous part of process.

14.3.3.9 Individual design project (IDP) (ST6902, or ST6903, or ST6904, 60 ECTS)

All PDEng trainees are carrying out an individual design project (IDP) during the second year of their PDEng traineeship. The necessary skills needed for this project are provided during their prior studies and during the first year of the traineeship.

The IDP aims for the (re)design of a (bio)chemical product (material, formulation and configured product device) and process or method, including attention for the imbedding in the supply chain.In each case, it should comprise a design problem that requires a novel design for a client, rather than applying already existing designs. Each individual design assignment generally contains design as well as research and development aspects. In all cases, the research and development must be beneficial to the design: e.g., acquiring required knowledge or experimental data. The design part should amount to at least 50%, leaving at most 50% for the research and development part.

The design task can be part of a larger project: e.g., a product and process development/design project. The designer will always interact with other project participants.

14.3.3.9.1 Confidentiality and working location

At the request of the principal, all knowledge obtained during the project will be treated as confidential and will not be disclosed to third parties without prior consent of the cooperating partner (principal). To this end, a cooperation agreement is signed by both the TU Delft and cooperating partner. This agreement also contains a nondisclosure agreement for the designer and TU Delft members of the supervising steering team. Detailed guidelines on how to act are given by the program management team.

For the IDP, the PDEng trainee will be mostly seconded to industry or to a research section at TU Delft. When working with the industrial partner, relocation may need to be considered.

14.3.3.9.2 Composition of steering group

The steering group of an IDP consists of a mentor from the principal engineer from the industrial party, a professor/mentor from university, a consultant with expert knowledge on the subject and an experienced design mentor from the PDEng program (the consultant and professor may be the same person). The PDEng program management will compose the steering team in close cooperation with the principal.

14.3.3.9.3 Milestones and implementation

During the IDP, at least four milestones can be identified: The kick off, basis of design (BOD), the Intermediate and the Final reviews (BOD, Intermediate and Final Reports are to be submitted). They consist of meetings with both the industry and university supervisors, and include a presentation by the PDEng trainee. The Kick-off meeting is scheduled within the first two to four weeks after the start of the project. The BOD review is scheduled two to three months after the start of the project, the Intermediate review is after 8 months, and the Final review is at the end of the project. All members of the steering group attend these meetings.

Generally, the Principal of the project will provide a project description, terms of reference, or proposal at the very beginning of the project. Often the description is relatively brief (concise) and not clear on all details. One of the first activities of the designer will be to lay down the detailed project description in close cooperation with the principal. The project description addresses the objective of the project, deliverables and a (time) plan how to obtain the deliverables (project approach). The project content will be formally discussed during the Kick-off meeting, which is to be organized preferably during week two and four of the project.

The main purpose of the Kick-off meeting is to make clear the objectives, and to agree upon them. The definition stage of a project leaves much room for misunderstandings between parties. Therefore, the description must be as explicit as possible. In order to avoid any misunderstandings, it is helpful to provide a description of the different project activities and the duration thereof. In this way, the parties involved will get a clear idea about on what subject and how time is spent, and whether parties' expectations are satisfied. Focus points during the Kick-off Meeting are:

- Project objective
- Scope of the project, delineation
- Approach
- Project planning, time table
- First results

Several other meetings with the supervisors and/or experienced design engineers are organized throughout the year the design project takes. The meeting frequency depends on the need and the opportunity. It is also the PDEng-trainee's responsibility to organize these meetings. Minimum will be a monthly meeting with the project manager/design mentor from the PDEng program (bi-weekly contact is advised).

Nontechnical skills needed for the execution of the project are to a large extent obtained during the first year, with dedicated courses on the social-economical aspects, and personal and professional skills as part of the core program. The designer is responsible to manage the project and to communicate adequately with all stakeholders involved. The designer has to organize and plan progress meetings. Moreover, the designer will also chair the meetings, and ensure to distribute agenda and minutes of meeting.

14.3.3.9.4 Reporting and oral presentations

The purpose of reporting is to document and transfer obtained knowledge from the designer to the project principal. To give evidence of skills, the designer is requested to prepare a BoD report and an intermediate report. In this way, the reporting effort is started at an early stage and is adequately spread over the project span.

Alongside the written report, the designer also will report orally during the steering group review meetings. At the end of the project two, final presentations are required. One presentation is to be held at the principal and one at the TU Delft. During the presentation at the principal, generally, representatives from both management and coworkers from the principal will be attending.

At the university, a colloquium must be given. As a rule, the colloquium is open to the public. Care must be taken to have the contents cleared by the principal beforehand, so as not to disclose any confidential information.

14.3.3.9.5 Assessment

After the completion of the IDP, the work will be reviewed with respect to theory, implementation and communication in accordance with the TU Delft PDEng assessment criteria. For all three aspects a mark will be given. The appraisal of the work will be carried out by the steering committee and under the responsibility of the professor/mentor.

14.3.4 OSPT courses (PhD, PDEng and participants from industry)

The *OSPT* (which is the Dutch abbreviation for the *Research School Process Technology*) organizes advanced two to five days long courses and workshops for PhD and PDEng students and for participants from industry.

The courses take place at TU Delft, TU Eindhoven or University Twente. The Chemical Product Centric Sustainable Process Design course (by R. Gani) held at TU Delft is described below in detail. The other courses are listed below with a reference to the OSPT course plan website [12].

14.3.4.1 Chemical product centric sustainable process design (PhD/PDEng course)

The objective of this PhD/PDEng course is to give the participants a view of chemical product design and the important process design issues related to their development (product centric process design). The course will highlight how to define the needs of a chemical product, how to identify the candidate chemicals and/or mixtures of chemicals and how to quickly evaluate the important process design issues so that decisions related to product development can be made in the early stages of product development. The objective is also to highlight the currently available methods and tools that can be applied to solve various types of problems associated with product process design in a systematic and integrated manner. Different case studies will be used as application examples. The ICAS software developed by the Computer Aided Process Engineering Center (CAPEC) at the Technical University of Denmark (DTU) will be used. Also, two new software tools, *ProCAFD* (a tool for synthesis-design of chemical processes) and *ProCAPD* (a tool for synthesis-design of chemical products), made available through the PSE for SPEED Company Ltd [13], will be used in the course.

In chemical product design and development, one first tries to find a product that exhibits certain desirable or targeted behavior and then tries to find a process that can manufacture it with the specified qualities. The product may be a single chemical, a mixture, or a formulation. For the later product type, additives are usually added to an identified active ingredient (molecule or mixture) to significantly enhance its desirable (target) properties. Examples of chemical products, such as *functional chemicals* (solvents, refrigerants, lubricants, etc.), agrochemicals (pesticides, insecticides, etc.), pharmaceuticals and drugs, cosmetics and personal care products, home and office products, etc., can be found everywhere. The workshop will only cover chemical based products and issues related to their design, development (including process design also) and analysis (product performance). Even though it is possible to identify many chemicals and/or their formulations as potential chemical products, only a small percentage actually becomes commercial products. Finding a suitable process that can sustainably (that is, reliably, efficiently and economically) manufacture the identified chemical with the desired product qualities, as well as evaluating product performance during application and analyzing market trends, play important roles in product design and development. From a process point of view there are products where the reliability of the quality of the manufactured chemical may be the deciding factor (for example, drugs and agrochemicals), while there are others where the cost of manufacturing the product is at least as important as the reliability of the product quality (such as solvents, refrigerants, lubricants). This means that product centered sustainable process design is important because identifying a feasible chemical product is not enough. To make it sustainable, the process needs to be efficient, reliable, economically feasible and environmentally acceptable. Also, while in the case of functional chemicals the identified molecule or mixture is the final product, in the case of chemicals based consumer products (drugs, cosmetics and personal care products, etc.), they are intermediate products from which the final products are obtained through additional processing. Therefore, the performance of the manufactured product, when applied, needs to be tested and validated. For some functional chemical products (such as solvents and refrigerants) this may be straight forward, but for some consumer products (such as drugs and food products), it may not be so straight forward.

14.3.4.2 Other OSPT courses (see [12])

- Computational Fluid Dynamics for Chemical Engineers
- Advanced Thermodynamics
- Fundamentals and Practice of Process Intensification
- Molecular Affinity Separation
- Numerical Methods for Chemical Engineers

Abbreviations

3TU.SAI	Federation of three Dutch technical Universities – Stan Ackermans Institute
4TU.SAI	Federation of four Dutch technical Universities – Stan Ackermans Institute
BSc	Bachelor of Science
BOD	Basis of design
CAPEC	Computer Aided Process Engineering Center
CB&I	Chicago Bridge & Iron Company
CCTO	Certification Committee for (post MSc) Technological Designer Programs
CO_2	Carbon dioxide
DCF	Discounted cash flow
DDM	Delft Design Map
DPTI	Delft Process Technology Institute
DTU	Denmark's Technical University
ECTS	European Credit Transfer System
GDP	Group design project
ICAS	Software tool developed at CAPEC/DTU
IDP	Individual design project
ISD	Inherent safer design
KIVI	Koninlijk Instituut Van Ingenieurs (Royal Dutch Society of Engineers)
LCA	Lifecycle assessment
LOPA	Layer of protection analysis
LST	Life Science & Technology (BSc and MSc programs)
MSc	Master of Science
MST	Molecular Science & Technology (BSc program)
OSPT	Onderzoeksschool Procestechnologie (Dutch for: Research School Process Technology)
PDEng	Professional Doctorate in Engineering
PhD	Doctor of Philosophy
PIN-NL	Process Intensification Network – Netherlands
PPP	Public/private partnerships

ProCAFD	Software tool for computer aided process synthesis design
ProCAPD	Software tool for computer aided product synthesis design
PSE-NL	Process Systems Engineering – Netherlands (network)
RUG	Rijksuniversiteit Groningen
TU	Technical university
TU/e	Technical University Eindhoven
UT	University Twente

References and further reading

[1] Molecular Science and Technology BSc curriculum, TU Delft, 2017. Accessed October 8, 2017, at https://www.tudelft.nl/en/education/programmes/bachelors/mst/bsc-molecular-science-technology/curriculum/.

[2] Life Science and Technology BSc curriculum, TU Delft, 2017. Accessed October 8, 2017, at: https://www.tudelft.nl/en/education/programmes/bachelors/lst/bsc-life-science-technology/curriculum/.

[3] TU Delft Process Technology Institute (DPTI), TU Delft, 2017. Accessed October 8, 2017, at http://dpti.tudelft.nl.

[4] QS world universities ranking by subject – chemical engineering. QS Top Universities, 2014, 2015, 2016). Accessed October 10, 2017, at http://www.topuniversities.com/university-rankings/university-subject-rankings/2014/engineering-chemical, https://www.topuniversities.com/university-rankings/university-subject-rankings/2015/engineering-chemical, https://www.topuniversities.com/university-rankings/university-subject-rankings/2016/engineering-chemical.

[5] Swinkels PLJ. Post MSc technological design (PDEng) traineeships by Dutch universities of technology catalyze industrial innovation, QScience Proceedings (Engineering Leaders Conference 2014) 2015: 21. Accessed October 8, 2017, at http://dx.doi.org/10.5339/qproc.2015.elc2014.21.

[6] Swinkels P, Grievink J, Hamersma P. Technical and human factors in chemical process design courses. In: Tu S-D, Wang Z-D (eds.), Total engineering education, Shanghai, East China University of Science and Technology Press, 2006, 59–65.

[7] Traineeships Engineering Design – PDEng programmes, TU Delft, 2017. Accessed October 8 2017, at http://www.pdeng.tudelft.nl.

[8] Netherlands Process Technology Networks – Process Systems Engineering, 2017. Accessed October 8, 2017, at http://www.processinnovation.nl/pse.

[9] Netherlands Process Technology Networks – Process Intensification Network (PIN-NL), 2017. Accessed October 8, 2017, at http://www.processinnovation.nl//homepinnl.

[10] Dorst CH. Design problems and design paradoxes. Design Issues, 22(3), 2006, 4–17.

[11] Jonker G and Harmsen J. Engineering for sustainability – a practical guide for sustainable design. Amsterdam, Netherlands, Elsevier, 2012.

[12] OSPT Courses, ISPT Innovation Academy. Accessed October 8, 2017, at http://www.ispt-innovationacademy.eu/course-agenda.html.

[13] PSE for SPEED – Sustainable Product Process Engineering, Evaluation & Design. Accessed October 13, 2017, at http://www.pseforspeed.com.

A3 Appendix to Chapter 3

Group design with Belbin team roles

Project groups are mostly constructed based on the required disciplines only. However, if groups have to cooperate closely for the desired results then they should get along with each other. In addition to disciplinary design, Belbin team roles can be used to populate the group or team [1].

Belbin role descriptions

- The **Plant** role is to be highly creative and good at solving problems in unconventional ways. He may be the outsider planted in the team.
- The **Monitor** provides a logical eye to make impartial judgments about options.
- The **Coordinato**r focuses on the team's objectives and delegates work appropriately.
- The **Resources Investigator** provides knowledge from outside the team.
- **The Implementer** plans a practical, workable strategy and carries it out as efficiently as possible.
- **The Finisher** makes sure at the end of a task, to remove errors, subjecting it to the highest standards of quality control.
- **The Teamworker** helps the team to gel. He may also fill in gaps left by others.
- **The Shaper** provides the necessary drive to keep the team moving to reach the goal.
- **The Specialist** provides in-depth knowledge of a key area. His weakness is to focus narrowly on the problem element and prioritize it over the team's progress. In the authors' view it is in general better to consult the specialist rather than making him a team member.

The preferred role of an individual can be determined by one's self using the Belbin self-perception method. This method can be found hereafter.

Arittzeta [2] provides a review of a large number of experimental studies on teams designed with Belbin team roles. His conclusion is that Belbin's team role theory has been validated. Some roles, however, show intercorrelation. From the intercorrelation information, the following major style groupings can be made:
Completer style: Completer-Finisher, Implementer, Monitor Evaluator
Bridge style: Teamworker, Monitor Evaluator, Coordinator
Innovative style: Resource Investigator, Plant, Shaper

Therefore, small teams should have one of each of these three major styles.

The authors have also had a powerful experience with groups designed using the Belbin self-perception team roles. The most complementary team indeed produced

https://doi.org/10.1515/9783110467741-015

the best results, while the team with members of only one role (specialists) performed worse by far.

Making inventory for Belbin team roles using self perception

Teams with complementary Belbin roles can be easily composed as follows. Each potential team member fills in a questionnaire and concludes their preferred role(s). Then a team or several teams can be composed with complementary roles. The questionnaire can be obtained from:
http://www.belbin.com/about{/\penalty\exhyphenpenalty}belbin-team-roles/

Creativity methods

The literature on creativity methods for innovation is vast. We limit ourselves here to methods for which the authors have experience in industrial product-process practice. A good introduction to further information can be found in Nijstad [3] and Hermann [4].

Brainwriting 6-3-5

Creating ideas within the product-process industries is probably slightly different from that within other industries in that process engineers as a majority are introverted people, who feel uncomfortable in directly expressing ideas. Brainwriting is sometimes a better method than brainstorming, in particular for introverts. On average brainwriting generates twice as many ideas as brainstorming [3]. For process engineers that factor could be even higher.

The brainwriting 6-3-5 method is very simple to execute and needs no trained facilitator.

Step 1: Plenary, a clear problem to solve is defined and stated clearly to all participants.

Step 2: Groups of preferably 6 people are formed.
Each member gets a writing chart with at least 3 columns and a number of rows at least equal to the number of members.

Step 3: In Round 1 each group member writes down silently 3 ideas in the first row on his writing chart.

Step 4: After 5 minutes the writing charts are rotated to the neighboring members and Round 2 is executed in which each member enriches the 3 ideas of the first row by putting additional ideas in the second row

Step 5: This 5 minute cycle is repeated until the members are exhausted of ideas. Of course additional clean sheets of writing charts can be provided to start new sessions and rounds.

Step 6: All filled in sheets are gathered by the facilitator and recorded outside the brainwriting session.

Potential problem analysis for start up

Potential problem analysis for the start-up phase is very useful for generating a robust start-up plan. Here is a template that the author has prepared and used for making robust start-up plans. It is derived from a problem analysis method [5] and turned into a potential problem analysis method.

Potential problem analysis template:

Question	Reference product/process	New design	Difference in new design	Risk
What				
Where				
When				
Size				

References and further reading

[1] Belbin, 2017. Sourced from: http://www.belbin.com/rte.asp.
[2] Aritzeta A, et.al. Belbin's team role model: Development, validity and applications for team building, Journal of Management Studies, 44(1), 2007, 96–118.
[3] Nijstad BA, Paulus PB. Group creativity: Innovation through collaboration. 2003, 326–339.
[4] Herrmann D, and Felfe J. Effects of leadership style, Creativity technique and personal initiative on employee creativity, Brit J Manage, 25, 2014, 209–227.
[5] Kepner CH, Tregoe BE. The new rational manager, New Jersey, USA, Kepner-Tregoe Inc., 1981.

A4 Appendix to Chapter 4

History of (chemical) product and process design leading to Delft Design Map

(Chemical) products and their manufacturing processes have been around for a long time, long before formal product and process design methodologies were developed and taught at universities of technologies since the end of the 1960s. Industry managed to do this based on heuristic knowledge of and operational experience with these products and processes. Experienced designers transferred their knowledge and experience through heuristic rules to novice designers. Design improvements were mostly incremental.

Design courses at universities in the 1970s and 1980s focused heavily on the (very powerful) process "unit operations" design, with attention toward shortcut methods for process integration, and performance analysis mainly based on cost estimation and techno-economic measures. The really good text books "Plant design and economics for chemical engineers" [1] and "Strategy of process engineering" [2] were used in many curricula in this period. The conceptual process design course taught at TU Delft by Prof. A. Montfoort (TU Delft 1970–1990) was based on this approach.

Process design methodology development for education started to take off in the 1980s. In this period, steady state process flowsheet simulatorsprocess flowsheet simulator (like ASPEN™, ASPEN-PLUS™ (AspenTech) and PRO-II™ (Simulation Sciences Inc.) emerged from research work at universities (like Massachusetts Institute of Technology (MIT)), and were launched as commercial tools for analyzing mass and energy balances of complete processes. Through these computational tools – including key thermodynamic models – a 'systems engineering approach', built up from unit operations and including process analysis and optimization became possible. The development of the process integration methodologies (thermal pinch analysis) at University of Manchester Institute of Technology (UMIST) by Linnhoff in 1968 [3], with applications in Imperial Chemical Industries (ICI) in the 1980s also gave process systems engineering thinking a clear boost. Here the systems engineering approach was used to arrive at considerable energy conservation.

After this attention to process design methods took off (see Figure A4.1 below).

In the period 1985–1995, dynamic flowsheet simulators were also developed for modeling process dynamics, startup, and control: SPEEDUP and gPROMS® were both developed at Imperial College London (ICL), whereas Jacobian was developed at MIT. The use of process flowsheet simulators allowed process systems engineering research and education to focus again on the fundamental process-synthesis problem: what unit operations and process structures were best suited for the required process task? The widely used textbook "Conceptual Design of Chemical Processes" by Douglas, issued in 1988 [4] successfully attempted to address these questions. In this book,

https://doi.org/10.1515/9783110467741-016

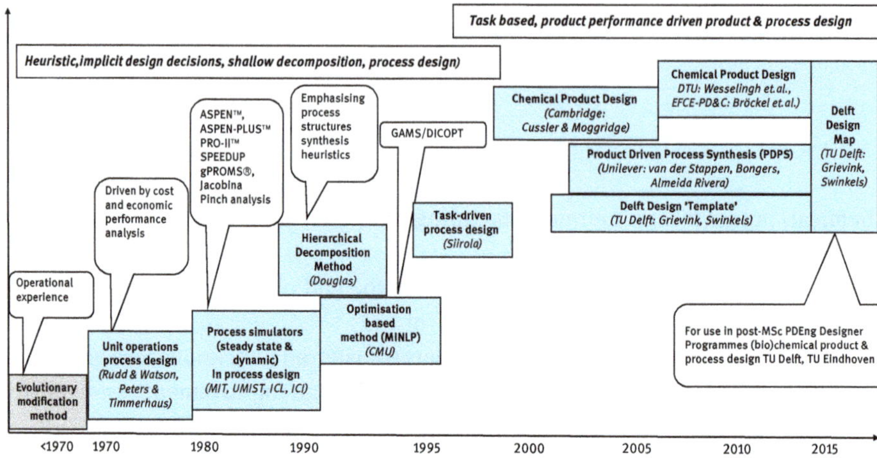

Fig. A4.1: (Chemical) product and process design methods – a historical perspective.

he successfully translated his vast wealth of industrial design experience into a hierarchical design decomposition scheme. Process structure options were generated/ synthesized using plausible heuristic rules in combination with the unit operations concept. In 1999 Seider and Lewin published the first edition of their student textbook on process design principles [5], taking a similar approach, but with more details on process structure, equipment design and cost. In the new millennium, new editions of this textbook have been upgraded over the years.

In the late eighties, computer-based optimization based methods were also developed to solve process design problems. Based on predefined superstructures encompassing all process flowsheet structure options, mixed integer nonlinear (MINLP) programming techniques were developed and applied to find the optimal flowsheet given the given optimization function. The GAMS/DICOPT MINLP algorithm was developed by Kocis and Grossman at Carnegie Mellon University [6].

In the mid-nineties, a novel task-driven process design approach was published by Siirola [7, 8]. For the first time, unit operations were no longer the building blocks for process design, but rather individual tasks or functions, such as adsorption, reaction, heat and mass transfer, and diffusion. These tasks could be combined in single columns generating novel design solutions, such as reactive distillation, in-situ crystallizer fermentation, etc. Inside Eastman Chemical Company, this task driven design approach had led to the design and implementation of large scale reactive distillation columns in which many tasks were combined, the well-known methyl acetate production route [9]. This multifunctional design approach is now part of the process intensification design tool box.

For product design, let alone concurrent product-process design, no generic design methods were yet developed and published in this period. For homogeneous products, such as food products and chemical products, designs were carried out inside companies, based on experience and tacit knowledge. In the nineties, the product design approach was at the phase where process design was in the late 1960s. Each industry has many heuristic rules, mostly limited to the existing product portfolio and based on in-house experimental validation. Industrial researchers are reaching out to academia where attempts were made to develop and understand application and manufacturing of new molecules, structures, and materials.

With an increasing focus in the chemical industry and related academia on product innovation, tools were becoming available and, once the desired product composition and structure is known, product driven and task based process design methods become available to tackle these design challenges. However, the product design phase – how to arrive from a desired function/performance of the product to its composition and structure – is another matter. This is a field where chemical engineers have historically not been educated and involved in, and is largely the domain of the chemistry and material sciences scientists.

In the first years of the millennium, the first (chemical) product design courses were initiated at several universities. Also, the first student textbooks were published at Cambridge by Cussler & Moggridge [10] at RU Groningen and DTU Copenhagen by Wesselingh et al. in 2007 [11]. In 2010 the third edition of the textbook by Seider et al. [12] was for the first time focused on product design aspects.

The approach adopted by Cussler et al. and Wesselingh et al. starts with describing and explaining how chemical products work in performing their task, and how changes can be made to improve them. Similar to the unit operation approach, this product related knowledge was first presented and taught in a rather compartmentalized way, with little connection between different classes of products and applications. Wesselingh (with his transport phenomena background) tried to put more emphasis on the importance of an engineering (quantification) approach with transport phenomena as a key part of product performance. He attempted to capture product application properties in dimensionless numbers, and advocated the use of time constant estimations of the ruling phenomena. This is a powerful tool process designer use in the 'scale-up by scale-down' approach for processes, but is equally suited for the characterization of product in-use processes.

Frens and Appel (both TU Delft) have also stressed in their product design courses, the importance of physical chemistry and complex rheology in the formulation and functioning of products in product design. Process systems engineering (PSE) scientists/engineers with a focus on applying chemical thermodynamics in process synthesis and engineering (Gani at Technical University of Denmark's [DTU] Computer Aided Process-Product Engineering Center (CAPEC) and K. Ng at Hong Kong University of Science and Technology [HKUST]) started to extend their work to product design, development, and simulation in the first decade of the new millennium.

In 2003 the European Federation of Chemical Engineering (EFCE) Section on Product Design and Engineering (PD&E) was founded in Granada, Spain. This happened at the first EFCE-sponsored PD&E Conference. In 2007 members of this section published two volumes of a book on "Product Design and Engineering – Best Practices" edited by Bröckel et al. [13, 14].

Starting in 2001 the cooperation between Grievink (ex-Shell) and Swinkels (ex-Unilever) at TU Delft led to the combination of their complementary process and product design knowledge and experience. The oil, gas, and chemicals company Shell has a long history of designing and implementing large scale continuous processes for homogeneous products. Unilever, the fast-moving consumer goods company has a long history of designing and developing novel structured and multiphase products for its world markets. Grievink and Swinkels developed in this period at that time were called hierarchically decomposed and task oriented "Delft Design Template for Conceptual Process Design." In 2014 a simplified version of the Delft Design Template for Conceptual Process Design was published in 2014 [15].

In cooperation with TU Delft, Unilever developed and tuned the Delft Design Template approach to fast moving consumer products (mainly foods) and proposed the "Product Driven Process Synthesis (PDPS)" methodology described by Meeuse in 2000 [16] and in 2017 by Almeida-Rivera [17].

Key developers van der Stappen, Bongers, and Almeida-Rivera at Unilever and Boom et al. (Wageningen University & Research [WUR]) also adopted this approach in their research into the design of new food products and their manufacturing.

A very similar approach in adopting a schematic conceptual model for chemical product design has been taken and published by Bernardo at University of Coimbra in 2017 [18], in which the same design levels/ functions are proposed (quality function, property function, and process function) as in the Delft method described in Chapter 7. Bernardo also proposes a mathematical approach to solve the concurrent/ simultaneous product-process design optimization as described in Chapter 8 and 9.

The design methods; Delft Design Template (DDT) (now restructured and called Delft Design Map, [DDM] and Unilever's Product Driven Process Synthesis [PDPS]) have been taught for many years at the post-MSc PDEng programs at TU Delft (Process & Equipment Design, BioProcess Engineering, Chemical Product Design) and TU Eindhoven (Process and Product Design) and applied in the PDEng design projects executed for and in industry. The advantage is that PDEng graduates learn a clear systematic approach to product oriented process design that is not dependent on product and process circumstances and can be applied in the product-process industry sectors, such as the chemical, biochemical, food, and others.

Delft Design Map for product and process design (DDM)

The Delft Design Map (DDM) is based on task-based design activity spaces, shaped in matrix form. The horizontal axis corresponds with innovation stages and the vertical axis corresponds with design steps, from problem definition to solution to reporting.

The empty blocks of the matrix are then filled during the design effort in a preplanned order until the matrix is filled. The design problem at hand (new product, new process and new supply chain, new reactor technology, heat integration) determines where the emphasis of the design work should lay, but also emphasizes that designers should be aware that implications of apparent small design changes (in reactor technology or catalyst type, for instance) may have profound effects (positive, negative) on the entire system. The design team can also use the DDM to map the route of the search for suitable design solutions, encompassing: formulating the design problem, planning the design tasks, and following through on these activities including reporting decisions and results. The DDM becomes then a planning, design management, and communication tool. Therefore, the authors (after consulting Grievink) renamed the 'Delft Template for Conceptual Process Design' into the 'Delft Design Map for Product and Process Design.'

This map will be even more helpful for students and industry designers than its predecessor. They get an overview of all design actions to be pursued to fill the matrix. This gives them confidence that they can make a design because the whole problem is cut up into small design action steps. The Delft Design Map is described in detail in Chapter 4.

References and further reading

[1] Peters MS, Timmerhaus KD. Plant design and economics for chemical engineers, 2nd edn, New York, NY, USA: McGraw-Hill, 1968.

[2] Rudd DF, Watson CC. Strategy of process engineering, New York, NY, USA: John Wiley & Sons Inc., 1968.

[3] Linnhoff B, Flower FR. Synthesis of heat exchanger networks: I. Systematic generation of energy optimal networks, AIChE Journal, 24(4), 1978, 633–642.

[4] Douglas JM. Conceptual design of chemical processes, New York: McGraw-Hill, 1988.

[5] Seider WD, Lewin DR. Process design principles: Synthesis, analysis and evaluation, New York, NY, USA, John Wiley & Sons Inc, 1999.

[6] Kocis GR, Grossman IE. Computational Experience with DICOPT: Solving MINLP problems in process systems engineering, Computers and Chemical Engineering, 13(3), 1989, 307–315.

[7] Siirola JJ. An industrial perspective on process synthesis, 4th International Conference on Foundations of Computer–Aided Process Design (FOCAPD), AIChE Symp Series USA 91(304), 1995, 222–233.

[8] Siirola JJ. Industrial applications of chemical process synthesis, In: Anderson JL, Bischoff KB (eds.), Advances in Chemical Engineering Volume 23: Process Synthesis. Academic Press, 1996, 1–92.

[9] Agreda, VH, Partin LR, Heise WH. High-purity methyl acetate via reactive distillation, Chemical Engineering Progress, 86(2), 1990, 40–46.

[10] EL, Moggridge GD. Chemical product design, Cambridge, UK: Cambridge University Press, 2001.

[11] Wesselingh JA, Kiil S, Vigild ME. Design & development of biological, chemical, food and pharmaceutical products, Hoboken: John Wiley & Sons Ltd, 2007.

[12] Seider WD, Seader JD, Lewin DR, Widagdo S. Process and process design principles: Synthesis, analysis and evaluation, 3rd edn, Hoboken, NJ, USA: John Wiley & Sons Inc, 2010.
[13] Bröckel U, Meier W, Wagner G. Product design and engineering – best practices, Volume 1: Basics and technologies, Weinheim, Germany: WILEY-VCH Verlag GmbH & Co. KGaA, 2007.
[14] Bröckel U, Meier W, Wagner G. Product design and engineering – best practices, Volume 2: Raw materials, additives and applications, Weinheim, Germany: WILEY-VCH Verlag GmbH & Co. KGaA, 2007.
[15] Grievink J, Swinkels PLJ, Van Ommen JR. Basics of Process Design. In: De Jong W, Van Ommen JR (eds.), Biomass as a sustainable energy source for the future: Fundamentals of conversion processes, 1st edn, Hoboken NJ, USA: Wiley & Sons Inc., 2014, 184–229.
[16] Meeuse M, Grievink J, Verheijen PJT, Vander Stappen MLM. Conceptual design of process for structured products, AIChE Symposium Series USA, 96, 2000, 324–328.
[17] Almeida-Rivera C, Bongers P, Zondervan E. A structured approach for product-driven process synthesis in foods manufacture. In: Martin M, Eden MR, Chemmangattuvalappil NG, eds. Computer Aided Chemical Engineering Vol. 39 Tools for chemical product design – from consumer products to biomedicine, Amsterdam, Netherlands: Elsevier BV, 2017, 417–441.
[18] Bernardo FP. Integrated process and product design optimization. In: Martin M, Eden MR, Chemmangattuvalappil NG, eds. Computer Aided Chemical Engineering Vol.39 Tools for chemical product design – from consumer products to biomedicine. Amsterdam, Netherlands: Elsevier BV, 2017, 347–372.

A9 Appendix to Chapter 9: Cases of process modeling for simulation and optimization

Johan Grievink

This Appendix presents thirteen application cases. An upfront remark about the restrictive use of some terms is in order. Behavior of a process or product relates to patterns of change in its chemical and physical quantities. The term performance is intentionally reserved for factors that express some kind of value and appreciation by society, including customers of a product and the operators and owners of a process. Such factors are related to safety, health, economy and ecology. The literature references appearing in this Appendix are listed at the end, independently of those in Chapter 9. Commonly used mathematical symbols are also listed at the end. However, several symbols are repeatedly in use in the various cases and sometimes have a different meaning. Therefore, to avoid ambiguity, care has been taken for each case to have some additional description of its symbols as well.

A9.1 Linear model applications for concept stage

Before starting the application cases, it is expedient to summarize five different mathematical forms of linear models, all of which occur in process modeling practice. A model is linear when each equation in the model is linear in all variables. Deterministic variables and models will be covered only, excluding stochastic and logical models. The five common linear deterministic model forms are:

	Symbols
(1) linear algebraic equations: $\mathbf{A} \cdot \underline{\mathbf{x}} = \underline{\mathbf{b}}$	$\mathbf{A}$: $m \cdot n$ matrix; $\underline{\mathbf{x}}$: n-vector; $\underline{\mathbf{b}}$: m-vector
(2) linear ordinary differential equations: $\mathrm{d}\underline{\mathbf{x}}(t)/\mathrm{d}t = \mathbf{A} \cdot \underline{\mathbf{x}}(t) + \mathbf{B} \cdot \underline{\mathbf{u}}(t)$	t: scalar coordinate; $\underline{\mathbf{x}}$: n-vector function; $\underline{\mathbf{u}}$: m-vector function; $\mathbf{A}$: $n \cdot n$ matrix; $\mathbf{B}$: $n \cdot m$ matrix
(3) linear partial differential equations: $\partial\underline{\mathbf{x}}(z,t)/\partial t + v \cdot \partial\underline{\mathbf{x}}(z,t)/\partial z = \mathbf{A} \cdot \underline{\mathbf{x}}(z,t) + \mathbf{B} \cdot \underline{\mathbf{u}}(z,t)$	t, z: scalar coordinates; v: scalar; $\underline{\mathbf{x}}$: n-vector function; $\underline{\mathbf{u}}$: m-vector function; $\mathbf{A}$: $n \cdot n$ matrix; $\mathbf{B}$: $n \cdot m$ matrix

https://doi.org/10.1515/9783110467741-017

(4) linear difference equations:

$\underline{\mathbf{x}}_{k+1} = \mathbf{A} \cdot \underline{\mathbf{x}}_k + \mathbf{B} \cdot \underline{\mathbf{u}}_k$

k: integer index, $k \in \{0, 1, 2, \ldots\}$
$\underline{\mathbf{x}}$: n-vector function;
$\underline{\mathbf{u}}$: m-vector function
$\mathbf{A}$: $n \cdot n$ matrix; $\mathbf{B}$: $n \cdot m$ matrix

(5) linear integer equations with binary variables:

$\mathbf{Z} \cdot \underline{\mathbf{Y}} \leq \underline{\boldsymbol{\omega}}$

$\mathbf{Z}$: $m \cdot n$ matrix; $\underline{\mathbf{Y}}$: n-vector;
$\underline{\boldsymbol{\omega}}$: m-vector

The matrices **A** and **B**, vectors $\underline{\mathbf{x}}$, and $\underline{\mathbf{u}}$, as well as the scalars t and z are all made up of real numbers. Matrix **Z** and vector $\underline{\boldsymbol{\omega}}$ take integer values. The entries in vector $\underline{\mathbf{Y}}$ take binary values {0, 1}. Matrices **A**, **B** and **Z** are constant, as they have fixed entries.

The application cases with linear models in this Appendix deal with algebraic equations and real variables. An exception to this is the modeling of the structure of a process block diagram/flow sheet involving a linear model with binary variables.

Case A9.1: Exploring the stoichiometric space for an innovative chemical conversion

Reaction stoichiometry offers freedom in the design of a process concept. A simple example will be given on a search over a stoichiometric space to find a suitable reaction that serves best a particular conversion goal. In this case the goal is the chemical conversion of carbon dioxide as a greenhouse gas into carbon monoxide as a chemical reactant in syngas. The reuse of carbon in chemical products helps in closing the carbon cycle.

In analogy to steam reforming [$CH_4 + H_2O \Rightarrow CO + 3H_2$], one can apply dry reforming with CO_2 as is done in the steel industry [$CH_4 + CO_2 \Rightarrow 2CO + 2H_2$]. Unfortunately, the conversion to CO by these reforming reactions is reduced by the water-gas-shift reaction [$CO + H_2O \Rightarrow CO_2 + H_2$], shifting CO partially back to CO_2. All reactions take place in the gas phase at high temperature. A disadvantage of the dry reforming reaction is the low molar ratio of $CO_2 : CH_4 = 1 : 1$. It would be nice to achieve a higher ratio of CO_2 to methane as a way of recycling more CO_2 molecules per methane molecule. This issue can be explored by means of modeling the possible reaction stoichiometry based on a given set of components. In this modeling exercise, the set {CH_4, CO_2, CO, H_2, H_2O} are the reacting species, comprising of {C, H, O} as elements.

Synthesis of model

The molar reaction stoichiometry is expressed in a generalized way:

$$\text{Generalized reaction:} \qquad a\,CH_4 + b\,CO_2 \Rightarrow x\,CO + y\,H_2 + z\,H_2O$$

There are three likely reaction products at the right hand side, excluding O_2 as a potential fourth candidate because the latter is not formed due to unfavorable thermodynamics.

The unknown stoichiometric coefficients {a, b, x, y, z} are constrained by three element balances:

$$\begin{array}{lrrrrrl} \text{C balance:} & a + & b - x & & & & = 0 \\ \text{H balance:} & 4a & & & - 2y & - 2z & = 0 \\ \text{O balance:} & & 2b - x & & & - z & = 0 \end{array}$$

The difference between the stoichiometric coefficients of CO_2 and CH_4 is introduced as a convenient, additional variable:

$$\delta = b - a \geq 0$$

These coefficients should be:

$$a \geq 1 \quad \text{and} \quad b \geq 1$$

The other coefficients must be nonnegative:

$$x \geq 0 \qquad y \geq 0 \qquad z \geq 0$$

When adopting the conventional form of writing chemical reaction equations, the stoichiometric coefficients take integer values only. From a mathematical point of view this integer condition is not strictly needed and one can take $a = 1$ as a pivot in the above reaction equation.

The objective is to find the reaction equation that maximizes the CO_2 consumption relative to methane consumption:

$$\text{Max}\forall\, F = b/a$$

Analysis of model

This model has six variables $\{a, b, x, y, z, \delta\}$ with four algebraic equality relations, resulting in two degrees of freedom. If a problem would have a much larger size, it can be better numerically solved and optimized by applying linear programming (LP). Because of the small size of this problem, an analytical approach with graphical solutions will be followed here. The stoichiometric coefficients of the reaction products $\{x, y, z\}$ can be expressed as linear functions of a and b:

$$\begin{array}{lll} x = a + b \geq 0 & \Rightarrow & a \geq 0 \text{ and } b \geq 0 \\ y = 3a - b \geq 0 & \Rightarrow & 3a \geq b \\ z = -a + b \geq 0 & \Rightarrow & b \geq a \end{array}$$

Now the problem has been reduced to a two dimensional search space {a, b} for optimization.

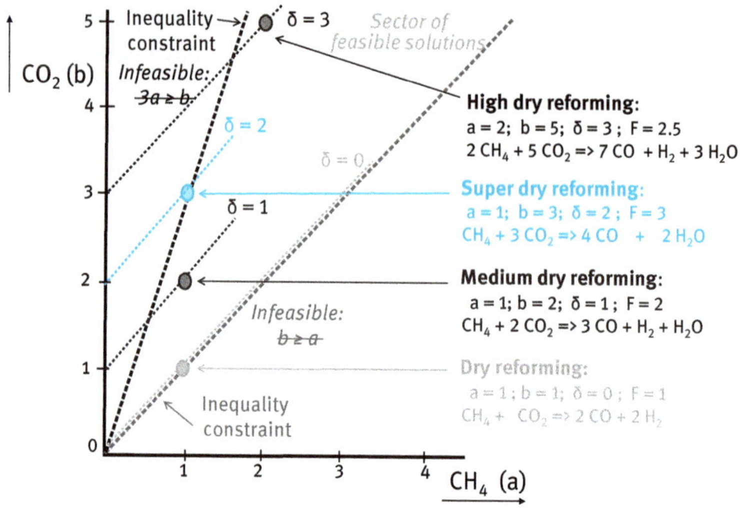

Feasible domain with three stoichiometric possibilities for CO_2 consumption.
The highest consumption (***super dry reforming***) arises for a=1 and b=3.

Fig. A9.1: Searching for optimized solutions in the stoichiometric reaction space.

Solutions of the model

Figure A9.1 depicts the $\{a, b\}$ solution space with contour lines of the objective function for (e.g., $\delta = 0, 1, 2$) and two inequality constraints $\{3a \geq b; b \geq a\}$. Three solutions are shown in this space:

$$\begin{aligned}
\text{Dry reforming} \quad & (a = 1\,; \quad b = 1\,; \quad \delta = 0) \\
\text{Semi dry reforming} \quad & (a = 2\,; \quad b = 3\,; \quad \delta = 1) \\
\text{Super dry reforming} \quad & (a = 1\,; \quad b = 3\,; \quad \delta = 2)
\end{aligned}$$

It is noted that the feasibility condition $3a \geq b$ imposes an upper bound on the objective function:

$$F = b/a \leq 3a/a \quad \Rightarrow \quad F \leq 3$$

One could wonder how many feasible solutions exist for $\delta = 1, 2, 3$ and so forth. It is convenient to take $\{a, \delta\}$ as the two independent variables. The remaining stoichiometric coefficients become:

$$b = a + \delta\,; \qquad x = 2a + \delta\,; \qquad y = 2a - \delta; z = \delta$$

Critical feasibility condition is:

$$3a \geq b \quad \Rightarrow \quad 2a \geq \delta$$

Lower and upper bounds on F:

$$1 < F < 3$$

Tab. A9.1: Possible ranges of stoichiometry of dry reforming reactions with CO_2 excess.

a	$b (= a + \delta)$	$x (= 2a + \delta)$	$y (= 2a - \delta)$	$z (= \delta)$	$a\,CH_4 + b\,CO_2 \Rightarrow x\,CO + y\,H_2 + z\,H_2O$	b/a
1	3	4	0	2	$1\,CH_4 + 3\,CO_2 \Rightarrow 4\,CO + 2\,H_2O$	3.00
2	5	7	1	3	$2\,CH_4 + 5\,CO_2 \Rightarrow 7\,CO + 1\,H_2 + 3\,H_2O$	2.50
1	2	3	1	1	$1\,CH_4 + 2\,CO_2 \Rightarrow 3\,CO + 1\,H_2 + 1\,H_2O$	2.00
5	9	14	6	4	$5\,CH_4 + 9\,CO_2 \Rightarrow 14\,CO + 6\,H_2 + 4\,H_2O$	1.80
4	7	11	5	3	$4\,CH_4 + 7\,CO_2 \Rightarrow 11\,CO + 5\,H_2 + 3\,H_2O$	1.75
3	5	8	4	2	$3\,CH_4 + 5\,CO_2 \Rightarrow 8\,CO + 4\,H_2 + 2\,H_2O$	1.67
5	8	13	7	3	$5\,CH_4 + 8\,CO_2 \Rightarrow 13\,CO + 7\,H_2 + 3\,H_2O$	1.60
7	11	18	10	4	$7\,CH_4 + 11\,CO_2 \Rightarrow 18\,CO + 10\,H_2 + 4\,H_2O$	1.56
2	3	5	3	1	$2\,CH_4 + 3\,CO_2 \Rightarrow 5\,CO + 3\,H_2 + 1\,H_2O$	1.50
1	1	2	2	0	$1\,CH_4 + 1\,CO_2 \Rightarrow 2\,CO + 2\,H_2$	1.00

What are the results of a systematic exploration of the $\{a, \delta\}$ space for $a = \{1, 2, 3, \ldots, 7\}$ with $2a \geq \delta$ and $\delta = \{0, 1, 2, 3, 4\}$? In fact, one finds a wide range of reaction equations. These reactions can be ordered in decreasing ratio of CO_2 consumption ($= b/a$). Table A9.1 shows the more striking reactions.

The upper one is super dry reforming ($b/a = 3$); the bottom one ($b/a = 1$) is normal-dry reforming. Theoretically, one can construct numerous reaction equations with a decreasing b/a ratio between 1.5 and 1.0. Being at the lower end of the performance spectrum, these reactions are of less interest.

In view of the many reactions in Table A9.1, one starts to wonder how many of these reactions are actually independent ones from a stoichiometric point of view, i.e., a reaction within a given set of reactions has an independent stoichiometry if the latter cannot be construed as a linear combination of the stoichiometry of some other reactions within this set. In this dry reforming case, the outcome is that there are, at most, two independent reactions (a mathematical example will be given below). To illustrate this outcome, let us take the dry reforming and super dry reforming reactions as the independent ones. It appears that all other reactions can be written as different linear combinations of the two independent ones. For instance, the medium dry reforming reaction ($1\,CH_4 + 2\,CO_2 \Rightarrow 3\,CO + 1\,H_2 + 1\,H_2O$) is the sum of the super dry reforming ($1\,CH_4 + 3\,CO_2 \Rightarrow 4\,CO + 2\,H_2O$) and dry reforming reaction ($1\,CH_4 + 1\,CO_2 \Rightarrow 2\,CO + 2\,H_2$) divided by two. In fact, all reactions other than the super dry and dry reforming reactions are linear combinations of these two.

It is interesting to note that the water-gas-shift reaction [$CO + H_2O \Rightarrow CO_2 + H_2$] is also included in the set of dependent reactions. If one subtracts the super dry reforming stoichiometry from the regular dry reforming stoichiometry and divides this by two, one gets the water-gas-shift reaction.

Can it be mathematically proven there are, at most, two independent reactions?

Independence of reaction stoichiometry has its mathematical basis in linear algebra and vector spaces. The stoichiometry of a reaction can be treated as a single vector in a real valued vector space ($\mathbb{R}^N$), where N is the number of chemical species between which a specified set of reactions occurs. N is the dimension of the 'species space'. A vector is a dependent one if it is a linear combination of some other vectors in this space. In ($\mathbb{R}^N$), one can always choose a set of N independent vectors which span the entire space. Any other vector in this space is a linear combination of these N independent ones. To determine the number of reactions with an independent stoichiometry (or 'independent reactions'), one must also take into account the conservation of elements in the reactions. The coefficients in the element (C, H, O) balance equations also make up vectors of dimension N in the species space. Let E be the element species matrix for the dry reforming case:

	Elements		CH_4	CO_2	CO	H_2	H_2O		⇐ Species
	C	[	1	1	1	0	0	]	
Matrix **E** =	H	[	4	0	0	2	2	]	
	O	[	0	2	1	0	1	]	

Each row in this matrix corresponds to a vector in the five dimensional species space. This is called an 'element' vector. The above matrix E has three linearly independent vectors in the species space, which can be easily verified; i.e., none of the rows can be written as a linear combination of the two other rows. Because of the principle of conservation of elements in reactions, each of these element vectors must be orthogonal to the reaction 'stoichiometric' vectors. Be reminded that a row in stoichiometric matrix S is the stoichiometry of a reaction, while a column in S shows the occurrence of a particular chemical species in the reactions. Due to the conservation principle, the inner product of an element vector and a stoichiometric vector is always zero. In matrix notation, this is as follows:

$$\mathbf{E} * \mathbf{S}^{\mathrm{T}} = O \quad \text{with} \quad N_{\text{species}} \geq N_{\text{indep.elements}} + N_{\text{indep.reactions}}$$

This inequality relation between the number of species and the numbers of independent elements and reactions is a useful one in chemical reaction network analysis. In the case of the dry reforming reactions, there are three element vectors in a five dimensional species space. Only two other independent reaction vectors are left to fully span the species space. These two vectors can be arbitrarily taken from any set of reactions with independent stoichiometry. Here, the dry reforming and the super dry reforming reactions are selected as the independent ones. This result completes the proof of having only two reactions with independent stoichiometry. For the sake of clarity, it is mentioned that stoichiometric dependence between reactions also implies dependence between their reaction equilibrium equations through the law of mass action.

Evaluation of model and solution

The derived stoichiometric model enables the exploration of potentially interesting and useful reactions in a given species space. The super dry reforming reaction is the highest performing reaction for CO_2 to CO conversion, corresponding to the ($b/a = 3$) entry in Table A9.1. An experimental study of this super dry reforming reaction and its underlying innovative catalytic scheme was first reported in [1] by Buelens et al. Our stoichiometric analysis also revealed the high dry ($b/a = 2.5$) and medium dry reforming reactions ($b/a = 2$) as interesting side reaction. In principle, these reactions may occur simultaneously with super dry reforming reactions. The stoichiometric scheme also implicitly covers the water-gas-shift reaction. Buelens et al., [1] have developed a novel experimental conversion scheme. The shift reaction is avoided by means of clever sequencing of catalytic reactions. In the first reaction, the methane is fully oxidized to H_2O and CO_2, using an oxygen carrying solid state medium:

$$CH_4 + 4\,[O] \Rightarrow CO_2 + 2\,H_2O$$

All CO_2 generated by the reaction and in the feed is chemically adsorbed in another solid phase, while the gas phase reaction water is flushed out of the reaction system. In the next step, all adsorbed CO_2 is reduced to CO:

$$4\,CO_2 \Rightarrow 4\,CO + 4\,[O]$$

The oxygen is adsorbed by a solid medium for subsequent use in the next oxidation cycle of CH_4, as in a chemical looping fashion. Jointly, these reactions satisfy the stoichiometry of super dry reforming.

As a final step, it is useful to reflect on uncertainties and limitations playing a role in formulating and solving of a model. In this case the model structure and the outcome of the analysis are determined by the chosen set of chemical species and reactions which codetermine the set of independent reactions. In this case, one could possibly argue a case to add solid carbon (C_{solid}) along with the Boudouard reaction ($2\,CO \Rightarrow CO_2 + C_{solid}$) if the CO_2 would have been released as a gas at lower temperatures.

Case A9.2: A conversion–separation–recycle structure using lumped units

A linear algebraic model of a common reactor–separator–recycle processing structure operating at steady state is derived and solved. This case may serve as an example on how to tackle such cases in general.

Synthesis of process block diagram and model

The arrangement of the main units in a process and their connecting mass streams is shown by means of a process block diagram (Figure A9.2), where each main unit is represented as a block. The process block diagram shows:

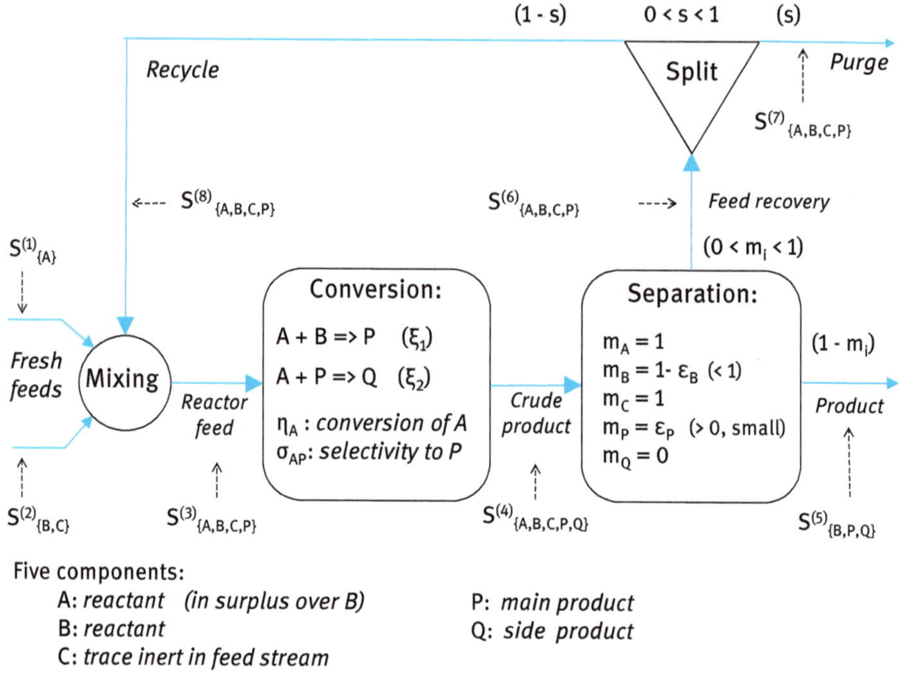

Fig. A9.2: Block diagram of a conversion–separation–recycle system.

- A '*conversion*' unit which turns, by two parallel reactions, the reactants A and B into the main product P (by reaction 1) and a side product Q (by reaction 2).

$$\text{Reactions:} \quad A + B \Rightarrow P$$
$$\text{and} \quad A + P \Rightarrow Q$$

- A '*separation*' unit recovers unconverted reactants from the products for recycling.
- A '*split*' unit purges an inert trace component (C) in the feeds from the process so as to avoid its accumulation through the recycle stream.
- A '*mixing*' unit combines the recycle stream with fresh feed into a feed to the conversion unit.
- The connecting '*streams*' to, in between and from the process units.

This process counts five chemical species {A, B, C, P, Q} occurring in varying amounts in seven streams $\{S^{(2)}, \ldots, S^{(7)}\}$. The labeled streams are indicated in the block diagram.

Synthesis of the model

At this stage of design and modeling, each process unit is defined by its steady state processing functionality (e.g., conversion) rather than as a piece of equipment (e.g., reactor). The units in a block diagram will be modeled in an input-output fashion. The transformations of the incoming streams (or mass) into the outgoing streams will be captured in a lumped way by means of some parameters, without expanding yet on detailed mechanistic workings inside. For example, a chemical conversion unit can be characterized by the degree of conversion of a feed component as well as by selectivity factors. The function of a separation unit can be simply represented by separation factors. Each factor indicates how much of a component in the feed to the separator is going to a particular exit stream of this unit.

The variables and equations for all streams and units will be specified. To avoid modeling errors, it is important to maintain modeling clarity by keeping the notation uniform. It is customary to work with a uniform representation of all streams and strive for similarity in the model representation of the process units. Furthermore, a deliberate attempt is made to keep the model equations linear in the variables to facilitate ease of computation and interpretation of solutions.

Streams

Each stream is characterized by a vector of molar flows ($\underline{\mathbf{f}}$) of the five species. When a component is absent in a particular stream, its molar flow is set to zero.

$f_i^{(j)}$: molar flow of species i [lower index] in stream j [upper index] j; $j = 1, \ldots, 8$

Note that this process model assumes fixed stream connections of the units, i.e., the process unit models contain the flows of these preassigned stream connections. There is no separate flow sheet structure model in this case. Such a structure model would allow for adjustable connections between units by being able to switch stream connections on or off.

Conversion unit

One wants to favor the conversion to main product P over the formation of the side product Q. The reaction stoichiometry shows that A is consumed by both reactions. Therefore, it should not be in short supply relative to B, which is involved in the first reaction only. The molar balances are given in a matrix vector notation by:

$$\underline{\mathbf{f}}^{(3)} - \underline{\mathbf{f}}^{(4)} + \mathbf{S}^{\mathrm{T}} \cdot \xi = 0$$

S is the stoichiometric matrix: $[-1\ -10+10]$ and extents of reactions ξ: $[\xi_1]$

$$[-100\ -1+1] \qquad [\xi_2]$$

The conversion of a component and the selectivity to products can be related to the extents (ξ) of reactions in which it participates. Choosing limiting reactant A as a reference, its degree of conversion of A (η_A) and the selectivity of reactant A to product P ($\sigma_{A\,P}$) are related to the two extents of reaction:

$$\text{Conversion:} \qquad f_A^{(3)} \cdot \eta_A - (\xi_1 + \xi_2) = 0$$
$$\text{Selectivity:} \qquad (\xi_1 - \xi_2) - \sigma_{A\,P} \cdot (\xi_1 + \xi_2) = 0$$

At this stage of design, one can use the degree of conversion and the selectivity as specified model parameters. The required numerical parameter values (or ranges for these values) can be derived from the literature or experimental conversion data. Having specified the parameter values, the two equations remain linear in the variables ($f_A^{(3)}, \xi_1, \xi_2$).

Separation unit

The amounts of chemical species entering the unit are distributed (by diffusional type of separation) over two exit streams, according to their thermodynamic preferences for either of the exit streams. These distributions are characterized by means of a single distribution parameter per species: $0 \le m_i \le 1$. This parameter m_i indicates the fraction of the incoming molar flow going to the recycle stream $S^{(5)}$, while the remaining fraction $(1 - m_i)$ goes to the product stream $S^{(4)}$. In this process the components A and C will go fully to the recycle stream, along with the bulk of B and a small fraction of P. A bit of B, the bulk of P and all of Q go to the product stream.

$$f_i^{(6)} - m_i \cdot f_i^{(4)} = 0 \quad \text{and} \quad f_i^{5)} - (1 - m_i) \cdot f_i^{(4)} = 0 \qquad i = \{A, B, C, P, Q\}$$

The numerical order of magnitude (or the numerical range) of the distribution coefficients must be derived from literature or experimental data.

Splitting and mixing

The entrance stream is split in two streams of identical composition. All components are distributed by the same fraction(s) over the two exit streams. In this case:

$$f_i^{(7)} - s \cdot f_i^{(6)} = 0 \quad \text{and} \quad f_i^{(8)} - (1 - s) \cdot f_i^{(6)} = 0 \qquad i = \{A, B, C, P, Q\}$$

The fraction(s) is a (specified) design decision parameter.

In a mixer, two or more incoming streams are added together to the exit stream. In this case:

$$f_i^{(8)} + f_i^{(1)} + f_i^{(2)} - f_i^{(3)} = 0 \qquad i = \{A, B, C, P, Q\}$$

Analysis and computation of model solution

Having achieved a compact linear form of the model, one should apply a degree of freedom analysis. How many unknowns do occur in this model and how many equations have been specified? The difference between the two is the degree of freedom in the model. Regarding the unknowns, there are five species flows in eight streams plus two extents of reactions, resulting in 42 variables. These variables can be ordered into a single vector of unknowns $\underline{\mathbf{x}}$ of dimension 42 in a vector space of real variables. The number of equations adds up to: 5 + 2 (conversion unit) + 2 · 5 (separator) + 2 · 5 (splitter) + 5 (mixing) = 32 equations. These equations are linearly independent, containing no redundancies. Thus, the degree of freedom of this model is 42−32 = 10. In order to make the model compact for a convenient solution procedure, the coefficients of the variables in these linear equations are put into a matrix $\mathbf{A}_\mathrm{M}$ of dimension 32 (rows) · 42 (columns). The right hand sides of the equations can be put into the right hand side vector $\underline{\mathbf{b}}_\mathrm{M}$ of dimension 32. In this case all right hand sides are zero and the vector $\underline{\mathbf{b}}_\mathrm{M}$ is the null vector. The resulting matrix vector form is nonsquare [$\mathbf{A}_\mathrm{M}$ (32 · 42); $\underline{\mathbf{x}}$ (42); $\underline{\mathbf{b}}_\mathrm{M}$ (32)], falling short by ten degrees of freedom to finding a solution to:

$$\mathbf{A}_\mathrm{M} \cdot \underline{\mathbf{x}} = \underline{\mathbf{b}}_\mathrm{M}$$

To fill the gap, one needs an external scenario about how this process is supposed to run with target feed and/or product flows. Such an external scenario must specify the remaining ten degrees of freedom.

Specification of an external scenario

The first step is to specify the feed streams. These streams come from elsewhere and usually one has no direct control over its composition. Having two fresh feed streams for the two main reactants, A and B, one may readily assume that feed stream (1) is rich in A and stream (2) is rich in B. The molar compositions of these streams will be given by known molar ratios $\{r_{i|\mathrm{A}}^{(1)}, r_{i|\mathrm{B}}^{(2)}\}$. Here $r_{i|\mathrm{A}}^{(1)}$ and is the given molar ratio of component i relative to main component A in stream (1), while $r_{i|\mathrm{B}}^{(2)}$ is the given molar ratio of component i relative to main component B in stream (2). Knowing these ratios, one can express the species feed rates relative to the feed rate of the main component:

$$f_i^{(1)} - r_{i|\mathrm{A}}^{(1)} \cdot f_\mathrm{A}^{(1)} = 0 \qquad i = \{\mathrm{B, C, P, Q}\}$$

$$f_i^{(2)} - r_{i|\mathrm{B}}^{(2)} \cdot f_\mathrm{B}^{(2)} = 0 \qquad i = \{\mathrm{A, C, P, Q}\}$$

These specifications of the two feed streams take away eight degrees of freedom, leaving two degrees of freedom to be determined.

Now there can be several alternative scenarios. For instance:

Specify the fresh feed streams:

$$f_\mathrm{A}^{(1)} = b_\mathrm{A}^{(1)} = \text{given flow specification of component A in stream (1)}$$

$$f_\mathrm{B}^{(2)} = b_\mathrm{B}^{(2)} = \text{given flow specification of component B in stream (2)}$$

Set targets for the output flows of the main and side products:

$$f_\mathrm{P}^{(5)} = b_\mathrm{P}^{(5)} = \text{target rate of product P in stream (5)}$$

$$f_\mathrm{Q}^{(5)} = b_\mathrm{Q}^{(5)} = \text{target rate of side product Q in stream (5).}$$

Set a target rate for the main product with a ratio for the side product rate:

$$f_{P}^{(5)} = b_{P}^{(5)} \quad = \text{target rate of product P in stream (5)}$$

$$f_{Q}^{(5)} - r_{Q|P}^{(5)} \cdot f_{P}^{(5)} = 0 \qquad r_{Q|P}^{(5)}. \text{ is molar ratio of Q over P in stream (5)}$$

These scenario specifications can be cast into a matrix vector form as well:

$$\mathbf{A}_{\text{Scenario}} \cdot \underline{\mathbf{x}} = \underline{\mathbf{b}}_{\text{Scenario}}$$

$$\text{Dimensions: } \mathbf{A}_{\text{Scenario}} : 10 \text{ (rows)}^{*}\ 42 \text{ and } \underline{\mathbf{b}}_{\text{Scenario}} : 10 \text{ (rows)}$$

The molar ratios $r_{i|j}^{(*)}$ become elements in the $\mathbf{A}_{\text{Scenario}}$ matrix while the flow specifications $b_{*}^{(*)}$ are put in the right hand side $\underline{\mathbf{b}}_{\text{Scenario}}$ vector.

Comments on model solution procedure

The 32 equations for the process units are combined with ten equations for the scenarios. Together they form a square problem $\mathbf{A} \cdot \underline{\mathbf{x}} = \underline{\mathbf{b}}$. The matrix $\mathbf{A}$ should be regular. If it is singular, the most likely cause is a modeling error such as having dependent mass balances in the set of equations.

A square linear problem $\mathbf{A} \cdot \underline{\mathbf{x}} = \underline{\mathbf{b}}$ can be quickly solved by any numerical linear equation solver. It may be tempting to solve this model analytically and computing the variables in a sequential manner. For instance, one can start by eliminating the extents of reactions from the variable and equation sets and work only with the species flows as unknowns. Then, one can solve for the flows of species A in the process as well as for the flows of the other species. However, such model reductions and transformations are not effective in terms of the engineering work efforts. That is, unless the derivations from the linear full model are reproduced again as an expensive repeat. Maintaining the transparency of the original process model is more important to effective engineering work than sacrificing it for small potential gains in computational speed. In engineering practice, the application of analytical model transformations to enhance speed of computation are often counterproductive in the long run.

Evaluation of the solution and the model

The model parameters (degree of conversion, selectivity, separation factors and split ratio), as well as the external scenario specifications can be assigned numerical values. These values will be called the reference values. One can solve the linear $\mathbf{A} \cdot \underline{\mathbf{x}} = \underline{\mathbf{b}}$ model and find the corresponding molar flows of the species ($\underline{\mathbf{x}}$) in the process. However, some model parameters may not be very accurate yet. At this stage of design these may vary over a range. Similarly, the external specifications (e.g., production rates) may show variability. As such, it makes sense to analyze the model solutions for a wider range than only the reference values of the model parameters. However, how does one select the parameter perturbations within their specified ranges of uncertainties? There are at least three options:

(1) To apply local sensitivity analysis by means of small parameter perturbations around the reference values.
(2) To apply a nonlocal sensitivity analysis by tracking a trajectory of parameters in the feasible parameter space.
(3) Nonlocal analysis by statistical sampling in the parameter space, governed by the probability density functions of the parameters.

Apart from the local sensitivity analysis, the two other methods are too involved to be discussed here. A suitable reference for further reading is found in [2] and in a reactor design application [3]. One can easily perform a numerical local sensitivity analysis. Given a parameter reference case, one can introduce a sequence of small perturbations in model parameters, $\{\underline{\mathbf{p}}^{(q)}, q = 0, 1, 2, \ldots, Q\}$, where $q = 0$ is the unperturbed reference. The model expands into multiple instances of matrix $\mathbf{A}$ and vector $\underline{\mathbf{b}}$:

$$\{\mathbf{A}(\mathbf{p}^{(q)}), \underline{\mathbf{b}}(\underline{\mathbf{p}}^{(q)})\}\,, \quad q = 0, 1, 2, \ldots, Q\,.$$

One can generate a sequence of dimension free sensitivity matrices:

$$\Omega^{(q)} = [(\underline{\mathbf{x}}^{(q)} - \underline{\mathbf{x}}^{(0)})/\underline{\mathbf{x}}^{(0)}]/[(\underline{\mathbf{p}}^{(q)} - \underline{\mathbf{p}}^{(0)})/\underline{\mathbf{p}}^{(0)}] \qquad q = 1, 2, \ldots, Q\,.$$

In view of the linear differences between $\mathbf{A}^{(q)}$, $\mathbf{A}^{(0)}$, $\underline{\mathbf{b}}^{(q)}$ and $\underline{\mathbf{b}}^{(0)}$, one would expect to have to avail a computational shortcut to easily compute $\underline{\mathbf{x}}^{(q)}$ from $\underline{\mathbf{x}}^{(0)}$. This is not the case. For every perturbation (q) the full set $\mathbf{A}^{(q)} \cdot \underline{\mathbf{x}}^{(q)} = \underline{\mathbf{b}}^{(q)}$ must be solved. However, it is not a burden given the efficiency of numerical solvers for linear equation sets. For further analysis, one may vertically stack all sensitivity matrices in a main sensitivity matrix:

$$\Omega^{(\text{main})} = [\Omega^{(1}/\Omega^{(2)}/\ldots/\Omega^{(q)}/\ldots\Omega^{(Q)}]$$

Application of singular value decomposition to $\Omega^{(\text{main})}$ yields the dominant singular value(s). The corresponding right singular vectors indicate which parameters or parameter combinations are the more influential ones.

Maintaining model linearity when adding an enthalpy balance

The enthalpy flow of a stream $(F_H^{(j)})$ is computationally the product of a mass flow and a temperature. It spoils the desired linear structure of the model equations. Yet, it is important for process design to get some quantitative understanding of the main enthalpy flows and the main enthalpy sources and sinks. By means of some reasonable assumptions about temperature levels, one can induce a linear structure for the enthalpy balance equations. This is done in the following way:

An enthalpy flow is obtained by taking the sum over all species in the stream of the product of the specific enthalpy, $h_i(T)$ and the flow rate of each component $f_i^{(j)}$:

$$F_H^{(j)} = \sum_{i=1,\ldots,N} h_i(T) \cdot f_i^{(j)}$$

The specific enthalpy $h_i(T)$ accounts only for the sensible heat effects to get to the temperature T from the thermodynamic reference condition. It excludes the enthalpy of formation of a component from its elements at reference conditions. The product of two process variables, temperature (T) and flow rate ($f_i^{(j)}$) destroys linearity. However, if one would be able to fix the temperatures of the streams and process units, and so the specific enthalpy, one can retain the linear form of the enthalpy flow variables. It is also presumed that the linear mass balance model has been solved and the local flows $\{ f_I^{(j)} \}$ are all available for the computation of the enthalpy flows. The enthalpy balance over the conversion unit to which a heat flow is added (Q_{exch}) for cooling or heating would then read:

$$F_H^{(3)} - \Delta H_{R,1} \cdot \xi_1 - \Delta H_{R,2} \cdot \xi_2 - F_H^{(4)} + Q_{\text{exch.}} = 0$$

$$F_H^{(3)} - \sum_{i=1,\ldots,N} h_i(T^{(3)}) \cdot f_i^{(3)} = 0$$

$$F_H^{(4)} - \sum_{i=1,\ldots,N} h_i(T^{(4)}) \cdot f_i^{(4)} = 0$$

When accounting for the reaction enthalpies $\{\Delta H_{R,1}, \Delta H_{R,2}\}$ as being given, these three equations match with three additional unknowns $\{F_H^{(3)}, F_H^{(4)}, Q_{\text{exch}}\}$.

This approach of using preset temperatures is justified when the effects of the enthalpy sources and sinks (such as a reaction enthalpy $\Delta H_R \cdot \xi$ or a phase change enthalpy) is much larger than the change in a stream enthalpy $\Delta F_H^{(j)}$ due to a variation in temperature.

Obviously, this linear approach by means of use of guesstimated temperatures will break down for an adiabatic operation of a process unit because of the additional specification $Q_{\text{exch}} = 0$. A new variable needs to be introduced which is the temperature of the unit outlet stream $T^{(4)}$ within this enthalpy balance. By adjusting this temperature, the right amount of enthalpy is carried away to ensure adiabatic operation.

Case A9.3: Alternative connectivity in a block diagram by use of a structure model

The units in the preceding block diagram in Figure A9.2 are directly connected by a stream. The behavior models of the process units include the connecting streams. There is no distinction between the behavioral models for the units and a structural model for describing how units are connected by streams. The behavioral and structural model are lumped together. This is disadvantageous if a process block diagram needs to be rearranged with rearranged connections between process units. A unit may have got into different streams. Then, the model equations of a unit involved in a reshuffle needs to be rewritten because ingoing and outgoing streams are relabeled. Is it logical that a change in connectivity of units is reflected in having to change the unit models? No, because unit models are the more complex ones in the overall pro-

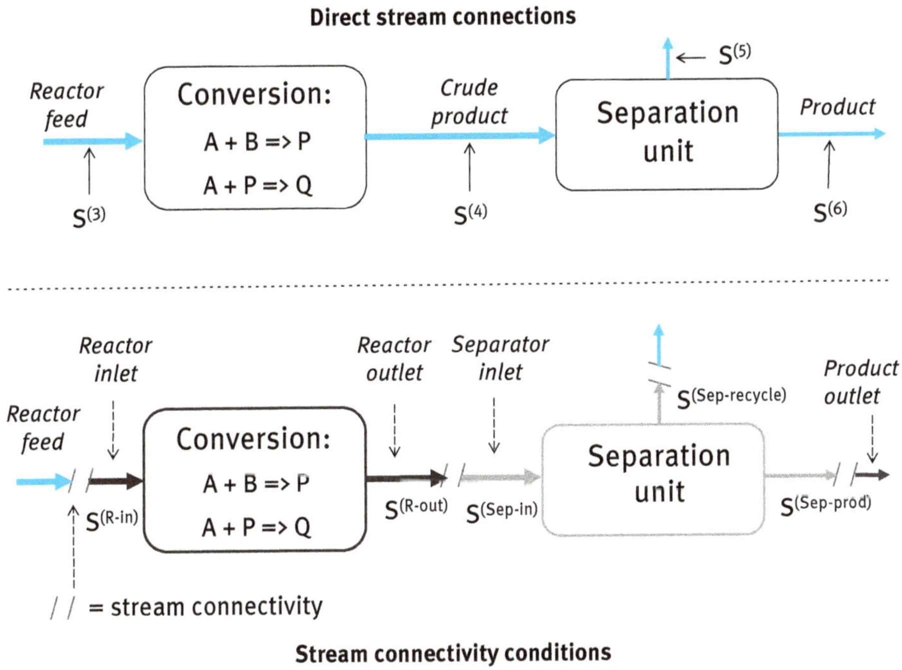

Fig. A9.3: Stream connectivity conditions between process units.

cess model and such model rewriting is error prone and may involve quite some effort. It would be better instead to have some changes in the stream connections only. This preferred way of modeling can be realized at a slight expense of introducing more linear connectivity equations in the model. The principle of creating flexible connections is explained in Figure A9.3.

The top part of this diagram shows a direct connection between units. For instance, stream $S^{(4)}$ directly connects the conversion and separation units. The lower part assigns generic slots to pass inlet and outlet streams to each unit. These slots can be labeled as pertaining to a particular unit. For instance, the generic entrance slot for a feed stream to the conversion unit is now labeled as $S^{(\text{R-in})}$, while its generic outlet slot is $S^{(\text{R-out})}$. Similarly, the generic inlet slot of the separator is $S^{(\text{Sep-in})}$.

The model equations for the conversion unit are written in terms of streams pertaining to its slots:

$$\begin{aligned}
&\text{Species balances:} && \underline{\mathbf{f}}^{(\text{R-in})} - \underline{\mathbf{f}}^{(\text{R-out})} + \mathbf{S}^{T} \cdot \xi && = \underline{\mathbf{0}} \\
&\text{Conversion:} && f_A^{(\text{R-in})} \cdot \eta_A - (\xi_1 + \xi_2) && = 0 \\
&\text{Selectivity:} && \xi_1 - \sigma_{A\,P} \cdot (\xi_1 + \xi_2) && = 0
\end{aligned}$$

Similarly, the separation equations in the separation unit can be written for all species (i) as:

$$\text{Recycle stream:} \quad f_i^{(\text{Sep-recycle})} - m_i \cdot f_i^{(\text{Sep-in})} = 0$$
$$\text{Product stream:} \quad f_i^{(\text{Sep-prod})} - (1 - m_i) \cdot f_i^{(\text{Sep-in})} = 0$$

The unit models have a generic form now; it is independent of their connectivity to other units. To connect units, the outlet slot of one unit is connected to the corresponding inlet slot of another unit. Making such a connection must be explicitly done by means of stream connectivity equations. Here is the example of connecting the stream through the outlet slot of the conversion unit with the stream through the inlet slot of the separation unit:

$$\text{Connectivity:} \quad \underline{\mathbf{f}}^{(\text{R-out})} - \underline{\mathbf{f}}^{(\text{Sep-in})} = \underline{\mathbf{0}}$$

The set of all connectivity equations forms a structure model of the block diagram.

Is there a price to pay for this convenient generalization by separating behavioral and structure models? The number of variables related to streams that connects two process units will double, and for every additional stream variable there is an additional connectivity equation. Fortunately, the price in terms of additional computing effort is very low. Modern numerical equation solvers are very efficient in dealing with such simple linear connectivity equations and variables. They are internally quickly eliminated by a solver, reducing the problem to its core structure of multi term model equations. Overall, one can say that keeping a process model transparent to its developer and users is much more important for efficient work processes than small amounts of additional computing costs.

Creating models of alternative block diagram structures by use of binary variables

The benefit of having explicit connectivity equations can also be exploited when inserting alternative process units. These alternatives exercise the same function but with different efficiencies and costs. Let us assume that the current separation block (now labeled as Sep-A) will be put in competition with an alternative separation unit (labeled Sep-B). It is now possible to flexibly connect the outlet stream of the conversion unit to the inlet stream of either of both separation units. This flexible connection is realized with the help of binary variables $Y_{\text{sep,A}}$ and $Y_{\text{sep,B}}$. A binary variable takes a value as 0 or 1 only.

$$\text{Connectivity:} \quad \underline{\mathbf{f}}^{(\text{R-out})} - Y_{\text{sep,A}} \cdot \underline{\mathbf{f}}^{(\text{Sep-A-in})} - Y_{\text{sep,B}} \cdot \underline{\mathbf{f}}^{(\text{Sep-B-in})} = \underline{\mathbf{0}}$$

Since only one of both types of separation units can be present, this exclusivity condition is expressed as a constraint on the binary variables:

$$\text{Exclusivity of units:} \quad Y_{\text{sep,A}} + Y_{\text{sep,B}} = 1$$

For $Y_{sep,A} = 1$, the conversion unit is connected with separation unit type A. For $Y_{sep,B} = 1$, a connection is made with separation unit type B. If the binary variables would be treated as variables along with continuous variables in the design computations, the model would become nonlinear due to the product form of a binary with a continuous variable in terms, such as: $Y_{sep,A} \cdot \underline{\mathbf{f}}^{(Sep-A-in)}$. Therefore, the explicit use of binary design variables will be postponed until the treatment of mixed integer, nonlinear programming approaches. The latter are applied to formulating and solving process synthesis and design problems in Chapter 9.4.

Case A9.4: Optimization of process performance by LP

For linear process design, optimization one has to do five things.

(a) To open up some degrees of freedom in the design model for optimization.
(b) To select a linear performance metric.
(c) To introduce linear inequality constraints on process variables to demarcate the physical domain in which a process design remains physically feasible.
(d) To select a reference production scenario and compute its performance.
(e) To compute the optimum solution and assess its relevance over the reference solution.

This will be applied to the conversion-separator-recycle (CSR) process shown in the block diagram in Figure A9.2. The species balances give rise to a linear model with some residual degrees of freedom.

Re (a): The feed rates of species A and B in streams (1) and (2) are set free for optimization. The internal process parameters {degree of conversion, selectivity, separation fraction and split factors} are kept fixed in order to maintain model linearity. The process performance can be optimized by varying an external production scenario (i.e., feed rates).

Re (b): The selected performance metric (E_{profit}) is an economic one; the profit rate. This rate is obtained by deducting the following from the proceeds of the products:

- costs of the feeds per unit time
- a cost term for cleaning of the purge stream

The resulting profit rate expression is:

$$E_{profit} = f_P^{(5)} \cdot p_P + f_Q^{(5)} \cdot p_Q - f_A^{(1)} \cdot c_A - f_B^{(2)} \cdot c_B - f_B^{(7)} \cdot c_{purge}$$

The economic parameters $\{p_P, p_Q, c_A, c_B, c_{purge}\}$ are, respectively, the unit sales prices of products P and Q, the unit costs of feed A and B and the purge cost of component B. Since a cost is associated with removing the excess of reactant B via the purge, one may expect that an optimization will push for reduced use of reactant B.

Tab. A9.2: Design, economic and feed stream data for CSR process.

Design parameters **(symbol)**	**Value (-)**	***Economic parameters*** **(symbol)**	**Value ($/kmol)**
Conversion of feed A (η_A)	0.98	Sale price of product P (p_P)	75.0
Selectivity of A to P (σ_{AP})	0.80	Sale price of product Q (p_Q)	50.0
Separation fraction A (m_A)	1	Cost of feed A (c_A)	15.0
Separation fraction B (m_B)	0.95	Cost of feed B (c_B)	30.0
Separation fraction C (m_C)	1	Cost of purging B (c_{purge})	5.0
Separation fraction P (m_P)	0.05	*Production parameters*	(kmol/s)
Separation fraction Q (m_Q)	0.0	*Nominal* production rate P ($f_P^{(5)}$)	0.10
Split factor (s)	0.05	Maximum sale of Q ($F_{Q,max}$)	0.02
Feeding ratios in stream (1)	Ratio (–)	*Feeding ratios in stream (2)*	Ratio (–)
A ($r_{A\|A}^{(1)}$)	1.0	A ($r_{A\|B}^{(2)}$)	0.0
B ($r_{B\|A}^{(1)}$)	0	B ($r_{B\|B}^{(2)}$)	1.0
C ($r_{C\|A}^{(1)}$)	0	C ($r_{C\|B}^{(2)}$)	0.02
P ($r_{P\|A}^{(1)}$)	0	P ($r_{P\|B}^{(2)}$)	0.0
Q ($r_{Q\|A}^{(1)}$)	0	Q ($r_{Q\|B}^{(2)}$)	0.0

Re (c): Two inequalities are imposed:
- a market constraint on the amount of Q that can be sold: $F_{Q,max} - f_Q^{(5)} \geq 0$
- the exit flow of B from the reactor cannot go below zero: $f_P^{(4)} \geq 0$

Re (d): The base case (reference) scenario is:
- $f_A^{(1)} = 0.9 * f_B^{(2)}$
- $f_P^{(5)} = 0.1$ [kmol/s]

Re (e): Comparing the results of the base case and the optimized case.

Table A9.2 lists numerical values of the design and economic and feed parameters in the CSR process model.

Given this input data, the 'mass' balances and economic performance function can be computed for the nominal and the optimized case. The mass balances are expressed by means of the *molar* flows of the five species in the process streams. The mass balance results are presented in Tables A9.3 and the key performance data in Table A9.4, covering both the nominal and optimized case.

The results in Table A9.3 show a substantial increase in the output rate of product P, while side product Q is pushed to its maximum level for attainable sale. It is also remarkable that reactant B is reacted nearly to extinction in the optimum case, thus reducing the cost of purging excess B. Also the recycle flows (stream 8) have become much smaller in the optimized case by better balancing the feed intakes, relative to the base case. In the optimized case there is a slight surplus of intake of A over B in order to push the production of side product Q to its sales limit.

Tab. A9.3: Molar flows in CSR process at nominal and optimized cases.

Species	Stream 1	Stream 2	Stream 3	Stream 4	Stream 5	Stream 6	Stream 7	Stream 8
Nominal	*case*							
A (kmol/s)	*0.12546*	0.00000	0.12789	0.00256	0.00000	0.00256	0.00013	0.00243
B (kmol/s)	0.00000	*0.13940*	0.38562	0.27282	0.01364	0.25918	0.01296	0.24622
C (kmol/s)	0.00000	0.00279	0.05576	0.05576	0.00000	0.05576	0.00279	0.05297
P (kmol/s)	0.00000	0.00000	0.00500	0.10526	*0.10000*	0.00526	0.00026	0.00500
Q (kmol/s)	0.00000	0.00000	0.00000	0.01253	*0.01253*	0.00000	0.00000	0.00000
Optimum	*case*							
A (kmol/s)	*0.20020*	0.00000	0.20408	0.00408	0.00000	0.00408	0.00020	0.00388
B (kmol/s)	0.00000	*0.18001*	0.18007	*0.00007*	0.00000	0.00007	*0.00000*	0.00007
C (kmol/s)	0.00000	0.00360	0.07200	0.07200	0.00000	0.07200	0.00360	0.06840
P (kmol/s)	0.00000	0.00000	0.00798	0.16798	*0.15958*	0.00840	0.00042	0.00798
Q (kmol/s)	0.00000	0.00000	0.00000	0.02000	*0.02000*	0.00000	0.00000	0.00000

Tab. A9.4: Comparison of nominal and optimum cases in CSR process.

	Profit ($/s)	$f_P^{(5)}$ (kmol/s)	$f_Q^{(5)} \leq 0.02$ (kmol/s)	$f_B^{(5)} \geq 0$ (kmol/s)	Feed A (kmol/s)	Feed B (kmol/s)	$\xi_{\text{reaction 1}}$ (kmol/s)	$\xi_{\text{reaction 2}}$ (kmol/s)
Nominal	1.99810	0.10000	0.01253	0.01364	0.12546	0.13940	0.11280	0.01253
Optimal	4.56521	0.15958	0.02000	0.00000	0.20020	0.18001	0.18000	0.02000

The results in Table A9.4 show that the economic performance has been substantially improved by optimization. For instance, the profit per unit of product P increased from 19.98 $/kmol P to 28.61 $/kmol. It is also seen that the optimum is indeed bounded by two active inequality constraints. The production of Q reaches its maximum (sales) level, while the component B is completely reacted away and has zero flow in the effluent.

Figure A9.4 shows the shift from the nominal case to the optimum point in the plane of the two (independent) feed flows, A and B. In the nominal case, the production of the maximum sales volume of Q is not reached yet. The optimum shifts along the profit gradient towards the intersection of two constraints, where profit takes its achievable maximum value.

Analysis of sensitivities of process performance for parameter changes

Suppose one would like to further improve the performance of the process. The question is where to have the most effective of further technological or commercial developments. To address this question a parametric sensitivity analysis will be made. The following three parameters have been chosen for further analysis of their critical role in process performance:

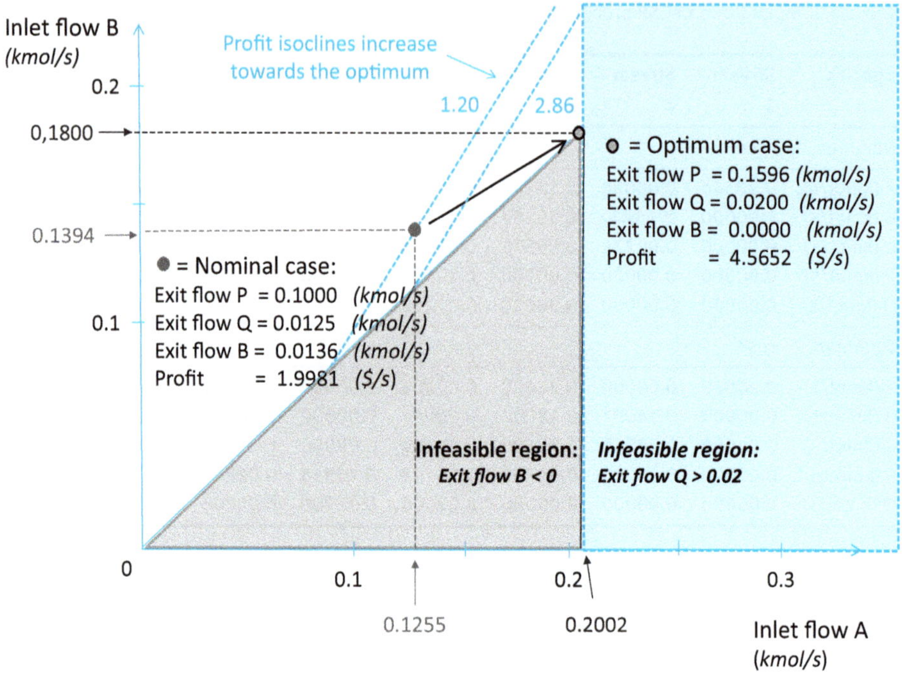

Fig. A9.4: Graph of outcome of optimization of reactor-separator-recycle process.

Technological

Selectivity factor of reactions from A to P (σ_{AP}) because it will enhance production of P.

Separation factor of component P (m_P) because it will also enhance output of P.

Commercial

An increase of the maximum sales volume of Q.

Starting from the optimized situation, the values of these three model parameters are sequentially changed by 2% relative to their nominal values. With each parameter change, the process design is reoptimized for maximum profit. The effect of the parameter change on the profit is expressed as a scaled sensitivity of the profit with respect to that parameter (p):

$$S_{par} = P_{ref} * \{\Delta\text{Profit}/\Delta p\}/\text{Profit}_{ref}$$

Selectivity factor of reactions from A to P (σ_{AP}):	0.11109
Separation factor P (m_P):	−0.00009
Larger sales volume of side product Q:	0.01948

The conclusion from these sensitivity data is that further performance enhancement, from among these three options, should focus on improving the selectivity of the reactions. However, this priority may be reversed when the cost of further technological development well exceeds the benefits of exercising the commercial option.

In a concluding reflection on this optimization case one could argue that it is a hypothetical example with nice results to show. Indeed, it is a hypothetical case with illustrative results. Yet the current way of presenting and analyzing the results between a base case and an optimized case is representative of practical applications.

Case A9.5: Inverse use of linear models for process targeting and estimation

Consider a linear algebraic process model, $\mathbf{A} \cdot \underline{\mathbf{u}} = \underline{\mathbf{y}}$, where $\underline{\mathbf{u}}$ is the N dimensional vector of process inputs and $\underline{\mathbf{y}}$ the M dimensional vector of process outputs. Matrix $\mathbf{A}$ $(M \cdot N)$ contains the internal process parameters. The matrix is nonsquare, making direct algebraic inversion impossible. Now there are three different ways to proceed in using such a model:

(1) input forward: given A and u, compute $\underline{\mathbf{y}}$. This case is straightforward and will be skipped
(2) output targeting: given A and targets for y, compute $\underline{\mathbf{u}}$.
(3) estimation: given a sequence of experimental observations of inputs $\{\underline{\mathbf{u}}_1, \underline{\mathbf{u}}_2, \ldots, \underline{\mathbf{u}}_K\}$ and of outputs $\{\underline{\mathbf{y}}_1, \underline{\mathbf{y}}_2, \ldots, \underline{\mathbf{y}}_K\}$ find the unknown process parameters which are the elements of matrix $\mathbf{A}$.

Both the targeting and the estimation are inverse problems. These are hard(er) to solve if the numbers of inputs and outputs are different. Then one has to deal with a nonsquare matrix $\mathbf{A}$ ($M > N$: more inputs than outputs or $M < N$, more outputs than inputs). One wants to find solutions match in some 'best possible' way with the linearity of the model. The best possible way has to be defined: here it will be done by means of an additional quadratic function that aims for minimization of deviations from model linearity. The fitting function is a quadratic function over the model residuals: $\underline{\boldsymbol{\varepsilon}} = \mathbf{A} \cdot \underline{\mathbf{u}} - \underline{\mathbf{y}}$. The benefit of using quadratic expressions as fitting functions in both targeting and estimation is that closed from linear analytical expressions result in answers. By using quadratic functions, linearity returns immediately when the necessary conditions for an extremum (minimum or maximum) are formulated in terms of the first order derivatives being set equal to zero. These derivative expressions are linear in all variables.

Targeting

One can set targets for the outputs and ask what are the inputs needed to realize these targets. When the number of output targets exactly matches the number of adjustable inputs, this inverse problem can be solved if the matrix A is regular (= nonsingular; $\mathbf{A}^{-1}$ exists):

$$\underline{\mathbf{u}} = \mathbf{A}^{-1} \cdot \underline{\mathbf{y}}_t \quad \text{with t = target}$$

What to do when there are more targets M than adjustable inputs N, $(M > N)$

Obviously, it will be impossible to exactly match the targets through the few inputs available. Then a strategy could be to minimize the differences between the targets and what the inputs can achieve on the outputs. If one takes a quadratic criterion with a (diagonal) weighting matrix **W** for scaling the different outputs one obtains the following minimization problem. It has a closed analytical expression for the best matching $\underline{\mathbf{u}}^*$:

Minimize over $\{\underline{\mathbf{u}}\}$ the matching function

$$\Omega = \frac{1}{2}(\mathbf{A}\underline{\mathbf{u}} - \underline{\mathbf{y}}_t)^T \cdot W_y^{-1} \cdot (\mathbf{A}\underline{\mathbf{u}} - \underline{\mathbf{y}}_t)$$

Result:

$$\underline{\mathbf{u}}^* = [\mathbf{A}^T \cdot \mathbf{W}_y^{-1} \cdot \mathbf{A}]^{-1} \cdot (\mathbf{W}_y^{-1} \cdot \mathbf{A}^T) \cdot \underline{\mathbf{y}}_t$$

The condition for $(\mathbf{A}^T \cdot \mathbf{W} \cdot \mathbf{A})^{-1}$ to exist is that matrix **A** must have full-row rank, meaning no linear dependencies between the outputs. The residual errors $(\underline{e}^*)$ of the matching are:

$$\underline{\mathbf{e}}^* = \mathbf{A} \cdot \underline{\mathbf{u}}^* - \underline{\mathbf{y}}_t$$

What to do when there are more adjustable inputs N than targets M, $(M < N)$

In such a situation, a number of inputs will remain undetermined by the target outputs. That is, unless one follows a policy of minimizing the changes in the input vector $\underline{\mathbf{u}}$, taking zero as the reference value. The result is again a mathematical optimization problem. One may scale the various inputs by means of the diagonal matrix $\mathbf{W}_u$.

Minimize the Lagrangian function over $\{\underline{\mathbf{u}}\}$

$$L = \frac{1}{2}(\underline{\mathbf{u}})^T \cdot \mathbf{W}_u^{-1} \cdot (\underline{\mathbf{u}}) - \underline{\boldsymbol{\lambda}}^T \cdot (\mathbf{A} \cdot \underline{\mathbf{u}} - \underline{\mathbf{y}}_t)$$

subject to:

$$(\mathbf{A} \cdot \underline{\mathbf{u}} - \underline{\mathbf{y}}_t) = \underline{\mathbf{0}}$$

The outcome is in a closed form for $\underline{\mathbf{u}}^*$:

$$\underline{\mathbf{u}}^* = \mathbf{W}_u \cdot \mathbf{A}^T \cdot [\mathbf{A} \cdot \mathbf{W}_u \cdot \mathbf{A}^T]^{-1} \cdot \underline{\mathbf{y}}_t$$

The Lagrange multipliers are:

$$\underline{\boldsymbol{\lambda}} = [\mathbf{A} \cdot \mathbf{W}_u \cdot \mathbf{A}^T]^{-1} \cdot \underline{\mathbf{y}}_t \qquad (\underline{\boldsymbol{\lambda}} \text{ has dim. } M) .$$

A Lagrangian multiplier indicates the sensitivity of the objective function L with respect to a change in the corresponding output target $y_{t,m}$: $\underline{\boldsymbol{\lambda}}_m = dL/dy_{t,m}$.

The interpretation of the solution is that the optimized inputs take minimized sizes while jointly meeting the output targets.

Estimation

Let be given a sequence of K experimental observations of inputs $\{\underline{\mathbf{u}}_1, \underline{\mathbf{u}}_2, \ldots, \underline{\mathbf{u}}_K\}$ and outputs $\{\underline{\mathbf{y}}_1, \underline{\mathbf{y}}_2, \ldots, \underline{\mathbf{y}}_K\}$. Experiment k yields the combined measurements of inputs and outputs $\{\underline{\mathbf{u}}_{k,e}, \underline{\mathbf{y}}_{k,e}\}$. The goal is to obtain statistical estimates of the process parameters. These parameters are in this case the elements of matrix $\mathbf{A}$, satisfying the linear model. The usual approach to obtain parameter estimates is a weighted least squares fitting to the experimental data. The first step is always to look into the scaling of the output variables. These variables may take very different magnitudes, e.g., the temperature of a stream is 300 K, while the molar fraction of a trace component in a stream can be $< 10^{-4}$. To ensure more robust solutions, it is prudent to scale all output variables to the same order of magnitude of unity. Let the scaling factors(s) be ordered in a diagonal matrix $\mathbf{D}_y$. Similarly one is advised to scale the inputs by means of another diagonal scaling matrix $\mathbf{D}_u$. Then the scaled outputs and parameter matrix become:

$$\underline{\mathbf{y}}' = \mathbf{D}_y^{-1} \cdot \underline{\mathbf{y}}\,; \qquad \underline{\mathbf{u}}' = \mathbf{D}_u^{-1} \cdot \underline{\mathbf{u}} \qquad \mathbf{A}' = \mathbf{D}_y^{-1} \cdot \mathbf{A} \cdot \mathbf{D}_u^{-1}$$

The goal is to find the parameters in the linear equation that match the experimental observations best according to some criterion.

Selecting a suitable criterion

The selection of a proper criterion is related to the nature of the probabilistic distribution of the measurement errors. Here, a *simplified approach* is taken by assuming that the error distribution of the residuals of the model equation using experimental input and output data is a multivariate, normal distribution with a mean of zero and uncorrelated spreads (i.e., the covariances of the residuals are zero). That is, the error in the residual of one equation is probabilistic independent of the error in another equation. Due to the scaling of the variables one can also assume that the spreads of all residuals are the same. A quadratic fitting criterion can be used, corresponding to the quadratic argument of a normal error distribution. If there are reasons to treat the measurement error distributions of outputs and inputs separately, one can use a more refined total least squares approach (see below for a formulation). For now a weighted least squares approach is applied to the residuals of the equations, working with scaled quantities. The objective function is defined as the sum over all experiments of the squared residuals of the equations, using the (scaled) experimental data for the inputs and outputs. One can also assign a weight per experiment $\{w_k, k = 1, \ldots K\}$ if a different degree of trust is put in the quality of the data per experiment. If one would put $w_k = 1/K$, the total sum (Ω) is scaled as the averaged squared residual over all experiments.

$$\text{Error sum:} \quad \Omega = \frac{1}{2} \sum_{k=1}^{K} \{\underline{\mathbf{y}}'_{k,e} - \mathbf{A}' \cdot \underline{\mathbf{u}}'_{k,e}\}^{\mathrm{T}} \cdot w_k \cdot \{\underline{\mathbf{y}}'_{k,e} - \mathbf{A}' \cdot \underline{\mathbf{u}}'_{k,e}\}$$

Minimization of this sum function with respect to the elements of matrix $\mathbf{A}$ results in an analytical closed expression for matrix $\mathbf{A}$. This least squares solution can be written in a compact form if the measured observations are ordered in data matrices:

$$\mathbf{Y}'_K = [\underline{\mathbf{y}}'_{1,e}, \underline{\mathbf{y}}'_{2,e}, \ldots, \underline{\mathbf{y}}'_{K,e}] \qquad \{M \cdot K\} \text{ data matrix}$$

$$\mathbf{U}'_K = [\underline{\mathbf{u}}'_{1,e}, \underline{\mathbf{u}}'_{2,e}, \ldots, \underline{\mathbf{u}}'_{K,e}] \qquad \{N \cdot K\} \text{ data matrix}$$

$$\mathbf{W}_K = \text{diagonal}[w_k, k = 1, \ldots, K] \quad \{K \cdot K\} \text{ weights matrix}$$

$$\text{Solution:} \quad \mathbf{A}'_{\text{estim}} = [\mathbf{Y}'_K \cdot \mathbf{W}_K \cdot \mathbf{U}'^{\,\mathrm{T}}_K] \cdot [\mathbf{U}'_K \cdot \mathbf{W}_K \cdot \mathbf{U}^{\mathrm{T}}_K]^{-1}$$

$$\text{Matrix dimensions:} \quad (M \cdot N) \qquad (M \cdot N) \qquad (N \cdot N)$$

This solution can be computed as long as the input information matrix $[\mathbf{U}'_K \cdot \mathbf{W}_K \cdot \mathbf{U}'^{\,\mathrm{T}}_K]$ is regular; meaning that matrix $\mathbf{U}'_K$ must have rank N. This requirement translates into the following practical conditions:
- The measurements of the components of $\underline{\mathbf{u}}$ should be linearly independent, or in other words: no measurement of a component should be a linear combination of measurements of some other components.
- There should be at least N experiments ($K \geq N$).
- No experiment should be a perfect linear combination of other experiments.

Having stated these requirements, the problem of matrix parameter estimation based on squaring the residuals of the equations is resolved.

Refined estimation by total least squares

With the conventional approach to estimating the model parameters, it is common to lump all measurement errors of input and output variables occurring in an equation into one remaining residual error per model equation. This is conceptually a bit rough, although computationally convenient. For those liking a more refined statistical approach, the following improvements are made:

The measurement errors of inputs and outputs are accounted for separately. Normal distributions with zero mean and covariances Σ_u and Σ_y are assumed.

A distinction is made between the *experimental* values of variables $\{\underline{\mathbf{y}}'_e; \underline{\mathbf{u}}'_e\}$ and their *theoretical* counterparts, $\{\underline{\mathbf{y}}'_m; \underline{\mathbf{u}}'_m\}$. The latter satisfy the linear model equations exactly:

$$\underline{\mathbf{y}}'_m - \mathbf{A}' \cdot \underline{\mathbf{u}}'_m = \underline{\mathbf{0}} \,.$$

The relation between the experimental and model variables is given by the presence of experimental and modeling errors lumped in the gross error vectors $\underline{\boldsymbol{\varepsilon}}_y$ and $\underline{\boldsymbol{\varepsilon}}_u$:

$$\underline{\mathbf{y}}'_e = \underline{\mathbf{y}}'_m + \underline{\boldsymbol{\varepsilon}}_y$$

$$\underline{\mathbf{u}}'_e = \underline{\mathbf{u}}'_m + \underline{\boldsymbol{\varepsilon}}_y$$

Given these relations, one can formulate the total least squares estimation problem involving a minimization of the sum over all experiments of the squared errors between experimental and model variables, subject to the model equations: Minimize (over A')

$$\Omega_{\mathrm{TLS}} = \frac{1}{2} \cdot \sum_{k=1}^{K} \{\underline{\boldsymbol{\varepsilon}}_{y,k}^{\mathrm{T}} \cdot \Sigma_y^{-1} \cdot \underline{\boldsymbol{\varepsilon}}_{y,k} + \underline{\boldsymbol{\varepsilon}}_{u,k}^{\mathrm{T}} \cdot \Sigma_u^{-1} \cdot \underline{\boldsymbol{\varepsilon}}_{u,k}\}$$

Subject to:

$$\begin{aligned} \underline{\mathbf{y}}'_{\mathrm{e},k} - \underline{\mathbf{y}}'_{\mathrm{m},k} - \underline{\boldsymbol{\varepsilon}}_{y,k} &= \underline{\mathbf{0}} \qquad k = 1, \ldots, K \\ \underline{\mathbf{u}}'_{\mathrm{e},k} - \underline{\mathbf{u}}'_{\mathrm{m},k} - \underline{\boldsymbol{\varepsilon}}_{u,k} &= \underline{\mathbf{0}} \qquad k = 1, \ldots, K \\ \underline{\mathbf{y}}'_{\mathrm{m},k} - \mathbf{A}' \cdot \underline{\mathbf{u}}'_{\mathrm{m},k} &= \underline{\mathbf{0}} \qquad k = 1, \ldots, K \end{aligned}$$

This problem objective function can be reformulated as a Lagrangian function with Lagrangian multipliers dealing with each of the associated algebraic equations.

$$L = \Omega_{\mathrm{TLS}} - \sum_{k=1}^{K} [\underline{\boldsymbol{\lambda}}_{y,k}^{\mathrm{T}} \cdot (\underline{\mathbf{y}}'_{\mathrm{e},k} - \underline{\mathbf{y}}'_{\mathrm{m},k} - \underline{\boldsymbol{\varepsilon}}_{y,k}) - \underline{\boldsymbol{\lambda}}_{u,k}^{\mathrm{T}} \cdot (\underline{\mathbf{u}}'_{\mathrm{e},k} - \underline{\mathbf{u}}'_{\mathrm{m},k} - \underline{\boldsymbol{\varepsilon}}_{u,k}) - \underline{\boldsymbol{\lambda}}_{u,k}^{\mathrm{T}} \cdot (\underline{\mathbf{y}}'_{\mathrm{m},k} - \mathbf{A}' \cdot \underline{\mathbf{u}}'_{\mathrm{m},k})]$$

The solution of this (large scale) optimization problem is more quickly determined by numerical means. A closed form, analytical solution exists but it contains many matrix and vector operations with matrix inversions. This will require numerical support anyway. It is also possible to find estimates for the covariance matrices of the unknown quantities, although the expressions become very involved.

In any case, consulting a book on engineering statistics is recommended (e.g., [4]) or an expert statistician on the nature of suitable error distributions, the corresponding choice of a suitable fitting function and aspects of interpretation of the estimation results to avoid conceptual pitfalls.

Case A9.6: Notes on the use of other types of linear models

(a) *Computing dynamic responses of a process (unit) by means of dynamic state space* models

Inputs to process: $\underline{\mathbf{u}}(t)$; internal states of process: $\underline{\mathbf{x}}(t)$; process outputs: $\underline{\mathbf{y}}(t)$
Model equations:

$$\begin{aligned} \mathrm{d}\underline{\mathbf{x}}(t)/\mathrm{d}t &= \mathbf{A} \cdot \underline{\mathbf{x}}(t) + \mathbf{B} \cdot \underline{\mathbf{u}}(t) \quad \text{with} \quad \underline{\mathbf{x}}(t_0) = \underline{\mathbf{x}}_0 \text{ and } t > t_0 \\ \underline{\mathbf{y}}(t) &= \mathbf{C} \cdot \underline{\mathbf{x}}(t) \end{aligned}$$

The analytical solution can be expressed in a closed form:

$$\underline{\mathbf{x}}(t) = \exp\{\mathbf{A} \cdot (t - t_0)\} \cdot \underline{\mathbf{x}}_0 + \int_{t_0}^{t} \exp\{\mathbf{A} \cdot (t - \tau)\} \cdot \mathbf{B} \cdot u(\tau) \cdot \mathrm{d}\tau$$

In the concept stage, such dynamic modeling will be useful when changes in the feed supply to the process are expected to occur and its response is of interest.

(b) *Distributed process/product dynamics*

Inputs to process: $\underline{\mathbf{u}}(z, t)$; internal states of process: $\underline{\mathbf{x}}(z, t)$
Model equations:

$$\partial\underline{\mathbf{x}}(z, t)/\partial t + v \cdot \partial\underline{\mathbf{x}}(z, t)/\partial z = \mathbf{A} \cdot \underline{\mathbf{x}}(z, t) + \mathbf{B} \cdot \underline{\mathbf{u}}(z, t) \quad \text{for} \quad t > t_0 \text{ and } z < z_0$$

Initial condition:

$$\underline{\mathbf{x}}(z, t_0) = \underline{\mathbf{x}}_0(z) \quad \text{for} \quad t = t_0 \text{ and } z > z_0$$

Boundary condition:

$$\underline{\mathbf{x}}(z_0, t) = \underline{\mathbf{x}}_{\mathrm{IN}}(t) \quad \text{for} \quad t > t_0 \text{ and } z = z_0$$

This is a Cauchy type of problem and an analytical solution in a closed form exists. While working with the analytical solution will give more insight, direct numerical equation solving will be quicker. In the concept stage, this form of dynamic modeling is useful for describing the change in the mass (or size) distribution of a particulate product in a process unit (e.g., crystals and polymer). In these cases, vector $\underline{\mathbf{x}}$ becomes a scalar function $x(z, t)$. Quantity $x(z, t)$ is the number of particulates per unit volume of the process unit, having mass z at time t. The quantity v is a fixed mass growth rate. Similarly, matrix $\mathbf{A}$ reduces to a scalar and usually it takes a negative value ($A \Rightarrow -a$), representing the mass decay rate of particulates. The final term, $\mathbf{B} \cdot \underline{\mathbf{u}}(z, t)$, also becomes a scalar one, representing the distributed mass birth rate of particulates with mass z at time t. Commonly, only particulates with small mass stand a chance to be born, e.g., $u(z, t) = u(z_0, t) \cdot e^{\{-\lambda \cdot z/z0\}}$ in which $u(z_0, t)$ is the birth rate of particulates with minimum mass (z_0) while λ (> 0) is a decline factor for the mass birth rate of bigger mass. The boundary condition $x(z_0, t) = x_{\mathrm{IN}}(t)$ indicates how many particulates of minimum mass z_0 are introduced from an external source over time t. If there is a nonzero birth rate term, $u(z, t)$, one usually takes $x_{\mathrm{IN}}(t) = 0$. This is the same the other way around also.

This model enables the computation of the evolution of the mass size distribution of the particulates under different scenarios, such as different initial distributions and the impact of changes in the physical parameters (V, A, B, $u(z_0, t)$, λ) on the mass distribution. It is of practical importance to know if the mass distribution becomes a steady one over a long enough time, or that the distribution will show sustained oscillations.

(c) *Discrete time model for the dynamics of a* fast moving consumer goods *(FMCG) warehouse inventory*

Inputs to process at time k:

$$\underline{\mathbf{u}}_k\,; \quad \text{internal states at time } k\text{: } \underline{\mathbf{x}}_k\,; \quad \text{Outputs: } \underline{\mathbf{y}}_k$$

Model equations:

$$\underline{\mathbf{x}}_{k+1} = \mathbf{A} \cdot \underline{\mathbf{x}}_k + \mathbf{B} \cdot \underline{\mathbf{u}}_k - \mathbf{C} \cdot \underline{\mathbf{y}}_k \qquad k = 1, 2, 3, \ldots, K \Rightarrow \infty$$

$$\mathbf{A}: \; N \cdot N \text{ matrix}; \quad \mathbf{B}: \; N \cdot J \text{ matrix}; \quad \mathbf{C}: \; N \cdot M \text{ matrix}$$

Initial condition:

$$\underline{\mathbf{x}}_k = \underline{\mathbf{0}} \qquad \text{for } k = 1$$

This model can be applied in the concept stage to determine the dynamics of the inventory of a warehouse for FMCG. Knowing the dynamics is important for the design of the storage capacity of the warehouse. A closed analytical expression for the solution $\mathbf{x}_{k+1}$ can be derived but it is more cumbersome to use than doing a direct numerical simulation.

Suppose N types of products are kept in stock in the warehouse. For each type of product, the stored mass may be different. The mass (x_i) of each type of product n, $\{n = 1, \ldots, N\}$ that is stored at instant k is accounted for in the mass hold up vector $\underline{\mathbf{x}}_K = [x_{1,k}, x_{2,k}, \ldots, x_{n,k}, \ldots, x_{N,k}]^{\mathrm{T}}$. The N products are supplied to the warehouse in J different types of packages, at every instant k. All packages contain the same N products; however, the mass fraction of a particular product may vary over the input packages. Each type of input package $\{u_j\}$ has a fixed composition of the N products. This composition is stored in the columns of delivery matrix $\mathbf{B}$ (= N rows $*$ J columns). An entry u_j $\{j = 1, \ldots, J\}$ in the input vector $\underline{\mathbf{u}}$ represents the mass of a package of type J. The vector $\underline{\mathbf{u}}_k = [u_{1,k}, u_{2,k}, \ldots, u_{j,k}, \ldots, u_{J,k}]^{\mathrm{T}}$ shows the masses of all supplied packages at instant k.

In the warehouse, the products are rearranged in new packages of different compositions for sales purposes. The products are sold in multiple packages: there are M different sales packages to customers. The composition of sales package m is specified by the data in column m of matrix $\mathbf{C}$ (= N rows $*$ M columns). There are M types of packages leaving at any instant k. The mass of sales package m at instant k is $y_{m,k}$. The column vector $\underline{\mathbf{y}}_k = [y_{1,k}, y_{2,k}, \ldots, y_{m,k}, \ldots, y_{M,k}]^{\mathrm{T}}$ is made up of the masses of all sold packages at instant k.

Product storage and handling in the warehouse leads to loss of product quality. So, a small fraction of the products deteriorates and has to be discarded. The decay fraction can vary per type of product but it remains constant over time. This product decay is expressed as a fractional loss in matrix $\mathbf{A}$:

$$\mathbf{A} = \mathbf{I}_N - \mathbf{D}\,.$$

$\mathbf{I}_N$ is the identity matrix of dimension N. The diagonal matrix $\mathbf{D}$ ($0 < \mathbf{D} < \mathbf{I}_N$) of dimension N contains the decay fractions of the product mass stored.

It is a design question and is about how big the warehouse should be in order to store its hold up in a steady state $\langle \underline{\mathbf{x}} \rangle$, given time averaged masses of the input and output packages, $\langle \underline{\mathbf{u}} \rangle$ and $\langle \underline{\mathbf{y}} \rangle$. The answer can be given in a closed form:

$$\langle \underline{\mathbf{x}} \rangle = (\mathbf{I}_N - \mathbf{D}) \cdot \langle \underline{\mathbf{x}} \rangle + \mathbf{B} \cdot \langle \underline{\mathbf{u}} \rangle - \mathbf{C} \cdot \langle \underline{\mathbf{y}} \rangle$$

The steady state hold up is:

$$\langle \underline{\mathbf{x}} \rangle = \mathbf{D}^{-1} \cdot \{\mathbf{B} \cdot \langle \underline{\mathbf{u}} \rangle - \mathbf{C} \cdot \langle \underline{\mathbf{y}} \rangle\}$$

Analysis of this solution:

The first question is what happens when a product n does not decay: $D_{nn} = 0$. An infinitely large stored mass of product n can only be avoided when the total input of product n taken over all supply packages is exactly balanced by its total output over all the sales packages:

$$\mathbf{B}|_n \cdot \langle \underline{\mathbf{u}} \rangle - \mathbf{C}|_n \cdot \langle \underline{\mathbf{y}} \rangle = 0$$

When D_{nn} is small, the total supply of product n must slightly exceed the total sales of this product in order to keep a modest storage $\langle \underline{\mathbf{x}}_n \rangle$:

$$\mathbf{B}|_n \cdot \langle \underline{\mathbf{u}} \rangle = \mathbf{C}|_n \cdot \langle \underline{\mathbf{y}} \rangle + D_{nn} \cdot \langle \underline{\mathbf{x}}_n \rangle$$

The design question about the size of the warehouse can be addressed now the total averaged mass hold up $\langle M_{\text{warehouse}} \rangle$ can be derived from the typical supply $\langle \underline{\mathbf{u}} \rangle$ and demand $\langle \underline{\mathbf{y}} \rangle$ pattern:

$$\langle M_{\text{warehouse}} \rangle = \sum_{n=1}^{N} \langle x_n \rangle$$

The warehouse hold up should be designed larger than the average mass content because of the dynamic fluctuations in supply and demand. The dynamic model $\underline{\mathbf{x}}_{k+1} = \mathbf{A} \cdot \underline{\mathbf{x}}_k + \mathbf{B} \cdot \underline{\mathbf{u}}_k - \mathbf{C} \cdot \underline{\mathbf{y}}_k$ can assist in performing scenario analysis for expected varying supply and demand patterns around the prior estimated steady states values $\langle \underline{\mathbf{u}} \rangle$, $\langle \underline{\mathbf{y}} \rangle$ and starting from the steady state hold up $\langle \underline{\mathbf{x}} \rangle$ for $k = 0$.

One can do further analysis of this design problem with demand and supply treated as stochastic processes; this is beyond the scope of this section.

(d) *Linear integer equations with binary variables*

Model equations:

$$\mathbf{Z} \cdot \underline{\mathbf{Y}} \le \omega$$

Symbols:

$\mathbf{Z}$: $m \cdot n$ matrix of integers ; $\underline{\mathbf{Y}}$: n-vector of binaries ; ω: m-vector of integers.

This type of algebraic equation relates binary variables to each other. It plays an essential role in defining structure models that numerically represent superstructure flow sheets, harboring many alternative specific flow sheets. A small example is given as an illustration of the use of these equations. Suppose that in flow sheet synthesis one can apply either reactor type A or reactor type B. It is known that reactor B performs better than reactor A (by achieving higher yields) but B is also substantially more expensive than A. Yet reactor type A may still outperform reactor type B by placing two reactors A in cascade. The superstructure for the conversion section is one reactor B in parallel to two reactors A in cascade. One of the following structures has to be extracted in the computer assisted process synthesis:

(1) one reactor B, no reactor(s) A
(2) one reactor A, no reactor B
(3) two reactors A in cascade, no reactor B
(4) exclusion: Reactor A2 cannot be placed in cascade after reactor B

Binary variables are assigned for the presence of a reactor ($Y = 1$: present) in the flow sheet:

$$Y_{A1} \text{ belongs to reactor A1}; \quad Y_{A2} \text{ to reactor A2}; \quad Y_B \text{ to reactor B.}$$

The following conditions enforce the above selection options:

$$Y_{A1} + Y_B = 1 \qquad \text{(either reactor A or B)}$$
$$Y_{A1} + Y_{A2} \leq 2 \qquad \text{(maximum of two reactors A in cascade)}$$
$$Y_{A1} + Y_{A2} \leq 0 \qquad \text{(reactor A2 only possible when A1 present)}$$

The optional structures mentioned above are all covered as solutions to this structure model:

(1) one reactor B, no reactor(s) A: $Y_B = 1$, $Y_{A1} = 0$, $Y_{A2} = 0$
(2) one reactor A, no reactor B: $Y_B = 0$, $Y_{A1} = 1$, $Y_{A2} = 0$
(3) two reactors A in cascade, no reactor B: $Y_B = 0$, $Y_{A1} = 1$, $Y_{A2} = 1$
(4) reactor A2 not in cascade after reactor B: $Y_B = 1$, $Y_{A1} = 0$, $Y_{A2} = 1$ is not a solution.

It requires a process design optimization effort to find out which of the three options is the more attractive one in terms of overall process performance.

A9.1.1 Nonlinear models and process simulations for development stage

Case A9.7: Note on ill conditioned constitutive equations in design models

This case is a simple illustration of a locally ill conditioned model by the use of ill-conceived reaction kinetic equation. It violates the computational requirement of having bounded derivatives in the model equations everywhere in the domain of model application. The model involves a continuous flow, well mixed tank reactor (CSTR) with a simple isothermal reaction A $\Rightarrow$ B. The reaction rate law is a commonly occurring type of kinetic expression. The powers of the species concentrations in the rate expression are allowed to deviate from the stoichiometric coefficients (Law of mass action) and become empirical ones. Often, such a power is allowed to take any real value (negative, zero, positive) when fitting the rate parameters to experimental kinetic data. Such type of rate law is applied here in the design equation for a CSTR.

Variables:	x	exit concentration of reactant A	$x \geq 0$
	x_f	inlet concentration of reactant A	$x_f > 0$
	r	reaction rate	$r \geq 0$
	τ	residence time of reacting medium	$\tau > 0$
	k	reaction rate constant	$k > 0$
	a	exponent of concentration in rate law	$0 \leq a < 1$

It is assumed that exponent a takes a value less than one, but remains nonnegative. The inlet concentration (x_f), the residence time (τ), the rate constant (k) and exponent (a) are all known quantities. The outlet concentration (x) is the unknown to be determined as a function of the residence time.

Equations:
Reactor design: $m \equiv x_f - x - \tau \cdot r(x) = 0$
Rate law: $r(x) = k \cdot x^a$
Derivative of reactor equation: $\partial m/\partial x = -1 - \tau \cdot k \cdot [a \cdot x^{(a-1)}]$
For $x \Rightarrow 0$: $x^{(a-1)} \Rightarrow \infty$ (numerical singularity)

Indeed, if one of the powers of a species concentration in a rate law is below one, the derivative of the rate law with respect to that species concentration becomes unbounded if this species concentration goes to zero. The smaller the concentration x the smaller the reaction rate r gets, but it slows down disproportionally fast. In fact, the rate slows down when the concentration approaches zero. The rate itself stops at zero concentration. The derivative of the reaction rate, with respect to the concentration, goes to infinity as the concentration goes down to zero. A numerical solution of the reactor design equation $m(x) = 0$ will fail when x gets close to zero. Beyond a mathematical singularity it is also a physical singularity. This kind of modeling problem appears at the lower boundary of the physically feasible domain; e.g., the boundaries of zero concentrations. *This case illustrates that constitutive laws must be formulated with due regard to correct physical behavior in boundary situations. Otherwise, it will bring process engineering computations in unnecessary trouble.*

This particular singularity problem could have been easily avoided by a physically more realistic modeling of the reaction rate equation:

Refined rate law: $r(x) = k \cdot x/(b + x)^{(1-a)}$ with $a < 1$, $k > 0$, $b > 0$ (small)

Parameter a is the empirical exponent while parameter b is an empirical threshold concentration (small). When the concentration x gets below this threshold value b and would go to zero, both the reaction rate and its derivative start to level down to zero.

Derivative: $\mathrm{d}r(x)/\mathrm{d}x = \{k \cdot [b + x] - k \cdot x \cdot (1 - a) \cdot (b + x)^{-a}\}/\{(b + x)^{(2-a)}\}$
It is noted that for $b > 0$ the derivative remains finite in

$\lim .x \to 0$: $\mathrm{d}r(x)/\mathrm{d}x = k \cdot b^{(-1+a)}$

For the upper limit $a \Rightarrow 1$ one gets the normal exponential decay rate: $\mathrm{d}r(x)/\mathrm{d}x = k$.

Case A9.8: Consistency of design specifications with thermodynamics

It is rather easy to overlook the (implicit) limits imposed by thermodynamics when setting design targets for process units. If targets are set beyond the limit of thermodynamic equilibrium, the computations with the process model and the thermodynamically infeasible design specifications will always fail due to this underlying inconsistency. The failure can show itself by lack of convergence in the computations or yielding unphysical design parameters for the unit. It seems like the numerical solver is failing. Actually, an infeasible problem is impossible to solve and any solver will fail!

A very simple design example illustrates this point. In a continuous flow, well mixed tank reactor, a single, reversible first order reaction A $\Rightarrow$ B is carried out at isothermal conditions. The reactor design question refers to what residence time is needed in order to meet a target conversion of A. The required residence time will determine the volume of the reactor. The reactor design model is given by the following process design equations and process data:

$[M]$:

$$C_{A,f} - C_A - \tau \cdot \{k_f \cdot C_A - k_b \cdot C_B\} = 0$$
$$C_{B,f} - C_B - \tau \cdot \{k_f \cdot C_A - k_b \cdot C_B\} = 0$$

Feed concentrations are known: $C_{A,f} = 1$ and $C_{B,f} = 0$

Known reaction rate parameters: $k_f = 1$ and $k_b = 1/3$

The outlet concentrations $\{C_A, C_B\}$ and the residence time τ are the unknowns. Apart from the process model equations $[M]$, another equation is needed for computing three unknowns. A design target must be added. The design target is set at 90% conversion of reactant A, leaving 10% unconverted:

$$[S_{perf}]: \qquad C_A - 0.1 C_{A,f} = 0$$

The process design equations $[M]$ are reworked to render the residence time explicit:

$$\tau = (C_{A,f} - C_A)/\{k_f \cdot C_A - k_b \cdot (C_{f,A} - C_A)\}$$

Computing the residence time with the given data yields a *negative* value:

$$\tau = -0.5$$

This answer is a bizarre outcome; what went wrong?

Analysis of the design problem

The reaction is reversible; i.e., conversion cannot be pushed beyond reaction equilibrium.

The equilibrium condition is:

$$k_f/k_b = K = C_{B,eq}/C_{A.eq} \quad \text{with} \quad C_{A,eq} + C_{B,eq} = 1$$

The equilibrium concentration of A:

$$C_{A,eq} = C_{A,f} \cdot 1/(1 + K) = 1/(1 + 3) = 0.25$$

When approaching reaction equilibrium, the denominator $\{k_f \cdot C_A - k_b \cdot (C_{f,A} - C_A)\}$ goes to zero. As a result, the residence time goes to infinity and it turns negative beyond the equilibrium condition. This infinity situation may give rise to a declaration of a singularity in computations with a numerical tool.

The initial target of 90% conversion of A results in an exit concentration for A of 0.1. This is smaller than the equilibrium concentration of 0.25. A more realistic target for C_A at 110% of the equilibrium concentration gives a positive residence time:

$$C_A = 0.275 > 0.25 \quad \Rightarrow \quad \tau = 21.75$$

Setting a more lax conversion target (160%):

$$C_A = 0.400 > 0.25 \quad \Rightarrow \quad \tau = 3.0$$

Moving away from the equilibrium boundary reduces the residence time and vessel volume, although at a loss of having a lower conversion. This analysis illustrates an inherent tradeoff between size of reaction vessel and conversion, which is in fact quite general. The resulting heuristic is:

Physical design targets should be located within a thermodynamically feasible domain.

The distance between a target and the thermodynamic equilibrium boundary is negotiable in design.

Case A9.9: Analysis of operability aspects in process design

The operational success of a process plant depends, to a large extent, on the quality of its design. Therefore, operational characteristics of process should be taken into consideration in process design. This can be done by both experimental testing and by model based simulation of the behavior of a process design under various operational scenarios. The experimental testing of vulnerable equipment parts in a process often focuses on the effects of active operation over a long time horizon, such as the tear and wear of equipment over time under a range of processing conditions, as well as loss of functionality of active agents in a process (e.g., catalysts, membranes and solvents) by deactivation, fouling or decompositions. Such long term effects play out on a timescale that is much longer than the total pass through time of feed material through the entire process. These effects are often hard to observe, model and simulate. It involves phenomena having low amplitudes and slow rates occurring below the threshold level of detail in a process (unit) model. Yet the effects of such phenomena consistently push in the same direction. By accumulation, they cause deterioration of process performance over a long time horizon (months to a year).

Model based operability analysis is suitable to study the effect of the dominant phenomena on the behavior and performance of a process over smaller time scales, e.g., the time scale of daily production and below. Operability analysis is needed to assess the effects of deterministic and random changes outside and inside a process. The changes in the surroundings of the process (e.g., feed quality, production rates, product quality requirements and environmental conditions) can be a switch from one steady level to another as well as dynamic fluctuations. Inside a process (model) there will be some uncertainties in its physical parameters. The outcomes of an operability analysis may give reason to change a design so as to make it easier to operate and control a process plant. Conditions for proper operability are often included as conditions to be satisfied (inequality constraints) in an optimization of process economics. Without respecting such conditions, it will be hard to realize the computed economic performance in reality by coping with an unwilling process plant. Safety analysis of a design always ranks high in any operability analysis of the intended production practice.

Operability analysis

Operability analysis of a process has become an extensive subfield of process engineering, linking process design, control and operations. It is too extensive to cover it in any degree of detail in this section. The aim is to create awareness of the main aspects of operability while giving some references to pertaining literature for further study. Marlin [5] reviews eight topics of operability from the perspective of use in process design teaching:

Safety

Process units can fail and people operating a process make mistakes from time to time, resulting in incidents or even accidents. Layers of protection (LOPA) can be designed to prevent failures or to mitigate the effects of failures.

The *operating window of process units and of the entire process*

The operating window is the joint ranges of processing conditions (T, P, x, throughput) for which the process (unit equipment) is supposed to function normally, according to specifications. Are there any peculiar features of the operating window? Does it offer enough room for flexible operation?

Flexibility and *controllability* of the process

A process is flexible when there is enough room to adjust some manipulated process variables within their operating window to counteract the effects of external changes or disturbances to keep the production goals on target. Controllability implies that enough manipulated inputs (by means of actuators such as valves for flows) are in place in the process to effectively keep the controlled output variables close to their target values; e.g., set points for product quality, physical conditions (P, T) and the rate of production. Often a distinction is made between steady state (looking at steady state changes) and dynamic controllability (the nature of time responses by outputs to changes in inputs). A process that is both flexible and controllable is resilient to external changes and disturbances. Care must be taken that process integration and intensification do not result in a loss of controllability. Integration often takes away operational degrees of freedom in the process adversely affecting controllability, while intensification will reduce time constants and increase speed of response.

Reliability

Process unit equipment will malfunction from time to time and the production rate of a process and/or the product quality will be adversely affected. One can express the reliability of a process unit by means of a probability of failure. By putting identical process units in parallel, one has backups in case of failure and the reliability of the entire process increases. An increase in reliability of a process with more product output per year improves income from product sales. However, this requires higher investment costs due to the extra units, leading to a tradeoff between enhanced production from higher reliability and additional investment costs.

Transitions

During process operations there will be transitions from one steady level of operation to another level. This can involve startup and shutdown, regeneration or cleaning and sterilization, switches between two successive steady state production modes and load following. The process design should address such transitions, making sure these can be accomplished over a practical time interval and at a reasonable cost.

Dynamic behavior

This involves the time responses of process units from inputs to outputs and aspects of stability and runaway. These can all be affected by process design variables such as holdup volumes and surface areas for heat and mass transfers. It also involves an identification of multiple steady states if these would occur in the process (units) and making a decision at which state the process should be run.

Efficiency

This includes effective use of resources in a process, including its utilities. Are there ways to maintain product quality (avoiding production losses) and steering on profit (avoiding loss of economic opportunity)?

Monitoring and diagnosis

This involves the continuous monitoring of processing and equipment conditions by means of sensors. The purpose is to offer information to the process operators to assist them in running the process, both short term for rapid analysis and actions and long term for identifying slow changes in the process, such a fouling, plugging and deactivation.

In which respects can process models and simulations contribute to operability analysis?

Let us suppose that a process model covers steady state and dynamic situations within the specified operating window. Then this model is applicable to all of the above operability cases, except reliability and safety. Reliability is an exception as it requires probabilistic models for failure rates. Models for reliability engineering are covered in, among others, [6] and [7]. Safety is another exception because it often involves process events going beyond the normal operating window. The conventional process models are rarely tuned to such exceptional conditions beyond the normal operating window. Separate dedicated models are required to study the physical behavior of a process unit under abnormal conditions, such as flow reversal in pipelines, a thermal runaway event in a reactor or flows through pressure relieve valves.

A9.1.2 Nonlinear models and process optimization for feasibility stage

Case A9.10: Constraint analysis to identify options for technological innovations

For reasons of keeping optimization complexity at bay only a case of nonlinear programming (NLP) is presented here. If a process design model would contain both continuous and discrete decision variables, it is assumed the discrete variables are kept fixed during the optimization over the continuous variables. Prior to presenting a small chemical engineering example on constraint analysis to identify options for process improvements, some necessary optimization theory is presented to set the scene for the application case.

Necessity conditions for an extremum in optimization

The generic format of the optimization reads as:

$$\begin{aligned} &\text{Minimize} && F(\underline{\mathbf{x}}) \\ &\text{over} && \{\underline{\mathbf{x}}\} \\ &\text{Subject to} && \underline{\mathbf{g}}(\underline{\mathbf{x}}) = \underline{\mathbf{0}} \\ &\text{and} && \underline{\mathbf{h}}(\underline{\mathbf{x}}) \geq \underline{\mathbf{0}} \end{aligned}$$

All functions are algebraic ones, and are supposed to be continuous and differentiable at least once. When analyzing and solving optimization problems, the Lagrange function plays a key role; see [8, 9].

$$\text{Lagrange function:} \qquad L(\underline{\mathbf{x}}, u, v) = F(\underline{\mathbf{x}}) + \underline{\mathbf{g}}(\underline{\mathbf{x}})^{\mathrm{T}} \cdot \underline{\mathbf{u}} + \underline{\mathbf{h}}(\underline{\mathbf{x}})^{\mathrm{T}} \cdot \underline{\mathbf{v}}$$

$\underline{\mathbf{u}}$ is a vector of Lagrange multipliers, while $\underline{\mathbf{v}}$ is a vector of Kuhn–Tucker multipliers. In a NLP optimization, one searches for the optimum point(s) in the solution space $\{\underline{\mathbf{x}}\}$ where:

(a) the gradient of the Lagrange function vanishes:

$$\nabla_x L^* = \nabla_x F(\underline{\mathbf{x}}^*) + \nabla_x \underline{\mathbf{g}}(\underline{\mathbf{x}}^*)^{\mathrm{T}} \cdot \underline{\mathbf{u}}^* + \nabla_x \underline{\mathbf{h}}(\underline{\mathbf{x}}^*)^{\mathrm{T}} \cdot \underline{\mathbf{v}}^* = \underline{\mathbf{0}}\,,$$

(b) the equality and inequality equations are satisfied:

$$\underline{\mathbf{g}}(\underline{\mathbf{x}}*) = \underline{\mathbf{0}} \quad \text{and} \quad \underline{\mathbf{h}}(\underline{\mathbf{x}}*) \geq \underline{\mathbf{0}}\,,$$

(c) the *complementary slack* conditions apply:

$$\underline{\mathbf{h}}(\underline{\mathbf{x}}^*)^{\mathrm{T}} \cdot \underline{\mathbf{v}}^* = 0 \quad \text{and} \quad \underline{\mathbf{v}}^* \geq \underline{\mathbf{0}}\,,$$

(d) there are no linear dependencies between the equality and the active inequality constraints.

These four conditions are jointly called the Karush–Kuhn–Tucker necessity conditions [9]. These conditions need to be satisfied in solution point $\{\underline{\mathbf{x}}^*, \underline{\mathbf{u}}^*, \underline{\mathbf{v}}^*\}$. Vector $\underline{\mathbf{u}}$ contains the Lagrange multipliers associated with the equalities $\underline{\mathbf{g}}(\underline{\mathbf{x}}) = \underline{\mathbf{0}}$. Vector $\underline{\mathbf{v}}$ contains the Kuhn–Tucker multipliers associated with the inequality constraints $\underline{\mathbf{h}}(\underline{\mathbf{x}}^*) \geq \underline{\mathbf{0}}$. A Kuhn–Tucker multiplier can be interpreted as the derivative of the objective function F with respect to active inequality constraint h_i:

$$v_i^* = \partial L / \partial h_i \quad \text{for} \quad \underline{\mathbf{x}} = \underline{\mathbf{x}}^*$$

The complementary slack conditions state that (a) if an inequality constraint is not active (> 0), its corresponding Kuhn–Tucker multiplier v_i is zero. For example:

$$h_i(\underline{\mathbf{x}}^*) > 0 \quad \text{and} \quad v_i = 0\,.$$

And (b) the other way around: active inequality constraints have positive Kuhn–Tucker multipliers:

$$h_i(\underline{\mathbf{x}}^*) = 0 \quad \text{and} \quad v_i^* > 0\,.$$

In addition to these necessity conditions for an extremum, there exist sufficiency conditions for the existence of a (local) minimum. For detailed, fundamental explanations, see [9].

Constraint analysis

Suppose one has performed an optimization and a minimum is obtained:

$$\{\underline{\mathbf{x}}^*, \underline{\mathbf{u}}^*, \underline{\mathbf{v}}^*\} \quad \text{and} \quad \{F^*, \underline{\mathbf{g}}(\underline{\mathbf{x}}^*) = \underline{\mathbf{0}}, \underline{\mathbf{h}}(\underline{\mathbf{x}}^*) \geq \underline{\mathbf{0}}\}$$

A subset of the inequality constraints will be active in the optimum and can codetermine the position of the minimum. The associated Kuhn–Tucker multipliers indicate how much the objective function F^* would change if a pertaining inequality constraint would shift a bit: $h(x) \geq 0 \Rightarrow h(x) - \underline{\boldsymbol{\varepsilon}} \geq 0$. Why is it relevant to know the effects of inequality constraints on the objective function? Realistic process models will reflect current limitations of the technologies applied in the process. Such limitations are often expressed as inequality constraints in the optimization problem. These inequalities delineate the domain in which a technology can trustfully be applied. The technologies that are constraining an optimum by means of active inequality constraints can be identified by means of their associated Kuhn–Tucker multipliers. When analyzing the results from a successfully completed optimization, one can put the Kuhn–Tucker multipliers in order of decreasing magnitude. In that way, one obtains the order by which active inequality constraints are restraining the objective function. That is, the active inequality constraint with the largest Kuhn–Tucker multiplier has the biggest influence on the objective function of all active inequality constraints. If one wants to enhance future performance of a process, one may decide to push away some active inequalities and particularly those with a high impact on the performance function. One may want to improve those technologies that have one or more constraining inequalities with a high impact on the performance. One may also widen the domain of application and so relax tight inequality constraints and increase the performance. The message is that an analysis of the active inequality constraints and their Kuhn–Tucker multipliers in an optimization outcome can help to find incentives and clues for further technology development and process innovations.

Interpretation of the results at a computed optimum of a simple reactor design case

Chemical engineering textbooks on optimization mainly present small scale, mathematically motivated examples also dealing with multiplier values. However, no further use is made of such multipliers for a posterior analysis, probably because these mathematical examples have no application context. Therefore, a small scale chemical engineering example is presented here to illustrate the use of constraint analysis. The example involves the optimization of a continuous flow well mixed reactor with an isothermal reaction A ⇒ B. The density of the reaction mixture is assumed constant. The reactor has two free design variables, both with an upper bound: reaction volume and temperature. Therefore, there are two inequality constraints (upper bounds). The rate of production of B is fixed. The objective of the process optimization is to maximize the profit rate. The profit rate accounts for the proceeds from product sales, minus the cost of feed A, the cost of waste treatment of unconverted A and a capital charge term related to the volume dependent capital investment of the reactor. The capital investment is assumed to scale linearly with reactor volume. The reactor optimization problem will be cast in the standard NLP format assuming the objective function will be minimized. This can be achieved by changing the sign of the profit rate.

Optimization problem (NLP1):
Objective function ($F(\underline{\mathbf{x}})$):
Minimize $F = -P$ (= maximize profit rate P)

Equality constraints ($\underline{\mathbf{g}}(\underline{\mathbf{x}}) = \underline{\mathbf{0}}$):

Profit rate expression:

$$P = p_B \cdot f_{B,out} - c_A \cdot f_{A,feed} - c_{waste} \cdot f_{A,out} - cc.\text{Inv}_{ref} \cdot (V_{reac}/V_{ref})/(\theta^* 3600.0)$$

Reactor design equations:

$$f_{A,feed} - f_{A,out} - V \cdot r = 0$$
$$f_{B,feed} - f_{B,out} + V \cdot r = 0$$
$$r - k \cdot x_{A,out} = 0$$
$$k - k_{T=300} \cdot \exp\{-\gamma \cdot [(T_{ref}/T) - 1]\} = 0$$
$$f_{A,feed} - \{\Phi \cdot x_{A,feed}\} = 0$$
$$f_{A,out} - \{\Phi \cdot x_{A,out}\} = 0$$
$$f_{B,feed} - \{\Phi \cdot x_{B,feed}\} = 0$$
$$f_{B,out} - \{\Phi \cdot x_{B,out}\} = 0$$
$$f_{B,out} - B_{production} = 0$$

Inequality constraints ($\underline{\mathbf{h}}(\underline{\mathbf{x}}) \geq \underline{\mathbf{0}}$):

Reactor temperature: $310.0 - T \geq 0$
Reactor volume: $500.0 - V \geq 0$

Variables ($\underline{\mathbf{x}}$):

$f_{A,feed}$	Flow rate of A to reactor	[kg s^{-1}]
$f_{A,out}$	Flow rate of A out of reactor	[kg s^{-1}]
$f_{B,feed}$	Flow rate of B to reactor	[kg s^{-1}]
$f_{B,out}$	Flow rate of B out of reactor	[kg s^{-1}]
k	First order reaction rate constant	[s^{-1}]
r	Reaction rate	[kg m^{-3} s^{-1}]
P	Profit rate	[€ s^{-1}]:
T	Reactor temperature	[K]
$x_{A,out}$	Concentration of A in effluent	[kg A m^{-3}]
$x_{B,out}$	Concentration of B in effluent	[kg B m^{-3}]
V	Reactor design volume	[m^3]
Φ	Volume flow (in and out)	[m^3 s^{-1}]

Parameters with numerical values:

c_B	Price of product B	1.200	[€/kg B]
c_A	Cost of fresh feed A	1.000	[€/kg A]
c_{waste}	Cost of waste processing	0.500	[€/kg A]
cc	Capital charge factor	0.250	[€/(€ year)]
B_{prod}	Production rate of B	2.0	[kg B s^{-1}]
Inv_{ref}	Reference capital investment reactor	$6.000 \cdot 10^{+6}$	[€]
k_{Tref}	Reaction kinetic constant at T_{ref}	$2.000 \cdot 10^{-3}$	[s^{-1}]
T_{ref}	Reference temperature for kinetics	300.0	[K]
V_{ref}	Reference reactor volume	260.0	[m^3]
$x_{A,feed}$	Concentration of A in feed	10.0	[kg A m^{-3}]
$x_{B,feed}$	Concentration of B in feed	0.0	[kg B m^{-3}]
γ	Kinetic parameter $\{= E_{act}/(R_{gas} \cdot T_{ref})\}$	50.0	[–]
Θ	Hours of production per year	8000.0	[h/y]

This optimization problem has twelve variables and ten equalities, including one economic function and nine reactor design equations. The difference between number of variables and number of equalities leaves two degrees of freedom. These two degrees of freedom are reactor volume and temperature and both have an upper bound.

Results of optimization

With two degrees of freedom, one can take values for T and V and easily compute all reactor variables as well as the profit rate P. The outcome of an optimization in the {T, V} plane is that the profit rate is constrained by the upper bounds of the reactor temperature and volume. The numerical results are:

P^*	Profit rate	0.175	[€ s^{-1}]
$f^*_{A,feed}$	Flow rate of A to reactor	2.083	[kg s^{-1}]
$f^*_{A,out}$	Flow rate of A out of reactor	0.083	[kg s^{-1}]
$f^*_{B,feed}$	Flow rate of B to reactor	0.0	[kg s^{-1}]
$f^*_{B,out}$	Flow rate of B out of reactor	2.000	[kg s^{-1}]
k^*	First order reaction rate constant	$1.003 \cdot 10^{-2}$	[s-1]
r^*	Reaction rate	$4.012 \cdot 10^{-2}$	[kg m^{-3} s^{-1}]
T^*	Reactor temperature	310.0	[K]
x^*_A	Concentration of A in effluent	0.4	[kg A m^{-3}]
x^*_B	Concentration of B in effluent	9.6	[kg B m^{-3}]
V^*	Reactor design volume	500.0	[m^3]
Φ^*	Volume flow (in and out)	0.208	[m^3 s^{-1}]

The profit rate is higher in the infeasible area; e.g., $T = 310$ [K]; $V = 562$ [m^3]; $P = 0.177$ [€ s^{-1}]

The corresponding Kuhn–Tucker multipliers in this constrained optimum are:

ν_T^*	Temperature multiplier in optimum	0.0175	[€ K^{-1}]
ν_V^*	Volume multiplier in optimum	0.0001	[€ m^{-1}]

This result suggests one should aim for an increase of the upper bound of the temperature.

The constraint analysis for this small scale optimization example can be carried over to large scale optimization problems that are solved by means of numerical mathematical programming algorithms. Upon extracting the Kuhn–Tucker multipliers of the active inequality constraints from the computed solution and putting these in decreasing order of magnitude, one can identify the corresponding dominating inequality constraints. Relaxation of the bounds of the more dominating inequalities will improve performance. These dominating inequalities often express underlying process technological restrictions. Such restrictions may be removed by focused improvements in technology development, enhancing process performance.

Case A9.11: An example of a Pareto tradeoff between two performance metrics

A simple chemical engineering example is given of a Pareto set of design solutions for two performance criteria. Commonly, two criteria in the realms of economy and ecology are often used. The latter may also cover efficiency of transfer of a critical chemical element from reactants to products.

Example

The chemical reactor problem of case A9.10 is used as a vehicle to demonstrate the effects of optimizing two criteria simultaneously. The two criteria are the profit rate (P) and the mass efficiency (ME) of the process. The latter indicates which fraction of the intake mass is usefully transferred to the product. The reactor temperature and volume are the two design variables. The results are shown in Figure A9.5. There is a region to the left in the diagram (where the profit rate rises to its peak) where both the mass efficiency and the profit rate get better. To the right of the peak starts the Pareto set where an improvement in the mass efficiency is met by a decrease of the profit rate.

Case A9.12: Optimization of designs of industrial processes – gas-to-liquid (GTL) example

Due to production scale and complexity, GTL processes are prime candidates for design optimization.

The main conversion steps in the GTL process are:

(1) Syngas manufacturing using steam (by reforming) or pure oxygen by catalytic partial oxidation (CPOx) to turn methane into syngas (CO: $n\,H_2$; $2 < n < 4$).

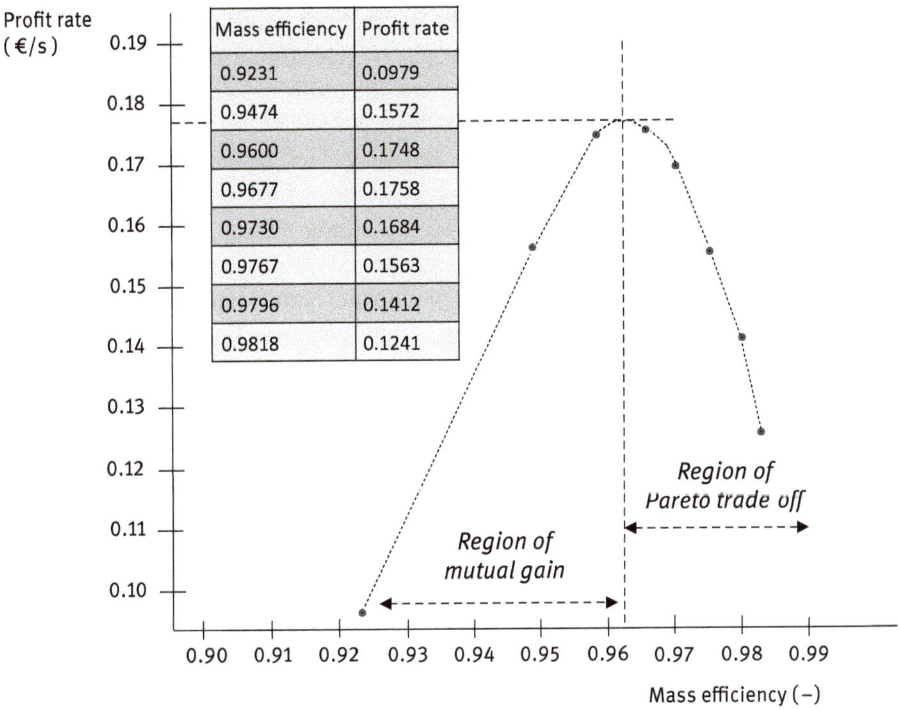

Fig. A9.5: Pareto tradeoff between profit and product mass efficiency of process.

(2) Fischer–Tropsch (FT) synthesis of wax (paraffinic hydrocarbons with some olefins and alcohols).

Reaction stoichiometry: $n\,CO + (2n+1)H_2 \Rightarrow C_nH_{2n+2} + n\,H_2O \quad n : 1 \Rightarrow\sim 300\,.$

Low temperature (200–240°C) conversion with cobalt catalyst is used for NG conversion.

(3) Hydrocracking of the wax into a slate of transportation liquids.

The FT reactors cannot fully convert the syngas due to technological limitations of catalyst and reactor equipment. The question arises of what to do with the unconverted syngas, being enriched with light hydrocarbons ($\leq C_4$) formed by FT synthesis. This is jointly called off gas. Four possibilities exist for off gas routing:

- Short recycle over the FT reactors to control the hydrogen to carbon ratio in reactor feeds.
- Long recycle to the syngas manufacturing section for reuse of carbon in CO and CO_2.
- Fuel to the utility system for steam and power generation.
- Stack (a fraction of the off gas must be flared to get rid of inerts in the feeds).

The design complexity is mainly due to the interactions between the syngas manufacturing, the FT conversion due to the syngas recycle structure as well as the heat exchanges between SGMU and FT sections and the utility system. The design of the hydrocracking is similar to existing refinery designs.

The list of process synthesis decision variables for syngas manufacturing units (SGMU) and FT sections is a long one:

- Type of SGMU, with choices between steam methane reforming (SMR), CPOx and autothermal reforming (ATR).
- Use of multiple SGMU technologies in parallel to control syngas composition.
- Oxygen to carbon ratio in feeds to SGMU's and their operating temperatures.
- Type of FT reactors (e.g., fixed bed, slurry and structured reactors) with cobalt catalyst.
- Amount of catalyst in a standardized FT reactor type.
- Number of sequential stages for FT reactors and number of parallel reactors per stage.
- Distribution of syngas from the SGMU's over the FT stages.
- Mixing ratio of fresh and recycle syngas to control $H_2 : CO$ ratio in the combined feed to a FT reactor stage (short recycle).
- FT reactor temperature and pressure, uniform per stage.
- Recycle fraction of off gas to the SGMU's with distribution over these units (long recycle).
- Fraction of off gas to the utility system for steam and power generation.
- Transfer rates of high quality thermal energy (high T) to lower quality levels in utility system (the energy quality levels have preassigned temperature and pressures).
- Production rate of liquid wax from FT section. If feed driven, intake of NG to SGMU section.
- Recycling of FT reaction water to SMR.

In addition, there will be multiple performance metrics:

- profitability, often expressed as net present value (NPV)
- carbon efficiency, expressed as amount of carbon in product over amount of carbon in feeds
- energy efficiency

In view of the complicated processing structure, it is also an option to place a lower bound on the availability and reliability of the process to assure a high enough annual product output.

An industrial case study has been published [10] on model based, optimization driven synthesis of a future GTL process, using profitability and carbon efficiency as performance measures. The process units are modeled in a parsimonious way, focusing on their main features of behavior. Each unit model makes use of one or more constitutive equations (kinetics and phase equilibria). For each unit, the domain of applicability of the constitutive equations has been outlined by means of inequality constraints. The resulting outcome of a GTL case study shows a design region in which both performance metrics can be improved simultaneously. However, after the peak of the profitability there is a Pareto curve with a tradeoff between the profitability and carbon efficiency. In a separate study [11], the operational reliability of a GTL process was modeled and optimized. One can go even one step further by jointly optimizing the design of a process along with planning of production and process maintenance. This has been successfully done for a multipurpose process [12].

Case A9.13: Process performance optimization by model and experiments

Data driven process model development is sometimes cheaper and quicker than developing a (semi)mechanistic model. However, it is required that one can indeed measure all input and output variables of a product/process, relevant to the characterization of its behavior and performance. This case study will focus on experimental process optimization, using a response surface approach. The procedure that is followed, applies equally well to the optimization of product performance.

The inputs to a process model must comprise the feed and key operating conditions of the process. The inputs may also include some quantitative geometric design decision variables as long as these can be implemented and adjusted in an experimental setup; e.g., by adjusting a volume or a contact surface between two different phases. The relevant measurable outputs are those that characterize the behavior and performance of a process and which can be measured. It is a key condition for data driven black box modeling that process inputs can be adjusted at will in a wide enough region to be able to collect information rich output data. Making a design of experiments will be instrumental to assure that measured output data is spread over the entire output space, rather than located in a subspace. Such a broad coverage is needed for good quality estimation of the unknown parameters in the black box model, which in turn will determine the descriptive accuracy of the model. Now, first a sequential, multistep procedure, represented in Figure A9.6, will be explained.

This work procedure combines modeling and experimentation in an optimization of the process performance. After having explained the procedure, the mathematical features of a black box model will be outlined. The objective of this work procedure is to closely approximate the point of optimum performance of the process through a sequence of model computations and experimentation. It is a strong feature of this approach that the point of optimum performance point is experimentally confirmed. The procedure gradually moves the performance of the process to the optimum in an alternating play between:

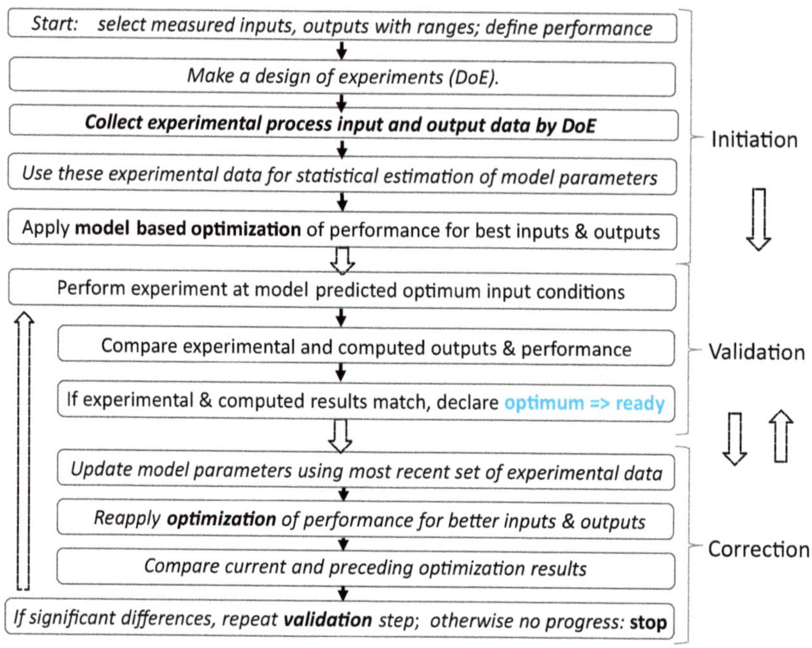

Fig. A9.6: A work procedure for an experimental approach to optimum process performance.

(a) *Prediction*: let the current model 'predict' at which inputs the optimum outputs are obtained.
(b) *Validation*: an experiment is performed at the model predicted optimum conditions. If the measured and predicted performances are close enough, the optimum is found and the search stops.
(c) *Correction*: if there is too much of a discrepancy between model and experimental performances the model needs adjustment. The newly generated experimental data under (b) will be used to improve the statistical estimates of the model parameter.

The sequence (prediction, validation, correction) is needed because any model, and specifically a black box one, remains an approximation to the real process with its unknown features. The structure of the procedure is shown in Figure A9.6.

The work procedure and its three phases are explained below.

Initiation phase:

Step 1: *Define process boundaries with measurable inputs and outputs and performance. The selected inputs and outputs must be relevant for performance.*

Step 2: *Make a (statistical) design of experiments (DoE).*
The objective of the DoE is to create a diverse, rich enough dataset which allows for statistical parameter estimation with good accuracy.

Step 3: *Perform experiments according to DoE plan and collect measured input and output data.*

Step 4: *Apply a statistical estimation procedure to derive the first set of parameter estimates.*
Check for poorly estimated parameters and collinearity of outputs.

Step 5: *Use the model with new parameter estimates to compute the optimum conditions.*
This computation is in fact a model based NLP optimization, requiring a model (available), bounds on the ranges of all inputs and outputs (given as inequality constraints) and a performance function in input and output variables.

Validation phase:

Step 6: *Perform an additional experiment with the computed (predicted) optimum input conditions.*

Step 7: *Compare the experimental and computational data {inputs, outputs and performance}.*
This can be done, for instance, by looking at the difference per variable over all variables.

Step 8: *Check how close the computed and experimental optimum outputs and performance are.*
If all outcomes are close (some quantitative criterion may be needed), then one reasonably assumes that the computed and the experimental results approximate the true but unknown optimum. The search is successfully finished.
If not close enough yet, a next corrective phase is entered.

Corrective phase:

The corrective phase acts on the difference between the model and the real process.

Step 9: *Update the model parameters.*
Use the most recent experimental data to update the model parameters by means of statistical estimation procedure updates, preferably in a Bayesian fashion: prior ⇒ posterior info.

Step 10: *Compute the next (new) optimum conditions, using the updated parameters.*

Step 11: *Check if the difference between the new and preceding optimum is significant.*
If so, start the next validation phase. If not so, further experimentation is likely not going to improve the optimization results. It is advised to terminate the procedure and reanalyze the model and the experimental approach.

It is likely that this sequence of validation corrections needs to be repeated a couple of times.

The outcome of this procedure is a set of input conditions for which the performance of the process is optimal and experimentally confirmed. As a side effect, a process model is obtained, having parameter estimates that are accurate enough to predict the outputs and the performance metric. The strength of this successive approach of gradual refinement of parameter estimates is that model predictions and experimental data should become consistently close in the region of optimum performance. Thus, a key benefit of this multistep approach is that it refines the accuracy of the computed model outcomes, particularly in a region that surrounds the optimum performance point of the process.

The process model in general will be a nonlinear multidimensional one, $\underline{\mathbf{Y}}^{(k)} = \underline{\mathbf{f}}(\underline{\mathbf{U}}^{(k)}; \underline{\mathbf{p}})$. Here, k is an index of an experimental situation. The structure of the equations is case dependent and should reflect as much as possible the basic understanding of the how the process works. In spite of its black box nature, the model (equations) should at least respect conservation principles of mass, the chemical elements and energy. Regarding accuracy of measurements, the conventional assumption that inputs ($\underline{\mathbf{U}}$) are set with 100% accuracy is often not realistic. There are always some fluctuations around the intended target settings. The assumption here is that multiple inputs ($\underline{\mathbf{U}}$) can be set with small known adjustment errors. These errors are characterized by means of an error covariance matrix (V_U). Usually the errors in the adjustments of the inputs are smaller than the measurement errors of the outputs ($\underline{\mathbf{Y}}$).

The experimental data and the associated model variables will be fitted by means of a quadratic norm (a weighted least squares approach), consistent with the common assumption of normally distributed measurement errors. The input variables in the model ($\underline{\mathbf{U}}$) are to be jointly fitted to the measured inputs as well as to model. Similarly, the process outputs ($\underline{\mathbf{Y}}$) in the model must be jointly fitted to measured data (experimental outputs) and obey the model equations. That is, the model variables must obey the model equations exactly, while the corresponding experimental data do not due to measurement errors and differences between model and the real process. While inputs and outputs are constrained by the model equations and their link to the experimental data, the remaining degree of freedoms in the fitting are the adjustable model parameters ($\underline{\mathbf{p}}$).

This approach falls into the category of response surface methods [13]. There are two extensions to the above chosen approach:

- The usual assumption in statistical textbooks that inputs have no errors is unrealistic for most chemical processes and must be relaxed. The result is a practical generalization of the least squares solution, commonly called total least squares or 'error in variables approach'.
- Application of a multistep sequential procedure to move to the optimum performance of the process in an alternating play between (a) doing an experiment at model predicted optimum conditions and (b) using the new experimental data to correct the model parameter estimates. The purpose of the multistep approach is to refine the accuracy of the model equations, particularly in a region that surrounds the optimum performance point of the process.

The final outcome is a process model having parameter estimates that are accurate enough to predict the optimum performance point. This point is also confirmed by experiments. The strength of this successive approach of gradual refinement of parameter estimates is that model outputs and experimental data should become consistent in the region of optimum performance.

List of symbols, subscripts and abbreviations:

Symbols

$\mathbf{A}$	real (non)square matrix (real values as entries)
$\mathbf{b}$	real vector
$\mathbf{B}$	real (non)square matrix, often mapping process inputs to states
c	cost related parameters in economic model expressions
c	molar concentration
d	symbol for design parameters (continuous) in a process model
E	set of quantitative performance metrics for a process design
$\mathbf{E}$	matrix of elemental composition of a set of chemical species
E	reaction activation energy
f	algebraic function
f	molar flow rate
F	objective function in an optimization problem
g	set of equality functions in an optimization problem
h	mole based specific enthalpy of a component
h	set of inequality functions in an optimization problem
ΔH_R	molar reaction enthalpy
i	generic index; e.g., for chemical species
j	generic index; e.g., for chemical reactions
k	index for discrete time stepping
k	reaction rate constant
K	reaction equilibrium constant
L	Lagrangian function
m	phase distribution factor
N	dimension of a vector
p	generic symbol for a set of physical parameters in a process model
p_P	proceeds from product sales, in economic model equations
P	performance function such as a profit rate
Q	heat exchange rate
r	reaction rate
R_{gas}	universal gas constant

s	split factor, distributing an incoming stream into two outgoing streams
$\mathbf{S}$	stoichiometric matrix
S	sensitivity factor
T	temperature
t	time coordinate in dynamic models
t	target value for a design quantity or performance function in process design
T	function expressing a metric (function) related to a target
u	generic symbol for inputs to a process model
v	Kuhn–Tucker multipliers in an optimization problem
V	volume
w	disturbance variables in a process model
$\mathbf{W}$	matrix of weight factor for scaling of process output variables
x	generic symbol for physical states in a process model
x	molar fraction of a chemical species
y	set of measured outputs of a process (model)
Y	generic symbol for structural design parameters (discrete) in a process model
z	generic symbol for a set of adjustable inputs to a process model
z	spatial coordinate in distributed models
$\mathbf{Z}$	matrix taking integer values only
$\{\alpha,\beta,\gamma\}$	indices for thermodynamic phases in models
γ	scaled reaction activation energy
$\underline{\boldsymbol{\varepsilon}}$	deviation between computed and experimental value of a process output variable
η	degree of conversion of a feed component in a reactor
ξ	extent of reaction
θ	number of production hours per year
$\underline{\boldsymbol{\lambda}}$	Lagrangian multiplier
Ω	sum of squared deviations between model and experimental outputs
τ	residence time
Φ	volume flow
σ	selectivity factor of a reaction network relative to a particular reaction product
$\mathbb{R}^N$	mathematical space of dimension N for real variables
∇	spatial derivative

Subscripts

average	averaged value of a quantity
design	design related
ecol	ecology related
ecom	economy related
oper	operations related
perf	performance related
ref	reference value of a quantity when making a comparison
tech	(process) technology related

Abbreviations

ATR	Autothermal reforming
CPOx	Catalytic partial oxidation (of natural gas)
CSR	Conversion separation recycle (process)
DAE	Differential algebraic equation set
GTL	Gas to liquid
FT	Fischer–Tropsch
LP	Linear programming
MILP	Mixed integer linear programming
MINLP	Mixed integer nonlinear programming
MTBE	Methyl *tertiary* butyl ether
NG	Natural gas
NLP	Nonlinear programming
SG	Syngas
SGMU	Syngas manufacturing units
SMR	Steam methane reforming
XTL	X = {biomass, coal, natural gas} to liquid fuels

References and further reading

[1] Buelens LC, Galvita VV, Poelman H, Detavernier C, Marin GB. Super-dry reforming of methane intensifies CO2 utilization via Le Chatelier's principle, Science 354(6311), 2016 449–452.
[2] Varma A, Morbidelli M, Wu H. Parametric sensitivity in Chemical Systems, Cambridge University Press, 1999.
[3] Haaker MPR, Verheijen PJT. Local and global sensitivity analysis for a reactor design with parameter uncertainty, Chemical Engineering Research & Design, 82, 2004, 591–598.
[4] Ogunnaike BA. Random Phenomena, fundamentals and engineering applications of probability and statistics, CRC Press, 2010.
[5] Marlin TE. Teaching operability in undergraduate chemical engineering design education, Computers & Chemical Engineering, 34, 2010, 1421–1431.
[6] S. Mannan. Lees process safety essentials. Hazard identification, assessment and control, Ch.5, Elsevier, 2014.

[7] Ogunnaike BA. Random phenomena, fundamentals and engineering applications of probability and statistics, Ch.21, CRC Press, 2010.
[8] Biegler LT. Grossmann IE, Westerberg AW. Systematic methods of chemical process design, McGraw Hill, 1997.
[9] Biegler LT. Nonlinear programming. Concepts, algorithms and applications to chemical processes, Ch. 4, SIAM, 2010.
[10] Ellepola JE, Thijssen N, Grievink J, Baak G, Avhale A, van Schijndel J. Development of a synthesis tool for Gas-To-Liquid complexes, Computers and Chemical Engineering, 42, 2012, 2–14.
[11] Blanco H, Grievink J, Thijssen N, Herder PM. Reliability integration to process synthesis applied to GTL processes, Computer-Aided-Process-Engineering, 33, 2014, 79–84.
[12] Goel HD, Grievink J, Weijnen MC. Integrated optimal reliable design, production, and maintenance planning for multipurpose process plants, Computers & Chemical Engineering, 27, 2013, 1543–1555.
[13] Ogunnaike BA. Random Phenomena, fundamentals and engineering applications of probability and statistics, Ch.19, CRC Press, 2010.
[14] Franceschini G, Macchietto S. Model-based design of experiments for parameter precision: State of the art, Chemical Engineering Science, 63, 2008, 4846–4872.
[15] Klebanov N, Georgakis C. Dynamic response surface models: A data driven approach for the analysis of time-varying process outputs, Industrial & Engineering Chemistry Research, 55, 2016, 4022–4034.

A13 Appendix to Chapter 13: Communicating

A13.1 Activity report, example

TU Delft	Design project: 'name & code' confidential	
Activity Report title: AR title: 'AR – PQ–PF/Knowledge – PSw'		Reference 'PQ–PF/Kn – PSw/20171015'

Version	Author	Date	Description
V1.0	Name	15/10/2017	Text

Summary

Text

Contents

- Objectives
- Approach, methods, tools
- Results & discussion
- Conclusions & recommendations
- References
- Appendices

Objectives

Text

Approach, methods, tools

Text

Results & discussion

Text

Conclusions & recommendations

Text

References

Appendices

https://doi.org/10.1515/9783110467741-018

A13.2 Stream specification (passing battery limit)

Table A13.1 shows the stream specification of the process stream passing the battery limit.

Tab. A13.1: Stream specification (passing battery limit) [2].

Stream name:	**Propylene/propane**				
Component	**Units**	**Specification**			**Additional information**
		Available	**Used in the Design**	**Notes**	**(also ref. note numbers)**
Propylene	%wt	50–70	69.0–70.0	(1)	(1) Values taken in consultation with principal.
Propane	%wt	50–30	29.0–30.0	(1)	
Light ends	%wt	<1.0		(2)	(2) As 'worst case' scenario, LEs and HEs are taken as ethane and butane respectively.
Heavy ends	%wt	<1.0		(2)	
Sulfur	ppm wt	<20		(3)	
Metals	ppm wt	<4		(3)	(3) Contaminants not harmful for the process. Compounds not included in mass balance.
Ethane	%wt		<1.0	(2)	
Butane	%wt		<1.0	(2)	
Total			100.0		(4) Not yet known. Will be provided by principal later.
	Process conditions and price				
Temp.	°C		25		Basis: Mixture of values for C3=/PG and C3o.
Press.	Bara		8		
Phase	V/L/S		V		Delivery per pipeline/rail car/ Tank truck/1000 l container/drum: indicate!
Price	Eu/ton		tbd	(4)	

A13.3 Concepts/criteria matrix

Table A13.2 shows an example of 'concepts/criteria matrix', or Pugh matrix [1]. This type of matrices support the design team's discussion on the design alternatives, stimulates creativity to further improve on the concepts and stimulates the decision making process. It should not be used as a quantitative decision making tool. Careful selection of the specifications (to avoid interrelations/overlap) and suppression of the tendency to quantify the subjective scores to a single score for each concept is needed.

Inferior and superior concepts can be easily identified, and the remaining concepts should be further developed and improved to get closer or surpass the superior concept(s) that fulfill and/or exceed the specifications.

Tab. A13.2: Concepts/criteria matrix.

Concepts/criteria matrix	Concept A	Concept B	Concept C
Specification 1	+	o	–
Specification 2	–	o	+
Specification 3	+	–	+
Specification 4	o	+	–
Specification 5	+	–	+

A13.4 Pure components properties table

Table A13.3 shows how to report pure component properties.

A13.5 Agenda and minutes of meeting (MOM) templates

Agenda template

Agenda

- Meeting information
  - Meeting title:
  - Project:
  - Date:
  - Starting time:
  - End time:
  - Location:
- Meeting preparation:
  - Please read:
  - Please bring:
- Invited attendees:
  - Names, affiliation:
- Agenda items
  - Attendance and introductions
  - Arrangement of agenda
  - Confirmation of MOM previous meeting
  - Matters arising from the MOM (action points)
  - Meeting topics
    - Topic 1, owner, allocated time
    - Topic 2, owner, allocated time
    - Etc.
  - Any other business
  - Next meeting

Minutes of meeting (MOM) template

Minutes of meeting

- Meeting information
  - Meeting title:
  - Project:
  - Date:
  - Starting time:
  - End time:
  - Location:

Tab. A13.3: Pure components properties table [2].

PURE COMPONENT PROPERTIES																		
Component name		Technological data								Health & safety data								
Design	Systematic	Formula	Mol. weight g/mol	Phase	Boiling point [1] °C	Melting point [1] °C	Flash point [1] °C	Liquid density [2] kg/m³	Vapour density [3] kg/m³	Auto-ignit temp. [1] °C	Flammable limits % by vol in air	Lower Explosion Limit (LEL) %	Upper Explosion Limit (UEL) %	LC_{50} in air/ water mg/m³	MAC value mg/m³	LD_{50} oral [4] g	Chemical reactivity	Notes
Propylene	Propylene	C_3H_6	42.08	V	-48.0	-185.2	-108	609.0	1.50	497.0	?	2.0	11.1	n.a.	n.a.	?	Extremely flamable,	(2),(5)
Propane	Propane	C_3H_8	44.09	V	-42.2	-187.6	-104	585.0	1.56	450.0	?	2.1	9.5	n.a.	1800	?	Explosive with air	(2),(6)

Notes:

[1] At 101.3 kPa

[2] Density at 25 °C, unless specified otherwise

[3] At 0 °C

[4] Oral ingestion in (g) for a male of 70 kg weight

[5] Density at -47 °C from H_2O at 4 °C

[6] Density at -45 °C from H_2O at 4 °C

Project ID number: GDP-PED-2017-1/1
Completion date: October 1st 2017

*Converting mg/m3 -->ppm & vice versa: mg/m3 to ppm calculator

- Agenda items
  - Attendance and introductions
    - Attendees: names, affiliations
    - Absentees: names, affiliations
  - Arrangement of agenda
  - Confirmation of MOM previous meeting
  - Matters arising from the MOM (action points)
  - Meeting topics
    - Topic 1, owner, decisions, action points and timing
    - Topic 2, owner, decisions, action points and timing
    - Etc.
  - Any other business
  - Next meeting
  - Reports, handouts presented, discussed during the meeting
  - List of decisions, action points and timing

A13.6 Frequently Occurring Opportunities For Improvement (FOOFI) list for design project presentations

This FOOFI list is composed in the form that frequently occurring errors/mistakes are indicated, based on [2].

General

- Page numbering missing
- Slides not readable (too small font size); minimum 16 pt
- University logo missing (use standard TU Delft or company format)
- Very few pictures, tables or figures are used: this makes the presentation boring
- Use of long sentences instead of keywords (use maximum seven lines of seven words)
- Presenter reads information from screen (mainly because information is in 'sentence form')
- Spelling mistakes (use spell checker)
- Presentation too long (aim at 1–1.5 slides/per minute)
- Too many colors used
- (Black) text invisible on colored background. Be very selective of colors; only use black text on white (or *very* light) backgrounds. White text works well on black/dark gray or (dark) blue backgrounds.

Title page

- Project principal (owner/customer) not mentioned
- Logo of Customer (in case of external project) missing
- Names of project team members missing
- Presentation date and location missing

Table of contents

- Missing table of contents
- Present presentation chapters at slide bottom, highlighting the current chapter

Introduction

- References to data, figures, tables missing

Project objectives

- Presentation has too much detail, without highlighting objectives.
- Conclusions & recommendations are not linked to objectives.

Basis of design (BOD)

- BOD section including all aspects agreed with customer is missing, e.g.:
  - Product performance specifications
  - Sales volume
  - Plant capacity (t/annum)
  - Number of operating hours/year (including shift pattern)
  - Location
  - Battery limit of system to be designed
  - Specifications (composition, delivery form & conditions, prices) at battery limit of:
    - products
    - byproducts
    - raw materials
    - available utilities
    - waste streams handling
  - Pure component properties table (including major SHE aspects)

Process block scheme (PBS)

- Missing
- No indication of conversion and split factors for separation blocks
- No indication of t/a, t/t values
- No indication of stream names

Process flow scheme (PFS)

- Standard symbols not used
- Not all raw materials enter at the left
- Not all products, byproducts and waste streams leave at the right
  - See separate information on making process flow diagrams
- PFS not readable: split into smaller (enlarged) parts

Mass balances

- No overall mass balance given (e.g., in the form of an input/output diagram showing all input and output streams, utilities and verification of *total in = total out*).

Tables

- Too many decimals behind decimal point
- Comma is used instead of decimal point
- Wrong units are specified: too many significant numbers are presented (copied from computing software)
- Numbers are not adjusted 'right' or at the decimal point (numbers should be easily viewed/compared)
- Units are not mentioned in the column title
- No SI-units are used ('kcal', 'j' instead of J, 'kJ/h' instead of 'kW' or 'MW', etc.)

Comparison of alternatives (conversion routes, separation methods, unit operations, etc.)

- Inappropriate comparison methods can be improved easily:
  - When comparing two different product and/or process design options the comparison criteria should be mentioned in the first column, and the performance of the alternatives should be scored for each criterion.
  - The criteria should be formulated in such a way that a score of '+' or '++' means that the performance for that criteria is good or very good.
  - The criteria should be ordered in order of importance.

- Use a 5-level scale: '– –', '–', '0', '+', '++'. Alternatively present actual performance data in the cells and use a 5-scale color code: '– –': red, '–': orange, '0': white, '+': light green, '++': dark green.
- This ensures that all criteria will be evaluated for each alternative and missing information is immediately evident.

Equipment design

- Design specifications per product component or equipment unit are not clear.
- The design variables to reach the design specifications for the product or process component are not made clear.
- It is unclear how the design variables are choosen/fixed to achieve the design specifications of each product or process component and the whole product or process.

Safety, health, environment, economy, technology, social (SHEETS) evaluation

- General: not a complete SHEETS evaluation given (mainly focus on economics).
  - Focus on all SHEETS aspects is required by stakeholders and evaluate the design accordingly.
- On economics:
  - Gross Income is not presented, or only at the very end; mention this upfront.
  - No distinction made between operating costs and investment costs
  - No conclusions formulated on the economical data presented and the objectives
  - No sensitivity analysis carried out: what are the financial risks for the project?

Conclusions and recommendations

- No reference is made to the project objectives: have they all been met?
- No presentation of the design's strengths and weaknesses: how can it be improved?
  - What would be possible next steps?
  - Difference between project objectives and achieved results should feed into the recommendations).

A13.7 Equipment summary and specification sheets

Tab. A13.4: Equipment summary sheet: reactors, columns & vessels – summary [2].

Equipment Nr.:	***C-01***	***V-01***
Name:	***P/P splitter***	***C-01 reflux accumulator***
–	Tray column	Horizontal
Pressure [bara]:	20.9 / 21.5	20.8
Temp. [°C]:	50.0 / 57.0	50.0
Volume [m^3]:		105.1 *(1)*
Diameter [m]:	4.5	3.5
L or H [m]:	69.0	11.0
Internals		
– Tray type:	Sieve Trays	n.a.
– Tray number:	130	n.a.
– Fixed packing		
Type:	n.a.	n.a.
Shape:	n.a.	n.a.
– Catalyst		
Type:	n.a.	n.a.
Shape:	n.a.	n.a.
Number		
– Series:	1	1
– Parallel:	–	–
Materials of construction *(2)*:	Trays: SS314 Column: CS	CS
Other:		

Remarks:
(1) V-01: effective volume = 68.8 m^3 for residence time of 5 minutes.
(2) SS = Stainless steel; CS = Carbon steel; n.a.: not applicable

Designers:	P.P. Kolom	W. Wisselaar	Project ID-Number:	*CDP3201*
	H.P. Pomp	R.G. Klep	Date:	January 1st 1998

Tab. A13.5: Equipment summary sheet: Heat exchangers & furnaces – Summary [2].

Equipment Nr.: ***Name:***	***E-01*** ***C-01 Ovhd condenser***	***E-02*** ***C-01 reboiler***	***E-03*** ***propylene cooler***	***E-04*** ***propane cooler***
	Finned tubes Air cooled	Thermosyphon	Fixed tube sheet	Fixed tube sheet
Substance				
– *Tubes:*	$C_3 = /n - C_3$	$n - C_3/C_3 =$	$C_3 = /n - C_3$	$n - C_3/C_3 =$
– *Shell:*	n.a.	L.P. Steam	Cooling Water	Cooling Water
Duty [kW]:	91,450	91,450	344	378
Heat Exchange area [m^2]:	12,760 *(1)*	2,850	75	62
Number				
– *Series:*	–	–	1	1
– *Parallel:*	–	6	–	–
Pressure [bara]				
– *Tubes:*	20.9	21.7	26.0	26.0
– *Shell:*	Atm.	4.0	4.0	4.0
Temperature In / Out [°C]				
– *Tubes:*	50.0 / 50.0	57.0 / 57.0	50.0 / 30.0	57.0 / 30.0
– *Shell:*	25.0 / 41.0	190.0 / 133.0	20.0 / 40.0	20.0 / 40.0
Special Materials of Construction (2):	Tubes: CS Shell: n.a.	Tubes: CS Shell: CS	Tubes: CS Shell: Al-Br	Tubes: CS Shell: Al-Br
Other:	Plot Area: 40 × 40 m			

Designers:	P.P. Kolom	W. Wisselaar	Project ID-Number:	*CDP3201*
	H.P. Pomp	R.G. Klep	Date:	January 1st 1998

Tab. A13.6: Equipment summary sheet: Pumps, blowers & compressors – Summary [2].

EQUIPMENT Nr.: ***NAME:***	***P-01*** ***C-01 Reflux:***	***P-02 A/B*** ***C-01 Bottom:***	***P-03 A/B*** ***Propylene Transfer:***
Type:	Centrifugal	Centrifugal	Centrifugal
Number:	2	2	2
Medium transferred:	$C_3 = /n - C_3$	$n - C_3/C_3 =$	$C_3 = /n - C_3$
Capacity [kg/s] [m^3/s]:	0.239	0.011	0.015
Density [kg/m^3]:	450	450	450
– *Pressure* [bara] *Suct. / Disch.:*	20.8 / 23.0	21.7 / 26.0	20.8 / 26.0
Temperature *In / Out* [°C]:	50.0 / 50.0	57.0 / 57.0	50.0 / 30.0
Power [kW]			
– *Theor.:*	53	5	8
– *Actual:*	75	8	13
Number			
– *Theor.:*			
– *Actual:*	2 *(1)*	2 *(1)*	2 *(1)*
Special Materials of Construction:	MS casing	MS casing	MS casing
Other:	Double mechanical seals	Double mechanical seals	Double mechanical seals

Remarks:
(1) Tray numbering from top to bottom

Designers:	P.P. Kolom H.P. Pomp	W. Wisselaar R.G. Klep	Project ID-Number: Date:	*CDP3201* January 1st 1998

Tab. A13.7: Equipment specification sheet: Distillation column [2].

EQUIPMENT NUMBER: ***C-01***
NAME: ***Propylene/Propane Splitter***

General Data

Service: – *distillation* / ~~*extraction*~~ / ~~*absorption*~~ /
Column Type: – ~~*packed*~~ / *tray* / ~~*spray*~~ /
Tray Type: – ~~*cap*~~ / *sieve* / ~~*valve*~~ /

Tray Number (1)

- *Theoretical:* 90
- *Actual:* 130
- *Feed (actual):* 85

Tray Distance (HETP) [m]*:* 0.500 *Tray Material: SS314* (2)
Column Diameter [m]*:* 4.500 *Column Material: CS* (2)
Column Height [m]*:* 69.000
Heating: ~~*none*~~ / ~~*open steam*~~ / *reboiler* / (3)

Process Conditions

Stream Details		*Feed*		*Top*		*Bottom*		*Reflux* / ~~*Absorbent*~~		~~*Extractant*~~ / ~~*side stream*~~	
Temp.	[°C]	: 53		: 50		: 57		: 50			
Pressure	[bara]	: 21.5		: 20.8		: 21.7		: 20.9			
Density	[kg/m^3]	: 450		: 450		: 450		: 450			
Mass Flow	[kg/s]	: 11.9		: 6.88		: 5.02		: 107.74			
Composition		mol%	wt%	mol%	wt%	mol%	wt%	mol%	wt%	mol%	wt%
Propylene		60.0	58.8	95.0	94.8	10.0	9.6	95.0	94.8		
Propane		40.0	41.2	5.0	5.2	90.0	90.4	5.0	5.2		

Column Internals (4)

Trays (5)		*Packing: Not Applicable*
Number of		*Type:*
~~*caps*~~ / *sieve holes* /*:*	...	*Material:*
Active Tray Area [m^2]*:*	...	*Volume* [m^3]*:*
Weir Length [mm]*:*	...	*Length* [m]*:*
Diameter of		*Width* [m]*:*
chute pipe / *hole* / [mm]*:*	...	*Height* [m]*:*

Remarks:
(1) Tray numbering from top to bottom.
(2) SS = Stainless Steel; CS = Carbon Steel.
(3) Reboiler is E-01; operates with LP steam.
(4) Sketch & measures of Column & Tray layout should have been provided.
(5) Tray layout valid for whole column.

Designers:	P.P. Kolom	W. Wisselaar	Project ID-Number:	*CDP3201*
	H.P. Pomp	R.G. Klep	Date:	January 1st 1998

A13.8 FOOFI (Frequently Occurring Opportunities For Improvement) list for design project reports

FOOFI list for Design Project Reports (Concept Stage, Feasibility Stage, Final Report)

This FOOFI list is composed in the form that frequently occurring errors/mistakes are indicated, and is based on [2].

General

- Page numbering missing
- Text not readable (too small font size); minimum 10 pt.
- University logo missing (use standard TU Delft or company format)
- Very few pictures, tables or figures are used: this makes the report very boring
- Use of very long sentences
- Spelling mistakes (use spell checker)
- Report main text too long (aim at 60 pages main text; remainder in appendices
- (Too) many colors used in figures
- (Black) text invisible on colored background: be very selective of colors; only use black text on white (or VERY light) backgrounds; white text works well on black/ dark gray or (dark) blue backgrounds.

Front page

- Design project stage (concept stage, feasibility stage, final report) omitted from front page
- Project principal (owner/customer) not mentioned
- Logo of customer (in case of external project) missing
- Names of project team members missing
- Report issue date missing
- Insufficient or nondescriptive key words
- Date on report review meeting missing on front page

Page numbering, table of contents, appendix table of content

- No page numbering before Chapter 1. (it should be: i, ii, iii, etc.)
- No page numbering for appendices (should be Appendix – 1 –, etc.)
- No page numbering of appendix table of contents, (it should be: Appendix – i –, Appendix – ii –, etc.)
- No information on appendices in (main) table of contents: list the detailed contents and pages of the appendices
- No ‘Appendix Table of Contents’ before Appendix 1

Chapters

- No symbol list and no list of abbreviations. These lists should be positioned before chapter references.
- Abbreviations are not explained the first time they appear.
- No chapter references present

Tables and figures

- Tables and figures do not have titles (above table, below figure).
- Tables and figures cannot be understood without the main text. Insufficient information is provided in the title and/or figure to be able to read/understand the table/figure.

Presentation of numbers in tables

- Too many decimals are present behind decimal point.
- Comma is used instead of decimal point.
- Wrong units are specified: too many significant numbers are presented (copied from computing software).
- Numbers are not adjusted 'right' or at the decimal point (numbers should be easily viewed/compared).
- Units are not mentioned in the column title.
- No SI-units are used ('kcal', 'j' instead of J, 'kJ/h' instead of 'kW' or 'MW' etc.).

Units

- SI units not used

Pure component properties table

- Table omitted from report
- Some data categories missing

Block flow diagram (BFD)

- Not all streams enter at the *left* and leave at the *righthand* side of the block scheme.
- No descriptive title of each individual stream. The name should already indicate whether this is a liquid or a solid. Indication of the phase and whether it is a solution or dispersion or solids is necessary, especially when dealing with multiple phases.
- Individual streams do not have numbers: <1>, etc. (For referral in process stream summary (PSS)).
- No use of bold lines for flows containing main raw materials/products
- Lines do cross; they shouldn't, use interrupted lines.

- No mass flow rate (t/a), no mass flow rate relative to main product or feed (t/t) for each individual stream present
- No indication of phase, T, P of entering and exiting streams (for completeness and later use for battery limit) indication
- Missing data on total *in* (t/a, t/t), total *out* (t/a, t/t)
- Blocks represent equipment instead of tasks (heat, cool, separate, recover).
- No indication of pressure, temperature, pH of the task blocks
- The block scheme should have a 1:1 relationship to the stream summary table.

Batch cycle diagram

- Batch cycle diagram is not included (for batch process).

Process stream summary, overall component mass balance and stream heat balance

- The process stream summary is not according to the prescribed format (listing for each stream: stream number, stream name, components names, incl. molecular weights (kg/kmol), enthalpy (kW), phase (V/L/S), pressure (bara), temperature (K), mass flow rate (kg/s) and moles flow rate (kmol/s). Show overall mass balance and stream heat balance.
- Pressure, temperature, phase and enthalpy have been forgotten.
- The stream summary table for the process block scheme or process flow scheme should have a 1:1 relationship to these schemes: they do not.
- No overall mass and energy balance is given, (e.g., in the form of an input/output diagram showing all input and output streams, utilities and verification of *total in* = *total out*.
- New formats are created/invented: easiest way is to download provided template files.

(Reaction) equations

- (Reaction) equations do not have numbers (no making it difficult to refer to). (E.g., per chapter: (Eq. 1.1), (Eq. 2.10).
- No phases (s), (l), (g), (aq.) indicated with species in reaction equations
- For complex molecular structure formulas, no component names are given.
- For complex molecules no molecular structure formula is given (stoichiometry cannot be verified).

Description of the design and design approach

- Principal not mentioned and what his/her interest in this design project is
- In what form will the product be delivered to what type of customers? What is/will be the supply pattern of the feedstocks and delivery of the product?
- No information given about competitor activities and patent situation.

- No mention of the main challenges you have/will need to overcome during the design: availability of data, uncertainties. How will you go about solving these issues?

Basic assumptions

- No clear definition of the battery limit (e.g., By drawing envelope in block scheme). Is storage and waste treatment included?
- No clear definition of streams entering and exiting the battery limit using the correct format (Table A13.1)
- No conclusions drawn on thermodynamic and kinetic literature data: region of validity, accuracy, demonstrated with experimental data.

Product and process designs

- Hierarchical decomposition method from Delft Design Map (DDM) not used. You should first view the overall supply chain, stakeholder and consumer needs and main conversion routes, and define rather broad tasks for the process blocks.
- Design choices (at different design levels) not clear or not clearly justified.
- No clear justification for the choice of the product composition & structure, process options. No use is made of comparing alternatives against predefined selection criteria.
- A lot of qualitative statements are made, without referring to the sources. Try to be as quantitative as possible to accept/reject choices, effectively using your estimation skills.

Comparison of alternatives (conversion routes, separation methods, unit operations, etc.)

- Inappropriate comparison methods are applied:
  - When comparing two different product and/or process design options, the comparison criteria should be mentioned in the first column, and the performance of the alternatives should be scored for each criterion.
  - The criteria should be formulated in such a way that a score of '+' or '++' means that the performance for that criteria is good or very good.
  - The criteria should be organized in order of importance.
  - Use a five level scale: '––', '–', '0', '+', '++'. Alternatively, present actual performance data in the cells and use a five scale color code: '––': red, '–': orange, '0': white, '+': light green, '++': dark green.
  - This ensures that all criteria will be evaluated for each alternative and missing information is immediately evident.

Equipment design

- Design specifications per product component or equipment units are not clear.
- The design variables to reach the design specifications for the product or process components are not made clear.
- It is unclear how the design variables are chosen/fixed to achieve the design specifications of each product or process component and the whole product or process.

Safety, health, environment, economy, technology, social (SHEETS) evaluation

- General: not a complete SHEETS evaluation given (mainly focus on economics).
  - Focus on all SHEETS aspects required by stakeholders and evaluate the design accordingly.
- On economics:
  - Gross income is not presented, or only at the very end; mention this upfront.
  - No distinction made between operating costs and investment costs.
  - No conclusions formulated on the economic data presented and the objectives.
  - No sensitivity analysis carried out: what are the financial risks for the project?

Conclusions & recommendations

- No reference is made to the project objectives: have they all been met?
- No presentation of the design's strengths and weaknesses: how can it be improved?
  - What would be possible for the next steps?
  - Difference between project objectives and achieved results should feed into the recommendations.

References

- Findings/assumptions/data, etc., are mentioned without referring to the sources. This makes it difficult for the future reader to check the validity and is considered plagiarism.

References and further reading

[1] Pugh S. Creating innovative products using total design, New York, NY, USA, Addison-Wesley-Longman, 1996.
[2] De Haan AB, Swinkels PLJ, de Koning PJ. Instruction manual conceptual design project CH3843 – from idea to design. Department of Chemical Engineering, Delft University of Technology, Delft, Netherlands, 2016.

Index

https://doi.org/10.1515/9783110467741-019

E

www.ingramcontent.com/pod-product-compliance
Ingram Content Group UK Ltd.
Pitfield, Milton Keynes, MK11 3LW, UK
UKHW052001041026
15590UKWH00008B/2080

9 783110 467727